National Aeronautics and Space Administration

July 2008

Space Shuttle Program Programmatic Environmental Assessment; Transition and Program Property Disposition

This page intentionally left blank.

Contents

Executive Summary ES-1
    ES.1   Introduction ES-1
    ES.2   Background ES-1
    ES.3   Purpose and Need for the Proposed Action ES-7
    ES.4   Proposed Action ES-9
        ES.4.1 Real Property ES-9
        ES.4.2 Personal Property ES-10
        ES.4.3 Proposed Action Schedule ES-11
    ES.5   No Action Alternative ES-11
    ES.6   Decision to be Made ES-11
    ES.7   Summary of Environmental Impacts ES-11
        ES.7.1 National Perspective on Socioeconomic Impacts ES-12
        ES.7.2 No Action Alternative ES-13
        ES.7.3 Proposed Action Alternative ES-14
    ES.8   Public and Agency Involvement ES-16

1. **Purpose and Need for the Proposed Action** 1-1
    1.1   Background 1-1
        1.1.1   Previous U.S. Human Space Exploration Programs 1-1
        1.1.2   Space Shuttle 1-2
        1.1.3   The Vision for Space Exploration 1-2
        1.1.4   NASA 2008 Budget Request 1-3
        1.1.5   Planning for SSP Transition and Retirement 1-4
    1.2   Need for the Proposed Action 1-5
    1.3   Purpose of the Proposed Action 1-8
        1.3.1   Decisions to be Made 1-8
        1.3.2   Public Involvement 1-8
        1.3.3   Issues Considered but Not Carried Forward 1-8
    1.4   Executive Order 12114 1-8

2. **Description of Proposed Action and Alternatives** 2-1
    2.1   Description of the Proposed Action and Preferred Alternative 2-2
        2.1.1   Disposition of Shuttle Assets 2-2
            2.1.1.1   Real Property 2-2
            2.1.1.2   Personal Property 2-4
            2.1.1.3   Property Disposition Schedule 2-5
        2.1.2   Space Shuttle Operations and Elements 2-8
            2.1.2.1   Space Shuttle Operations 2-8

		2.1.2.2 Space Shuttle Space Flight Hardware Elements 2-15	
		2.1.3 Proposed Action Schedule 2-22	
	2.2	Description of the No Action Alternative 2-22	
	2.3	Alternatives Considered but Not Carried Forward 2-22	
	2.4	Summary Comparison of Alternatives 2-22	
3.	**Affected Environment** 3-1		
	3.1	Overview of Resource Areas 3-1	
	3.2	Kennedy Space Center 3-3	
		3.2.1 Air Quality 3-9	
		3.2.1.1 Affected Environment 3-9	
		3.2.2 Biological Resources 3-10	
		3.2.2.1 Affected Environment 3-10	
		3.2.3 Cultural Resources 3-14	
		3.2.3.1 Affected Environment 3-14	
		3.2.4 Hazardous and Toxic Materials and Waste 3-16	
		3.2.4.1 Affected Environment 3-16	
		3.2.5 Health and Safety 3-19	
		3.2.5.1 Affected Environment 3-20	
		3.2.6 Hydrology and Water Quality 3-20	
		3.2.6.1 Affected Environment 3-20	
		3.2.7 Land Use 3-25	
		3.2.7.1 Affected Environment 3-25	
		3.2.8 Noise 3-28	
		3.2.8.1 Affected Environment 3-28	
		3.2.9 Site Infrastructure 3-30	
		3.2.9.1 Affected Environment 3-30	
		3.2.10 Socioeconomics 3-30	
		3.2.10.1 Region of Influence 3-30	
		3.2.10.2 Affected Environment 3-33	
		3.2.11 Solid Waste 3-34	
		3.2.11.1 Affected Environment 3-34	
		3.2.12 Traffic and Transportation 3-35	
		3.2.12.1 Region of Influence 3-35	
		3.2.12.2 Affected Environment 3-35	
	3.3	Johnson Space Center 3-36	
		3.3.1 Air Quality 3-41	
		3.3.1.1 Affected Environment 3-41	
		3.3.2 Biological Resources 3-41	
		3.3.2.1 Affected Environment 3-41	
		3.3.3 Cultural Resources 3-42	
		3.3.3.1 Affected Environment 3-42	

		3.3.4	Hazardous and Toxic Materials and Waste	3-47
			3.3.4.1 Affected Environment	3-47
		3.3.5	Health and Safety	3-51
			3.3.5.1 Affected Environment	3-51
		3.3.6	Hydrology and Water Quality	3-52
			3.3.6.1 Affected Environment	3-52
			3.3.6.2 Affected Environment	3-56
		3.3.7	Noise	3-59
			3.3.7.1 Affected Environment	3-59
		3.3.8	Site Infrastructure	3-60
			3.3.8.1 Affected Environment	3-60
		3.3.9	Socioeconomics	3-61
			3.3.9.1 Region of Influence	3-61
			3.3.9.2 Affected Environment	3-65
		3.3.10	Solid Waste	3-66
			3.3.10.1 Affected Environment	3-66
		3.3.11	Traffic and Transportation	3-66
			3.3.11.1 Region of Influence	3-66
			3.3.11.2 Affected Environment	3-66
	3.4	Ellington Field		3-67
		3.4.1	Air Quality	3-67
			3.4.1.1 Affected Environment	3-67
		3.4.2	Hazardous and Toxic Materials and Waste	3-71
			3.4.2.1 Affected Environment	3-71
		3.4.3	Health and Safety	3-72
			3.4.3.1 Affected Environment	3-72
		3.4.4	Hydrology and Water Quality	3-73
			3.4.4.1 Affected Environment	3-73
		3.4.5	Land Use	3-73
			3.4.5.1 Affected Environment	3-73
		3.4.6	Noise	3-74
			3.4.6.1 Affected Environment	3-74
		3.4.7	Site Infrastructure	3-75
			3.4.7.1 Region of Influence	3-75
			3.4.7.2 Affected Environment	3-75
		3.4.8	Solid Waste	3-76
			3.4.8.1 Affected Environment	3-76
		3.4.9	Traffic and Transportation	3-76
			3.4.9.1 Region of Influence	3-76
			3.4.9.2 Affected Environment	3-76
	3.5	El Paso Forward Operation Location		3-77
		3.5.1	Air Quality	3-81
			3.5.1.1 Affected Environment	3-81

CONTENTS (CONTINUED)

	3.5.2	Hazardous and Toxic Materials and Waste		3-82
		3.5.2.1	Affected Environment	3-82
	3.5.3	Health and Safety		3-82
		3.5.3.1	Affected Environment	3-82
	3.5.4	Hydrology and Water Quality		3-83
		3.5.4.1	Affected Environment	3-83
	3.5.5	Noise		3-84
		3.5.5.1	Affected Environment	3-84
	3.5.6	Site Infrastructure		3-84
		3.5.6.1	Region of Influence	3-84
		3.5.6.2	Affected Environment	3-84
	3.5.7	Solid Waste		3-85
		3.5.7.1	Affected Environment	3-85
	3.5.8	Transportation and Traffic		3-85
		3.5.8.1	Region of Influence	3-85
		3.5.8.2	Affected Environment	3-85
3.6	Stennis Space Center			3-89
	3.6.1	Air Quality		3-93
		3.6.1.1	Affected Environment	3-93
	3.6.2	Biological Resources		3-94
		3.6.2.1	Affected Environment	3-94
	3.6.3	Cultural Resources		3-102
		3.6.3.1	Affected Environment	3-102
	3.6.4	Hazardous and Toxic Materials and Waste		3-102
		3.6.4.1	Affected Environment	3-102
	3.6.5	Health and Safety		3-106
		3.6.5.1	Affected Environment	3-106
	3.6.6	Hydrology and Water Quality		3-107
		3.6.6.1	Affected Environment	3-107
	3.6.7	Land Use		3-112
		3.6.7.1	Region of Influence	3-112
		3.6.7.2	Affected Environment	3-112
	3.6.8	Noise		3-114
		3.6.8.1	Affected Environment	3-114
	3.6.9	Site Infrastructure		3-115
		3.6.9.1	Region of Influence	3-115
		3.6.9.2	Affected Environment	3-115
	3.6.10	Solid Waste		3-115
		3.6.10.1	Affected Environment	3-115
	3.6.11	Traffic and Transportation		3-116
		3.6.11.1	Region of Influence	3-116
		3.6.11.2	Affected Environment	3-116

3.7	Michoud Assembly Facility		3-116
	3.7.1	Air Quality	3-120
		3.7.1.1 Affected Environment	3-120
	3.7.2	Biological Resources	3-121
		3.7.2.1 Affected Environment	3-121
	3.7.3	Cultural Resources	3-126
		3.7.3.1 Affected Environment	3-126
	3.7.4	Hazardous and Toxic Materials and Waste	3-127
		3.7.4.1 Affected Environment	3-127
	3.7.5	Health and Safety	3-131
		3.7.5.1 Affected Environment	3-131
	3.7.6	Hydrology and Water Quality	3-132
		3.7.6.1 Affected Environment	3-132
	3.7.7	Land Use	3-136
		3.7.7.1 Affected Environment	3-136
	3.7.8	Noise	3-136
		3.7.8.1 Affected Environment	3-136
	3.7.9	Site Infrastructure	3-137
		3.7.9.1 Region of Influence	3-137
		3.7.9.2 Affected Environment	3-137
	3.7.10	Socioeconomics	3-138
		3.7.10.1 Region of Influence	3-138
		3.7.10.2 Affected Environment	3-138
	3.7.11	Solid Waste	3-142
		3.7.11.1 Affected Environment	3-142
	3.7.12	Traffic and Transportation	3-143
		3.7.12.1 Region of Influence	3-143
		3.7.12.2 Affected Environment	3-143
3.8	Marshall Space Flight Center		3-145
	3.8.1	Air Quality	3-149
		3.8.1.1 Affected Environment	3-149
	3.8.2	Biological Resources	3-150
		3.8.2.1 Affected Environment	3-150
	3.8.3	Cultural Resources	3-157
		3.8.3.1 Affected Environment	3-157
	3.8.4	Hazardous and Toxic Materials and Waste	3-161
		3.8.4.1 Affected Environment	3-161
	3.8.5	Health and Safety	3-165
		3.8.5.1 Affected Environment	3-165
	3.8.6	Hydrology and Water Quality	3-165
		3.8.6.1 Affected Environment	3-165
	3.8.7	Land Use	3-169
		3.8.7.1 Affected Environment	3-169

CONTENTS (CONTINUED)

		3.8.8	Noise	3-170
			3.8.8.1 Region of Influence	3-170
			3.8.8.2 Affected Environment	3-170
		3.8.9	Site Infrastructure	3-171
			3.8.9.1 Region of Influence	3-171
			3.8.9.2 Affected Environment	3-171
		3.8.10	Socioeconomics	3-172
			3.8.10.1 Region of Influence	3-172
			3.8.10.2 Affected Environment	3-175
		3.8.11	Solid Waste	3-176
			3.8.11.1 Affected Environment	3-176
		3.8.12	Traffic and Transportation	3-176
			3.8.12.1 Region of Influence	3-176
			3.8.12.2 Affected Environment	3-176
	3.9	White Sands Test Facility		3-177
		3.9.1	Air Quality	3-181
			3.9.1.1 Affected Environment	3-181
		3.9.2	Biological Resources	3-182
			3.9.2.1 Affected Environment	3-182
		3.9.3	Cultural Resources	3-183
			3.9.3.1 Affected Environment	3-183
		3.9.4	Hazardous and Toxic Materials and Waste	3-187
			3.9.4.1 Affected Environment	3-187
		3.9.5	Health and Safety	3-191
			3.9.5.1 Affected Environment	3-191
		3.9.6	Hydrology and Water Quality	3-193
			3.9.6.1 Affected Environment	3-193
		3.9.7	Land Use	3-194
			3.9.7.1 Region of Influence	3-194
			3.9.7.2 Affected Environment	3-194
		3.9.8	Noise	3-195
			3.9.8.1 Affected Environment	3-195
		3.9.9	Site Infrastructure	3-197
			3.9.9.1 Region of Influence	3-197
			3.9.9.2 Affected Environment	3-197
		3.9.10	Socioeconomics	3-198
			3.9.10.1 Region of Influence	3-198
			3.9.10.2 Affected Environment	3-198
		3.9.11	Solid Waste	3-201
			3.9.11.1 Affected Environment	3-201
		3.9.12	Traffic and Transportation	3-201
			3.9.12.1 Region of Influence	3-202
			3.9.12.2 Affected Environment	3-202

3.10	Dryden Flight Research Center		3-205
	3.10.1 Air Quality		3-205
		3.10.1.1 Affected Environment	3-205
	3.10.2 Cultural Resources		3-209
		3.10.2.1 Affected Environment	3-209
	3.10.3 Hazardous and Toxic Materials and Waste		3-210
		3.10.3.1 Affected Environment	3-210
	3.10.4 Health and Safety		3-213
		3.10.4.1 Affected Environment	3-213
	3.10.5 Hydrology and Water Quality		3-214
		3.10.5.1 Region of Influence	3-214
		3.10.5.2 Affected Environment	3-214
	3.10.6 Land Use		3-215
		3.10.6.1 Affected Environment	3-215
	3.10.7 Noise		3-216
		3.10.7.1 Affected Environment	3-216
	3.10.8 Site Infrastructure		3-216
		3.10.8.1 Affected Environment	3-216
	3.10.9 Solid Waste		3-217
		3.10.9.1 Affected Environment	3-217
	3.10.10 Traffic and Transportation		3-218
		3.10.10.1 Region of Influence	3-218
		3.10.10.2 Affected Environment	3-218
3.11	Palmdale		3-220
	3.11.1 Air Quality		3-220
		3.11.1.1 Affected Environment	3-220
	3.11.2 Cultural Resources		3-223
		3.11.2.1 Affected Environment	3-223
	3.11.3 Hazardous and Toxic Materials and Waste		3-223
		3.11.3.1 Affected Environment	3-223
	3.11.4 Health and Safety		3-224
		3.11.4.1 Affected Environment	3-224
	3.11.5 Noise		3-224
		3.11.5.1 Affected Environment	3-224
	3.11.6 Site Infrastructure		3-224
		3.11.6.1 Affected Environment	3-224
	3.11.7 Solid Waste		3-225
		3.11.7.1 Affected Environment	3-225
	3.11.8 Traffic and Transportation		3-225
		3.11.8.1 Region of Influence	3-225
		3.11.8.2 Affected Environment	3-225

CONTENTS (CONTINUED)

4.	**Environmental Consequences**		4-1
	4.1	Overview of Cultural Resources and Socioeconomics	4-2
		4.1.1 National Perspectives on Cultural Resources	4-2
		4.1.2 National Perspective on Socioeconomic Impacts	4-3
		4.1.2.1 Socioeconomic Effects of Federal Agency Actions	4-3
		4.1.2.2 NASA's Vision for Exploration Systems and Space Operations	4-6
		4.1.2.3 Overall Effects of the Proposed Action	4-8
		4.1.2.4 Overall Effects of No Action Alternative	4-8
	4.2	Kennedy Space Center	4-10
		4.2.1 Environmental Consequences for KSC	4-14
	4.3	Johnson Space Center	4-19
		4.3.1 Environmental Consequences for JSC	4-24
	4.4	Ellington Field	4-29
		4.4.1 Environmental Consequences for Ellington Field	4-30
	4.5	El Paso Forward Operation Location	4-34
		4.5.1 Environmental Consequences for EPFOL	4-35
	4.6	Stennis Space Center	4-38
		4.6.1 Environmental Consequences Summary for SSC	4-39
	4.7	Michoud Assembly Facility	4-45
		4.7.1 Environmental Consequences Summary for MAF	4-45
	4.8	Marshall Space Flight Center	4-50
		4.8.1 Environmental Consequences Summary for MSFC	4-52
	4.9	White Sands Test Facility	4-57
		4.9.1 Environmental Consequences Summary for WSTF	4-59
	4.10	Dryden Flight Research Center	4-64
		4.10.1 Environmental Consequences Summary for DFRC	4-64
	4.11	Palmdale	4-68
		4.11.1 Environmental Consequences Summary for Palmdale	4-68
	4.12	Environmental Justice	4-72
		4.12.1 Definitions	4-72
		4.12.1.1 Minority Individuals and Minority Populations	4-72
		4.12.1.2 Low-income Individuals and Low-income Populations	4-73
		4.12.1.3 Disproportionately High and Adverse Human Health Effects	4-73
		4.12.1.4 Disproportionately High and Adverse Environmental Effects	4-73
		4.12.2 Methodology	4-73
		4.12.3 Population Characterization and Impact Analysis	4-74
	4.13	Cumulative Impacts	4-74

5.	List of Preparers	5-1
6.	References	6-1

Appendixes

A: Acronyms and Abbreviations and Common Metric/British System Equivalents
B: Distribution List
    B.1: Responses to Draft EA Public Review Comments
C: Criteria Used to Determine Historic Property Eligibility for the Space Shuttle Assets
D: Applicable Regulations and Laws

Exhibits

ES-1	Summary Comparison of Alternatives	ES-2
1-1	Timeline of the United States' Human Exploration of Space	1-3
1-2	Issues Considered but Eliminated from Further Analysis	1-10
2-1	Property Excess Planned Burndown	2-6
2-2	Property Transfer Planned Burndown	2-6
2-3	Property Excess Planned Burndown by Location	2-7
2-4	Property Transfer Planned Burndown by Location	2-7
2-5	SSP Facilities	2-9
2-6	SSP Hardware Flow	2-11
2-7	Space Shuttle Elements Flow at SSP Related NASA Centers	2-13
2-8	Space Shuttle Configuration	2-15
2-9	Space Shuttle Orbiter	2-16
2-10	Space Shuttle Main Engine	2-18
2-11	Space Shuttle External Tank	2-19
2-12	Space Shuttle Reusable Solid Rocket Motor	2-20
2-13	Space Shuttle Solid Rocket Booster	2-21
2-14	Summary Comparison of Alternatives	2-23
3-1	SSP Facilities	3-2
3-2	Kennedy Space Center Location Map	3-4
3-3	Buildings Used by SSP at Kennedy Space Center	3-7
3-4	Kennedy Space Center Vegetation	3-11
3-5	State and Federally Protected Species at KSC	3-12
3-6	Buildings on or Eligible for the NRHP, Kennedy Space Center	3-17
3-7	Kennedy Space Center Water Resources	3-21

CONTENTS (CONTINUED)

3-8	Designated Uses for Surface Waters on KSC	3-24
3-9	KSC Administrative Areas	3-27
3-10	Noise Generated at KSC	3-29
3-11	Economic Region of Influence for Kennedy Space Center	3-31
3-12	Kennedy Space Center Transportation Map	3-37
3-13	Johnson Space Center Location Map	3-39
3-14	Johnson Space Center, Ellington Field, and Sonny Carter Training Facility Wetlands	3-43
3-15	Johnson Space Center, Ellington Field, and Sonny Carter Training Facility Floodplain	3-45
3-16	Buildings on or Eligible for the NRHP, Johnson Space Center	3-49
3-17	Johnson Space Center Water Resources	3-53
3-18	Johnson Space Center, Ellington Field, and Sonny Carter Training Facility Landcover	3-57
3-19	Economic Region of Influence for Johnson Space Center, Ellington Field, and Sonny Carter Training Facility	3-63
3-20	Transportation Features within Johnson Space Center	3-69
3-21	El Paso Forward Operation Location, Location Map	3-79
3-22	Typical Astronaut Training Mission Activities Conducted at EPFOL	3-81
3-23	El Paso Forward Operation Location Transportation Map	3-87
3-24	Stennis Space Center Location Map	3-91
3-25	Stennis Space Center Wetlands	3-97
3-26	Stennis Space Center Floodplain	3-99
3-27	Federal- and State-listed Wildlife Species with Ranges that Include SSC	3-101
3-28	Buildings on or Eligible for the NRHP, Stennis Space Center	3-103
3-29	Stennis Space Center Water Resources	3-109
3-30	SSC Groundwater Well Use Permits	3-111
3-31	Transportation Features Within Stennis Space Center	3-117
3-32	Michoud Assembly Facility Location Map	3-119
3-33	Nominal Six ETs-per-Year Production Cycle	3-121
3-34	Michoud Assembly Facility Vegetation	3-122
3-35	Michoud Assembly Facility Wetlands	3-124
3-36	Michoud Assembly Facility Floodplain	3-125
3-37	Buildings on or Eligible for the NRHP, Michoud Assembly Facility	3-129
3-38	Michoud Assembly Facility Water Resources	3-133
3-39	Economic Region of Influence for Michoud Assembly Facility and Stennis Space Center	3-139
3-40	Transportation Features within Michoud Assembly Facility	3-144
3-41	Marshall Space Flight Center Location Map	3-147
3-42	Vegetation and Land Cover, Marshall Space Flight Center	3-151
3-43	Wetlands, Marshall Space Flight Center	3-153
3-44	Floodplain, Marshall Space Flight Center	3-155

3-45	Federal- and State-Protected Species with Potential Habitat at MSFC	3-158
3-46	Buildings on or Eligible for the NRHP, Marshall Space Flight Center	3-163
3-47	Marshall Space Flight Center Water Resources	3-167
3-48	Economic Region of Influence for Marshall Space Flight Center	3-173
3-49	Transportation Features Within Marshall Space Flight Center	3-178
3-50	White Sands Test Facility Location Map	3-179
3-51	Vegetation Features Within White Sands Test Facility	3-185
3-52	Buildings on or Eligible for the NRHP, White Sands Test Facility	3-189
3-53	WSTF Discharge Plans	3-193
3-54	Major Sources of Noise at WSTF	3-196
3-55	Economic Region of Influence for White Sands Test Facility	3-199
3-56	Transportation Features within White Sands Test Facility	3-203
3-57	Dryden Flight Research Center Location Map	3-207
3-58	Buildings on or Eligible for the NRHP, Dryden Flight Research Center	3-211
3-59	Transportation Features Within Dryden Flight Research Center	3-219
3-60	Palmdale, CA, Location Map	3-221
4-1	Contribution of the Space Shuttle Program to Regional Economies in FY 2006	4-5
4-2	NASA FY 2008 Budget Request for Exploration Systems and Space Operations	4-7
4-3	Major SSP Real and Personal Property at KSC	4-10
4-4	Summary of Environmental Consequences for KSC	4-14
4-5	Major SSP Real and Personal Property at JSC	4-19
4-6	Summary of Environmental Consequences for JSC	4-25
4-7	Ellington Field SSP Property	4-29
4-8	Summary of Environmental Consequences for EF	4-30
4-9	Major SSP Property at EPFOL	4-34
4-10	Summary of Environmental Consequences for EPFOL	4-35
4-11	Major SSP Property at SSC	4-39
4-12	Summary of Environmental Consequences for SSC	4-40
4-13	Major Property at MAF	4-45
4-14	Summary of Environmental Consequences for MAF	4-45
4-15	Major SSP Property at MSFC	4-50
4-16	Summary of Environmental Consequences for MSFC	4-53
4-17	Major SSP Property at WSTF	4-57
4-18	Summary of Environmental Consequences for WSTF	4-60
4-19	SSP Property at DFRC	4-64

CONTENTS (CONTINUED)

4-20 Summary of Environmental Consequences for DFRC 4-64
4-21 SSP-related Property at Palmdale 4-68
4-22 Summary of Environmental Consequences for Palmdale 4-69
4-23 Minority and Low-income Population Characteristics and Potential Environmental Impacts 4-75

Executive Summary

ES.1 Introduction

This Programmatic Environmental Assessment (EA) has been prepared by the National Aeronautics and Space Administration (NASA). This Programmatic EA will assist in the decision-making process as required by the National Environmental Policy Act (NEPA) of 1969, as amended (42 United States Code [U.S.C.] 4321 et seq.), Council on Environmental Quality (CEQ) Regulations for implementing the provisions of NEPA (40 *Code of Federal Regulations* [CFR] Parts 1500 through 1508), NASA's policies and procedures at 14 CFR Subpart 1216.3, and Executive Order (EO) 12114, *Environmental Effects Abroad of Major federal Actions*.

This Programmatic EA provides information associated with the potential environmental impacts of the transition and retirement (T&R) of NASA's Space Shuttle Program (SSP). The T&R of the SSP would consist of the disposition of both real property (land, buildings and other structures and their associated built-in systems that cannot readily be moved without changing the essential character of the real property) and personal property (all assets not classified as real property owned by, leased to, or acquired by the government). Property disposition activities are the primary focus of this EA because this is the T&R activity with the greatest potential for environmental impacts. The Programmatic EA approach allows NASA to assess the overall T&R activities, although some specific options are not yet sufficiently developed to assess in detail.

This Executive Summary includes the background, purpose, and need for the Proposed Action; the No Action Alternative; the decisions to be made; the methodology of the EA; and a summary of the environmental impacts. Exhibit ES-1 (at the end of this section) summarizes the environmental impacts of implementing the Proposed Action by resource area.

ES.2 Background

When the United States (U.S.) began the space program in the late 1950s, missions were accomplished using expendable launch vehicles (ELVs). The Saturn vehicles provided the launch capabilities for the manned lunar exploration program (Apollo), and smaller vehicles such as Titan, Atlas, Delta, and Scout were used to launch a variety of automated spacecraft such as communications, weather, and science satellites.

EXHIBIT ES-1
Summary Comparison of Alternatives

Resource Area	Potential Impact of Proposed Action	Potential Impact of No Action Alternative
Kennedy Space Center		
Air Quality	minimal to no impact	minimal to no impact
Biological Resources	minimal impact	minimal impact
Cultural Resources	moderate impact	moderate impact
Hazardous/Toxic Materials and Waste	minimal impact	minimal impact
Health and Safety	minimal impact	minimal impact
Hydrology and Water Quality	minimal impact	minimal impact
Land Use	minimal impact	minimal impact
Noise	minimal impact	minimal impact
Site Infrastructure	minimal impact	minimal impact
Socioeconomics	minimal to no impact	minimal to no impact
Solid Waste	minimal impact	minimal impact
Traffic and Transportation	minimal impact	minimal impact
Johnson Space Center		
Air Quality	minimal to no impact	minimal to no impact
Biological Resources	no impact	no impact
Cultural Resources	moderate impact	moderate impact
Hazardous/Toxic Materials and Waste	minimal impact	minimal impact
Health and Safety	minimal impact	minimal impact
Hydrology and Water Quality	minimal impact	minimal impact
Land Use	minimal impact	minimal impact
Noise	minimal impact	minimal impact
Site Infrastructure	minimal impact	minimal impact
Socioeconomics	minimal to no impact	minimal to no impact
Solid Waste	minimal impact	minimal impact
Traffic and Transportation	minimal impact	minimal impact
Ellington Field		
Air Quality	minimal to no impact	minimal to no impact
Hazardous/Toxic Materials and Waste	minimal to no impact	minimal impact

EXHIBIT ES-1
Summary Comparison of Alternatives

Resource Area	Potential Impact of Proposed Action	Potential Impact of No Action Alternative
Health and Safety	minimal impact	minimal impact
Hydrology and Water Quality	minimal impact	minimal impact
Land Use	minimal impact	minimal impact
Noise	minimal impact	minimal impact
Site Infrastructure	minimal impact	minimal impact
Solid Waste	minimal impact	minimal impact
Traffic and Transportation	minimal impact	minimal impact
El Paso Forward Operating Location		
Air Quality	minimal to no impact	minimal to no impact
Hazardous/Toxic Materials and Waste	minimal to no impact	minimal impact
Health and Safety	minimal impact	minimal impact
Hydrology and Water Quality	minimal impact	minimal impact
Noise	minimal impact	minimal impact
Site Infrastructure	minimal impact	minimal impact
Solid Waste	minimal impact	minimal impact
Traffic and Transportation	minimal impact	minimal impact
Stennis Space Center		
Air Quality	minimal to no impact	minimal to no impact
Biological Resources	minimal impact	minimal impact
Cultural Resources	moderate impact	moderate impact
Hazardous/Toxic Materials and Waste	minimal impact	minimal impact
Health and Safety	minimal impact	minimal impact
Hydrology and Water Quality	minimal impact	minimal impact
Land Use	minimal impact	minimal impact
Noise	minimal impact	minimal impact
Site Infrastructure	minimal impact	minimal impact
Solid Waste	minimal impact	minimal impact
Traffic and Transportation	minimal impact	minimal impact

EXHIBIT ES-1
Summary Comparison of Alternatives

Resource Area	Potential Impact of Proposed Action	Potential Impact of No Action Alternative
Michoud Assembly Facility		
Air Quality	minimal to no impact	minimal to no impact
Biological Resources	minimal impact	minimal impact
Cultural Resources	moderate impact	moderate impact
Hazardous/Toxic Materials and Waste	minimal impact	minimal impact
Health and Safety	minimal impact	minimal impact
Hydrology and Water Quality	minimal impact	minimal impact
Land Use	minimal impact	minimal impact
Noise	minimal impact	minimal impact
Site Infrastructure	minimal impact	minimal impact
Socioeconomics	minimal to no impact	minimal to no impact
Solid Waste	minimal impact	minimal impact
Traffic and Transportation	minimal impact	minimal impact
Marshall Space Flight Center		
Air Quality	minimal to no impact	minimal to no impact
Biological Resources	minimal impact	minimal impact
Cultural Resources	moderate impact	moderate impact
Hazardous/Toxic Materials and Waste	minimal impact	minimal impact
Health and Safety	minimal impact	minimal impact
Hydrology and Water Quality	minimal impact	minimal impact
Land Use	minimal impact	minimal impact
Noise	minimal impact	minimal impact
Site Infrastructure	minimal impact	minimal impact
Socioeconomics	minimal to no impact	minimal to no impact
Solid Waste	minimal impact	minimal impact
Traffic and Transportation	minimal impact	minimal impact
White Sands Test Facility		
Air Quality	minimal to no impact	minimal to no impact
Biological Resources	minimal impact	minimal impact
Cultural Resources	moderate impact	moderate impact

EXHIBIT ES-1
Summary Comparison of Alternatives

Resource Area	Potential Impact of Proposed Action	Potential Impact of No Action Alternative
Hazardous/Toxic Materials and Waste	minimal impact	minimal impact
Health and Safety	minimal impact	minimal impact
Hydrology and Water Quality	minimal impact	minimal impact
Land Use	minimal impact	minimal impact
Noise	minimal impact	minimal impact
Site Infrastructure	minimal impact	minimal impact
Socioeconomics	minimal to no impact	minimal to no impact
Solid Waste	minimal impact	minimal impact
Traffic and Transportation	minimal impact	minimal impact
Dryden Flight Research Center		
Air Quality	minimal to no impact	minimal to no impact
Cultural Resources	moderate impact	moderate impact
Hazardous/Toxic Materials and Waste	minimal impact	minimal impact
Health and Safety	minimal impact	minimal impact
Hydrology and Water Quality	minimal impact	minimal impact
Land Use	minimal impact	minimal impact
Noise	minimal impact	minimal impact
Site Infrastructure	minimal impact	minimal impact
Solid Waste	minimal impact	minimal impact
Traffic and Transportation	minimal impact	minimal impact
Palmdale		
Air Quality	minimal to no impact	minimal to no impact
Cultural Resources	moderate impact	moderate impact
Hazardous/Toxic Materials and Waste	minimal impact	minimal impact
Health and Safety	minimal impact	minimal impact
Noise	minimal impact	minimal impact
Site Infrastructure	minimal impact	minimal impact

EXHIBIT ES-1
Summary Comparison of Alternatives

Resource Area	Potential Impact of Proposed Action	Potential Impact of No Action Alternative
Solid Waste	minimal impact	minimal impact
Traffic and Transportation	minimal impact	minimal impact

Notes:
No Impact–No impacts expected
Minimal–Impacts are not expected to be measurable, or are measurable but are too small to cause any change in the environment
Minor–Impacts that are measurable but are within the capacity of the affected system to absorb the change, or the impacts can be compensated for with little effort and few resources so that the impact is not substantial
Moderate–Impacts that are measurable but are within the capacity of the affected system to absorb the change, or the impacts can be compensated for with effort and resources so that the impact is not substantial
Major–Environmental impacts that, individually or cumulatively, could be substantial

Approved as a National program in 1972, the Shuttle is a unique design because, except for the External Tank (ET), all Shuttle components are reusable. The Shuttle's purpose is to deliver payloads into low Earth orbit and to dock with satellites and the International Space Station (ISS). However, the President and Congress have established new objectives and direction for the Nation's space exploration program. On January 14, 2004, President George W. Bush presented his Vision for U.S. Space Exploration to the nation. The fundamental goal of this vision is to advance U.S. scientific, security, and economic interests through a robust space exploration program. In support of this goal, the U.S. will do the following:

- Implement a sustained and affordable human and robotic program to explore the solar system and beyond.
- Extend the human presence across the solar system, starting with a human return to the moon by the year 2020, in preparation for human exploration of Mars and other destinations.
- Develop the innovative technologies, knowledge, and infrastructures to both explore and support decisions about the destinations for human exploration.
- Promote international and commercial participation in exploration to further U.S. scientific, security, and economic interests (NASA, 2004g).

Congress expressly endorsed the President's space exploration initiative and provided additional direction for the initiative in the NASA Authorization Act of 2005 (Public Law [P.L.] 109-155). Both Congress and the President have directed NASA to develop a "crew exploration vehicle" and associated systems to support the exploration initiative and provide U.S. human spaceflight capability after the retirement of the Shuttle. NASA is in the planning stages of T&R activities for the SSP that efficiently will address the reuse of critical skills, human capital, and property. NASA initiated and is in the early planning stages of the "Constellation

Program," which is intended to develop and operate the human space exploration systems necessary to implement the vision. NASA has evaluated the potential environmental impacts of its proposed Constellation Program and its various components in the *Final Constellation Programmatic Environmental Impact Statement* (Cx PEIS) (2007t).

ES.3 Purpose and Need for the Proposed Action

In announcing the Vision for Space Exploration, the President directed NASA to retire the Space Shuttle by 2010 (NASA, 2004g). Congress expressly endorsed the President's exploration initiative and provided additional direction for the initiative in the NASA Authorization Act of 2005, authorizing NASA to "...establish a program to develop a sustained human presence on the Moon, including a robust precursor program to promote exploration, science, commerce and U.S. preeminence in space, and as a stepping stone to future exploration of Mars and other destinations" (P.L. 109-155).

Under presidential direction, NASA will cease operations of its SSP at all locations, including those addressed in this EA: Kennedy Space Center (KSC), Johnson Space Center (JSC), Ellington Field (EF), El Paso Forward Operating Location (EPFOL), Stennis Space Center (SSC), Michoud Assembly Facility (MAF), Marshall Space Flight Center (MSFC), White Sands Test Facility (WSTF), Dryden Flight Research Center (DFRC), and Palmdale. The retirement of the program necessitates the disposition of all SSP assets.

DFRC is a tenant of Edwards Air Force Base (EAFB) in California. EPFOL is located on El Paso International Airport (EPIA), which is owned and operated by the City of El Paso, Texas, and NASA leases land from the City. Palmdale (also known as Air Force Plant 42 Site 1 [AFP 42]), located at EAFB, is owned by the U.S. Air Force (USAF), leased by NASA, and operated by Boeing Company. The White Sands Missile Range (WSMR) is a U.S. Department of Defense (DoD)-owned facility operated by the Department of the Army (DA) and is located at WSTF. All other facilities are owned and operated by NASA.

All NASA Centers and prime contractor facilities were considered for inclusion in this EA. The criteria used to screen out potential NASA Centers and prime contractor facilities were as follows:

- If SSP activities occur or occurred at the Center.
- If so, the scale and timeframe of the SSP operations that took or take place were considered.

- Centers with limited SSP operations or those that did conduct SSP operations at one time, but are no longer used for SSP support, were eliminated from this evaluation because there is minimal Shuttle-unique property to be disposed.
- Contractor-owned properties were not included because contractors are responsible for the disposition of their own properties. However, government-owned property at contractor sites is included in this EA.

The complete list of NASA Centers and prime contractor facilities considered for this EA is provided in Section 1.2. It was determined that the Sonny Carter Training Facility (SCTF), Ames Research Center, Glenn Research Center, Goddard Space Flight Center, Jet Propulsion Laboratory, Langley Research Center, and Wallops Flight Facility would not be included in this EA because their respective operations support multiple NASA programs and there is minimal Shuttle-unique property to be disposed. However, a few Centers have property that is eligible for listing under the National Register of Historic Places (NRHP) and will be disposed in accordance with applicable laws and regulations.

Santa Susana Field Laboratory (SSFL) is not included in the EA because SSP activities and property usage have been minimal for many years. The infrastructure in place has supported numerous NASA program activities. NASA environmental compliance and restoration activities are ongoing at SSFL and are being conducted by NASA Infrastructure and Administration Office. Consequently, the disposition of assets at SSFL will be addressed outside of the SSP T&R activities. NASA is currently assessing the future needs for SSFL. If NASA decides to excess the property at SSFL, the U.S. General Services Administration (GSA) would be responsible for disposal activities and would prepare the required NEPA documentation.

The prime contractor facilities that were considered for inclusion in this EA included ATK (Promontory, Utah), Boeing (Huntington Beach, California), Lockheed Martin (at MAF), Pratt Whitney Rocketdyne (West Palm Beach, Florida; and Canoga Park, California), and United Space Alliance (USA) (primarily KSC and JSC locations). These facilities were not included (except for MAF's NASA operations) because they are responsible for the disposition of their own properties. However, government-owned property at contractor sites is included in this EA as described in Section 1.2.

The purpose of the proposed action is the disposition of Shuttle assets, including real and personal property, in a manner that fully realizes any remaining value of those assets and that is compliant with applicable federal, state, and local laws and regulations.

ES.4 Proposed Action

Under presidential direction, NASA will cease operations of its SSP in 2010. A number of assets will be dispositioned during the T&R activities. SSP property disposition activities may extend several years beyond 2010.

NASA proposes to implement a centralized process, consisting of a coordinated series of actions, for the disposition of the SSP real and personal property. SSP real and personal property would be evaluated in accordance with NASA Procedural Requirements (NPR) 8800.15, "Real Estate Management Program Implementation Manual," and NPR 4300.1, "NASA Personal Property Disposal Procedural Requirements," to select the best option for disposition.

ES.4.1 Real Property

When the SSP disposes of real property, the responsible NASA Center will evaluate whether the property can be used by another NASA program (reutilization), or it may mothball or destroy the property. If NASA decides to convey the property to another federal, state, local, or private individual, NASA relinquishes the property to the U.S. General Services Administration (GSA). The GSA will convey the property according to federal laws and regulations. The property disposition options that will be evaluated for real property are as follows:

- **Reutilization:** The first option for disposal of government property is reutilization by another NASA program. Property is screened for reutilization by NASA's ongoing programs and for transfer and use by future programs.
- **Utilization:** If the property is not required by other NASA programs, it is made available to other federal agencies. The receiving federal agency would be responsible for the applicable NEPA analysis and documentation resulting from the use of the property.
- **Mothball**: Under this option, NASA would mothball particular SSP real property in place. Under this scenario, NASA would maintain these properties at some low level of support in the event that a Center or new program could use them in the future.
- **Destruction**: Under this option, the property would be demolished or otherwise removed from NASA property to an appropriate location, such as a landfill or hazardous waste treatment, storage, or disposal facility (TSDF).
- **Release to GSA**: If the property is no longer needed by NASA, it may be relinquished to the GSA for conveyance to other federal, state, local, or private individuals.

NASA real property is evaluated for historic significance per the National Historic Preservation Act (NHPA) to assess eligibility for listing in the NRHP. NASA's Historic Preservation Working Group (HPWG) drafted a set of standard criteria for the evaluation of SSP-related properties at all NASA Centers. If the evaluation recommends that the property meets the criteria for historic significance under the NHPA, it is submitted to the State Historic Preservation Officer (SHPO) for comment and concurrence of historic significance. For those properties determined eligible for listing in the NRHP, the undertakings involving the expenditure of federal funds will be submitted to the SHPO for review per the requirements of the NHPA.

ES.4.2 Personal Property

Shuttle-related personal property includes hundreds of thousands of items ranging from common parts, such as nuts and bolts, to complex tooling and flight hardware. The disposition of common parts has no potential for significant impacts to the environment. Consequently, disposition of personal properties such as complex tooling and flight hardware that may have the potential to adversely affect the environment are analyzed in this Programmatic EA. When personal property is no longer required by the SSP, it is disposed according to NASA's established procedures for disposal. The disposal procedure progresses through a series of options, as described below:

- **Reutilization:** The first option for disposal of government property is reutilization by another NASA program. Property is screened for reutilization by NASA's ongoing programs and for use by future programs.
- **Storage:** Under this option, NASA would relocate particular SSP personal property to appropriate storage locations (such as laydown yards or warehouses). At these locations, the property would be maintained at some minimum level of support in the event that a Center or new program could use it in the future. These locations would have an appropriate level of security provided by the location's owner, which would be NASA or some other federal agency. The storage locations could be located onsite or offsite, or be newly constructed areas or buildings. Because it is not currently known whether any new storage areas would be constructed to store SSP property, the information necessary to analyze the potential environmental impacts for constructing such areas does not exist at this time. Therefore, environmental analyses for the construction of new structures for storage of SSP property are deferred until the construction becomes less speculative, and the information necessary for analyses becomes available. Any additional NEPA analyses will be conducted by the responsible Center.
- **Utilization:** If the property is not required by other NASA programs, it is made available to other federal agencies. The receiving federal agency would be responsible for the applicable NEPA analysis and documentation resulting from the use of the property.
- **Donation:** If the property is not required by another federal agency, it is eligible for donation. Under this option, federal excess property can be provided to the state for screening and then to other eligible applicants, including nonprofit

educational and public health activities, nonprofit and public programs (such as museums) for the elderly, educational activities of special interest, public airports, or the homeless.
- **Sales:** Under this option, providing that efforts to reutilize and/or donate have been exhausted, NASA would dispose of the property by means of a competitive bid process such as an auction, sealed bid, or retail sales, in accordance with the guidelines.
- **Destruction:** Under this option, the property would be demolished or otherwise removed from NASA property to an appropriate location, such as a landfill or hazardous waste TSDF.

The evaluation criteria to assess the potential historic significance of personal property and preservation requirements are being developed by NASA. Once completed, these requirements will be applied to SSP personal property to determine what is historically significant.

ES.4.3 Proposed Action Schedule

The SSP is scheduled for retirement in 2010. Under the Proposed Action, once an asset is determined to no longer be needed by the SSP, it would become slated for disposition. Disposition could occur for some assets before SSP retirement in 2010. However, many assets will be needed until the final SSP mission is completed. Furthermore, the evaluation of the potential usefulness of some assets for other NASA programs may not be possible until those programs reach a certain level of maturity. Therefore, so that NASA may best use its SSP assets, final disposition of SSP assets under the Proposed Action may extend several years beyond 2010.

ES.5 No Action Alternative

Under the No Action alternative, NASA would not implement the proposed comprehensive and coordinated effort to disposition SSP property under a structured and centralized SSP process. The disposition of SSP property instead would occur on a Center-by-Center and item-by-item basis in the normal course of NASA's ongoing facility and program management.

ES.6 Decision to be Made

The primary decision to be made by NASA, supported in part by the information contained in this EA, is the manner of disposition of the Shuttle assets.

ES.7 Summary of Environmental Impacts

Twelve environmental areas were evaluated to provide a context for understanding the potential effects of the Proposed Action and a basis for assessing the significance of the potential impacts. These areas include air quality; biological resources; cultural resources; hazardous and toxic materials and waste; health and safety; hydrology and water quality; land use; noise; site infrastructure; socioeconomics; solid waste; and transportation. Lists of the activities necessary to accomplish the Proposed Action and No Action Alternative were developed. Those activities that have the potential to affect the environment were identified and analyzed to evaluate their potential impacts.

This subsection summarizes the conclusions of the analyses made for each of the environmental areas based on the application of the described methodology. Only those activities for which a potential environmental concern was determined at each location are described. Exhibit ES-1 summarizes this information. The impacts were evaluated as follows:

- No Impact–No impacts expected
- Minimal–Impacts are not expected to be measurable, or are measurable but are too small to cause any change in the environment
- Minor–Impacts are measurable but are within the capacity of the affected system to absorb the change, or the impacts can be compensated for with little effort and few resources so that the impact is not substantial
- Moderate–Impacts are measurable but are within the capacity of the affected system to absorb the change, or the impacts can be compensated for with effort and resources so that the impact is not substantial
- Major–Environmental impacts that, individually or cumulatively, could be substantial

ES.7.1 National Perspective on Socioeconomic Impacts

This Programmatic EA evaluates NASA's decision about how to disposition the SSP's real and personal property assets; therefore, the socioeconomic impact analysis addresses only the impacts of NASA's discretionary actions regarding disposition of the SSP's real and personal property. It does not address the broader socioeconomic impacts of the President's decision to discontinue the SSP, because the Presidential decision to discontinue the SSP has already been made and is not subject to NEPA analysis..

Nevertheless, to provide context for this EA's limited socioeconomic analysis, the EA provides information about the current and projected socioeconomic influence of the SSP and other NASA programs.

The President's Fiscal Year (FY) 2008 budget request for NASA shows a steadily increasing investment in exploration systems and space operations (the portion of the budget that covers the SSP, ISS, Constellation Programs, and other ongoing activities) over the budget period of FY 2006 through FY 2012. As the SSP transitions and retires, the Constellation Program will increase the pace of development and testing of the nation's new space vehicles, leading to an initial operating capability by 2015. Even with the new programs, there will be an approximate 4-year gap between the termination of the SSP and the operation of the new vehicles, during which employment and expenditures would be affected.

NASA will continue to invest in other space operations at existing Centers and will distribute the new work across NASA's existing Centers, aligning the work to be performed with the capabilities of the individual NASA Centers. New NASA programs and projects will help fill the void left by the SSP T&R activities; however, localities that host NASA Centers that are heavily involved in the SSP would experience adverse socioeconomic impacts.

The disposition of SPP assets would have little to no discernible effects on socioeconomics, in comparison to the potentially considerable, although temporary, changes in employment (especially at Centers such as KSC, JSC, and MAF) that could result from the Presidential decision to close down the SSP. As recognized in the Final Cx PEIS (NASA, 2007t), a detailed analysis of changes in employment and expenditures at each Center is precluded by the fact that the Constellation Program is at an early stage of development and would be subject to adjustments and changes as requirements become better defined.

NASA recognizes that a skilled NASA and contractor work force is an essential ingredient to successful implementation of the Constellation Program and that there will be challenges for retaining skilled personnel. NASA is examining a variety of personnel initiatives to effect a smooth transition to Constellation operations and is committed to preserving the critical and unique capabilities provided by each NASA Center.

ES.7.2 No Action Alternative

Under the No Action Alternative, NASA would not implement the proposed comprehensive and coordinated effort to disposition SSP property under a structured and centralized SSP process. Instead, the disposition of SSP property would occur on a Center-by-Center and item-by-item basis in the normal course of NASA's ongoing facility and program management.

Consequently, the environmental impact would be expected to be similar to that of the Proposed Action Alternative, which is described below. However, if a centralized process were not used to disposition assets (i.e., proposed action), the property disposal process could become overwhelmed with the volume of property to disposition. The volume of property to be processed could result in schedule and cost impacts if a structured disposal process were not implemented. Also, artifacts may not be properly identified and made available to museums for display. In addition, the amount of solid and hazardous waste that would require disposal could exceed landfill and less than 90-day hazardous waste storage yard capacities at some Centers.

ES.7.3 Proposed Action Alternative

ES.7.3.1 Kennedy Space Center

The specific disposition methods selected for SSP real and personal property are likely to have minimal to no or minimal discernible effects on air quality; biological resources; hazardous and toxic materials and waste; health and safety; hydrology and water quality; land use; noise; site infrastructure; socioeconomics; solid waste; and transportation. Moderate impacts to cultural resources could occur if the disposition of real property would require the demolition of an NRHP-listed or eligible building. This would be true even assuming the required consultation with the SHPO.

ES.7.3.2 Johnson Space Center

The specific disposition methods selected for SSP real and personal property are likely to have no discernible effects on biological resources and minimal to no or minimal discernible effects on air quality; hazardous and toxic materials and waste; health and safety; hydrology and water quality; land use; noise; site infrastructure; socioeconomics; solid waste; and transportation. Moderate impacts to cultural resources could occur if the disposition of real or personal property would require the demolition of an NRHP-listed or eligible building. This would be true even assuming the required consultation with the SHPO.

ES.7.3.3 Ellington Field

The specific disposition methods selected for SSP real and personal property are likely to have minimal to no or minimal discernible effects on air quality; hazardous and toxic materials and waste; health and safety; hydrology and water quality; land use; noise; site infrastructure; solid waste; and transportation.

ES.7.3.4 El Paso Forward Operating Location

The specific disposition methods selected for SSP real and personal property are likely to have minimal to no or minimal discernible effects on air quality; hazardous and toxic materials and waste; health and safety; hydrology and water quality; noise; site infrastructure; solid waste; and transportation.

ES.7.3.5 Stennis Space Center

The specific disposition methods selected for SSP real and personal property are likely to have minimal to no or minimal discernible effects on air quality; biological resources; hazardous and toxic materials and waste; health and safety; hydrology and water quality; land use; noise; site infrastructure; solid waste; and transportation. Moderate impacts to cultural resources could occur if the disposition of real or personal property would require the demolition of an NRHP-listed or eligible building. This would be true even assuming the required consultation with the SHPO.

ES.7.3.6 Michoud Assembly Facility

The specific disposition methods selected for SSP real and personal property are likely to have minimal to no or minimal discernible effects on air quality; biological resources; hazardous and toxic materials and waste; health and safety; hydrology and water quality; land use; noise; site infrastructure; socioeconomics; solid waste; and transportation. Moderate impacts to cultural resources could occur if the disposition of real or personal property would require the demolition of an NRHP-listed or eligible building. This would be true even assuming the required consultation with the SHPO.

ES.7.3.7 Marshall Space Flight Center

The specific disposition methods selected for SSP real and personal property are likely to have minimal to no or minimal discernible effects on air quality; biological resources; hazardous and toxic materials and waste; health and safety; hydrology and water quality; land use; noise; site infrastructure; socioeconomics; solid waste; and transportation. Moderate impacts to cultural resources could occur if the disposition of real or personal property would require the demolition of an NRHP-listed or eligible building. This would be true even assuming the required consultation with the SHPO.

ES.7.3.8 White Sands Test Facility

The specific disposition methods selected for SSP real and personal property are likely to have minimal to no or minimal discernible effects on air quality; biological resources; hazardous and toxic materials and waste; health and safety; hydrology and water quality; land use; noise; site infrastructure; socioeconomics; solid waste; and transportation. Moderate impacts to cultural resources could occur if the disposition of real or personal property would require the demolition of an NRHP-listed or eligible building. This would be true even assuming the required consultation with the SHPO.

ES.7.3.9 Dryden Flight Research Center

The specific disposition methods selected for SSP real and personal property are likely to have minimal to no or minimal discernible effects on air quality; hazardous and toxic materials and waste; health and safety; hydrology and water quality; land use; noise; site infrastructure; solid waste; and traffic and transportation. Moderate impacts to cultural resources could occur if the disposition of real or personal property would require the demolition of an NRHP-listed or eligible building. This would be true even assuming the required consultation with the SHPO.

ES.7.3.10 Palmdale

The specific disposition methods selected for SSP real and personal property are likely to have minimal to no or minimal discernible effects on air quality; hazardous and toxic materials and waste; health and safety; noise; site infrastructure; solid waste; and transportation. Moderate impacts to cultural resources could occur if the disposition of real or personal property would require the demolition of an NRHP-listed or eligible building. This would be true even assuming the required consultation with the SHPO.

ES.8   Public and Agency Involvement

The Notice of Availability of the Programmatic EA was announced in the *Federal Register* (FR) on 25 or 26 February 2008. Comments on the Programmatic EA were solicited through notices of availability published in newspapers in Alabama, California, Florida, Louisiana, Mississippi, New Mexico, Texas, and Washington, D.C., as well as in the FR. Public comments were encouraged by offering a variety of means by which to submit comments, including written comments sent through the postal system, electronic mail, and facsimile.

1. Purpose and Need for the Proposed Action

This Programmatic Environmental Assessment (EA) has been prepared by the National Aeronautics and Space Administration (NASA) to assist in the decision-making process as required by the National Environmental Policy Act (NEPA) of 1969, as amended (42 United States Code [U.S.C.] 4321 et seq.), Council on Environmental Quality (CEQ) regulations. [Note: A list of acronyms and abbreviations, and a metric and English conversion table, are provided in Appendix A.] This Programmatic EA implements the provisions of NEPA (40 *Code of Federal Regulations* [CFR] Parts 1500 to 1508), Executive Order (EO) 12114 ("Environmental Effects Abroad of Major Federal Actions"), and NASA policies and procedures at 14 CFR Subpart 1216.3.

This Programmatic EA provides information associated with the potential environmental impacts of the transition and retirement (T&R) of NASA's Space Shuttle Program (SSP). The T&R of the SSP would consist of the disposition of both real property (land, buildings and other structures and their associated built-in systems that cannot readily be moved without changing the essential character of the real property) and personal property (all assets not classified as real property owned by, leased to, or acquired by the government). Property disposition activities are the primary focus of this EA because this is the T&R activity with the greatest potential for environmental impacts. The Programmatic EA approach allows NASA to assess the overall T&R activities, although some specific options are not yet sufficiently developed to assess in detail.

1.1 Background

The SSP T&R includes both the transition of SSP important assets to new and current NASA Programs and the cost-effective retirement of assets and capabilities that will not be needed when the SSP retires. The capabilities held by the SSP include human capital, real property, and personal property.

1.1.1 Previous U.S. Human Space Exploration Programs

Beginning in the late 1950s, the United States (U.S.) embarked upon the ongoing effort of human exploration of space. The first human spaceflight initiative was Project Mercury, established in October 1958, with crewed spacecraft first launched from the Cape Canaveral Air Force Station (CCAFS) in the early 1960s. NASA's Launch Operations Center and the portions of CCAFS that were used by NASA were renamed the John F. Kennedy Space Center (KSC) in late 1963. Project Mercury was followed by Project Gemini and the Apollo Program. Project Gemini was announced in January 1962 and served to perfect maneuvers in Earth orbit. The

Apollo Program, initiated in 1961, successfully landed U.S. astronauts on the Moon and returned them safely to Earth.

1.1.2 Space Shuttle

Approved as a National program in 1972, the Space Transportation System (STS)—commonly known as the Space Shuttle—is a unique design because, except for the External Tank (ET), all parts are reusable. The Space Shuttle's purpose is to deliver payloads into lower Earth orbit and to dock with satellites and the International Space Station (ISS). Designed solely for missions to Earth orbit, the Space Shuttle was the first and is still the only winged U.S. spacecraft capable of launching crew vertically into orbit and landing horizontally upon returning to Earth. Over the past 25 years, the Space Shuttle fleet has supported more than 100 missions to Earth orbit.

1.1.3 The Vision for Space Exploration

On January 14, 2004, President George W. Bush presented his Vision for U.S. Space Exploration to the nation. The fundamental goal of this Vision is to advance U.S. scientific, security, and economic interests through a robust space exploration program. In support of this goal, the following steps will be taken:

- Implement a sustained and affordable human and robotic program to explore the solar system and beyond.
- Extend human presence across the solar system, starting with a human return to the moon by the year 2020, in preparation for human exploration of Mars and other destinations.
- Develop the innovative technologies, knowledge, and infrastructures to both explore and support decisions about the destinations for human exploration.
- Promote international and commercial participation in exploration to further the U.S. scientific, security, and economic interests (NASA, 2004f).

In announcing the Vision for Space Exploration, the President directed NASA to use the Space Shuttle to fulfill its obligation to complete assembly of the ISS and then to retire the Shuttle in 2010. Congress expressly endorsed the President's space exploration initiative and provided additional direction for the initiative in the NASA Authorization Act of 2005 (Public Law [P.L.] 109-155). Both Congress and the President have directed NASA to develop a "crew exploration vehicle" and associated systems to support the exploration initiative and to provide U.S. human spaceflight capability after the retirement of the Shuttle. NASA is in the planning stages of T&R activities for the SSP that will efficiently address the reuse of critical skills, human capital, and property. NASA initiated and is in the early planning stages of the "Constellation Program," which is intended to develop and operate the human space exploration systems necessary to implement the vision. NASA has evaluated the potential environmental impacts of its proposed Constellation Program and its various components under a separate *Final Constellation*

Programmatic Environmental Impact Statement (Cx PEIS) and tiered NEPA documentation, as appropriate (NASA, 2007t).

1.1.4 NASA 2008 Budget Request

Implementing the President's Vision requires the retirement of the Space Shuttle in 2010, while bringing new human spaceflight capabilities online shortly thereafter. NASA's Fiscal Year (FY) 2008 budget request reflects these two goals. Exhibit 1-1 is a timeline for the U.S. human exploration of space.

EXHIBIT 1-1
Timeline of the United States' Human Exploration of Space

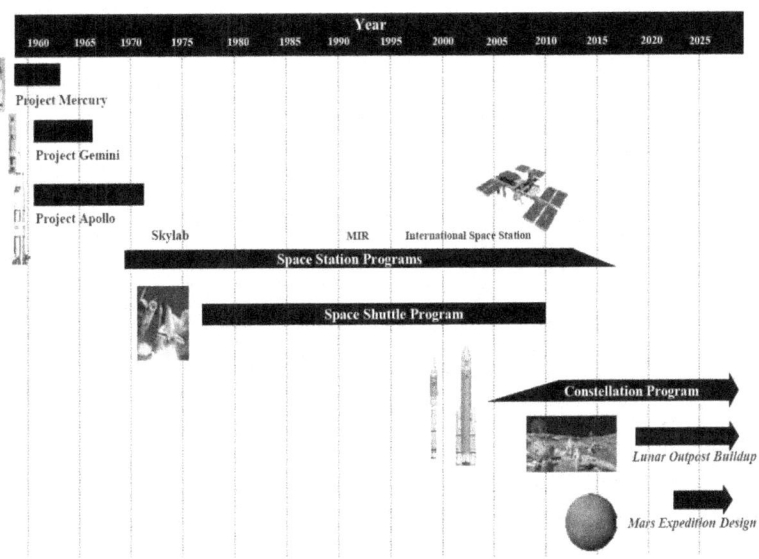

Over the budget period covering FY 2006 through FY 2012, as SSP annual budgets decrease, investment in other areas of NASA's Exploration Systems and Space Operations will increase steadily. This portion of NASA's budget covers the SSP, ISS, and Constellation Programs, as well as the ongoing activities supporting human space flight and advanced capabilities development (see Section 4.1.2 for more information). As the SSP T&R is carried out, the Constellation Program will increase the pace of development and testing of the nation's new space vehicles (NASA, 2007t).

The Constellation Program consists of new spacecraft, launchers, and associated hardware that would facilitate manned and unmanned missions. The new crew transportation system includes three elements: the Orion Crew and Service Modules, the Lunar Lander, and the Earth Departure Stage (EDS). The rockets to be used for launching the different components consist of the Ares V (for the EDS and either the Lunar Lander or cargo), and the Ares I for the Orion spacecraft. Several elements of the Constellation Program's hardware are derived from those originally developed for the SSP. The Orion Spacecraft is influenced by the Apollo spacecrafts, consisting of a two-part crew and service module system (NASA, 2007t).

The Cx PEIS (NASA, 2007t) provides additional information about that proposed program.

1.1.5 Planning for SSP Transition and Retirement

The goals and objectives of the SSP T&R were developed to implement the President's directive to retire the Shuttle in 2010 in a manner that also provides optimum support for all aspects of the Vision for space exploration. Specifically, the SSP T&R goals are as follows:

- Take no action that will impede the ability to safely and effectively complete the fly-out of the Shuttle Program.
- Perform T&R cost-effectively and as soon as possible.
- Provide an interface to other programs and institutional elements for capability transition.

The organizational structure begins at NASA Headquarters (HQ) with the Associate Administrator (AA) and the Space Operations Mission Directorate (SOMD) Transition Manager. At the program level, an SSP Transition Manager is assigned responsibility for the SSP T&R activities.

To accomplish the T&R functions, processes and tools have been developed to assess the capabilities of the SSP; to develop plans to retain, transfer, or excess these capabilities; and then to implement those plans.

1.1.5.1 Strategic Capabilities

The SSP is identifying strategic capabilities across the Program, which will allow decisions to be made relative to a capability–the human capital, real property, and personal property.

1.1.5.2 Human Capital Management

NASA's Number 1 priority is safe and successful mission execution through Space Shuttle fly-out and retirement no later than 2010. At the same time, the agency must plan for the smooth transition of much of the same workforce to other exploration programs during the timeframe between SSP retirement and the beginning of future space flight programs.

1.1.5.3 Property Management

The primary objective of SSP property management during the T&R is to maintain Program integrity while simultaneously implementing the divestiture of Program property no longer needed to meet the Program mission requirements. Prompt disposition of SSP property will make valuable assets available for follow-on programs and will minimize agency costs for storage and sustainment. In 2007, the SSP identified more than 900,000 line items that must be dispositioned.

1.1.5.4 Historic Properties

The SSP strives to identify historic properties and artifacts as early as possible in the T&R process to ensure that adequate time is available to resolve technical and funding issues and to minimize implementation delays. Historic preservation is an integral part of property management.

1.1.5.5 Environmental Management

The environmental objectives of the SSP T&R include the following:

- To enable mission success by managing environmental responsibilities, identifying and mitigating environmental risks, providing adequate resources and technical support, and working with the mission stakeholders.
- To comply with all applicable federal, state, and local laws and regulations, as well as all applicable NASA requirements.
- To honor all agreements with other agencies, industries, organizations, and entities that are relevant to NASA's ongoing environmental responsibilities.
- To include environmental considerations in the program and project management processes with emphasis on prevention, conservation, compliance, and restoration.

1.2 Need for the Proposed Action

To accomplish the Vision for U.S. Space Exploration, one of the steps mandated by the President is to retire the Space Shuttle in 2010 (NASA, 2007f). Under presidential direction, NASA will cease operations of its SSP activities at all locations, including those addressed in this EA:

- KSC
- Johnson Space Center (JSC)
- Ellington Field (EF)
- El Paso Forward Operating Location (EPFOL)
- Stennis Space Center (SSC)
- Michoud Assembly Facility (MAF)
- Marshall Space Flight Center (MSFC)
- White Sands Test Facility (WSTF)
- Dryden Flight Research Center (DFRC)
- Palmdale

The T&R of the Program necessitates the disposition of all SSP assets (NASA, 2004g).

DFRC is a tenant of Edwards Air Force Base (EAFB). EPFOL is located on El Paso International Airport (EPIA), which is owned and operated by the City of El Paso, and NASA leases land from the City. Palmdale (also known as Air Force Plant 42 Site 1 [AFP 42]), is located at EAFB, California. Palmdale is owned by the U.S. Air Force (USAF), leased by NASA, and operated by Boeing Company. The White Sands Missile Range (WSMR) is a U.S. Department of Defense (DoD)-owned facility operated by the Department of the Army (DA), located at WSTF. All other facilities are owned and operated by NASA.

The following NASA Centers and prime contractor facilities were considered for inclusion in this EA:

- Ames Research Center
- ATK Launch Systems (ATK) (Promontory, Utah)
- Boeing (Huntington Beach, California)
- DFRC
- EF
- EPFOL
- Glenn Research Center
- Goddard Space Flight Center (GSFC)
- Jet Propulsion Laboratory
- JSC
- KSC
- Langley Research Center
- Lockheed Martin (at MAF)
- MSFC
- MAF
- Palmdale (AFP 42, operated by Boeing)
- Pratt Whitney Rocketdyne (West Palm Beach, Florida; and Canoga Park, California)
- Santa Susana Field Laboratory (SSFL)
- Sonny Carter Training Facility (SCTF)
- SSC
- United Space Alliance (USA) (primarily KSC and JSC locations)
- Wallops Flight Facility
- WSTF

1. PURPOSE AND NEED FOR THE PROPOSED ACTION

A screening process was used to eliminate sites from the analysis based on the following criteria:

- If SSP activities occur or occurred at the Center
- If so, the scale and timeframe of the SSP operations that took or take place were considered
- Centers with limited SSP operations or those that did conduct SSP operations at one time but are no longer used for SSP support were eliminated from this evaluation because there is limited to no SSP property disposal.
- Contractor-owned properties were not included because they are responsible for the disposition of their own properties.

It was determined that SCTF would not be included in this EA because the operations there support multiple NASA programs and there is minimal SSP-unique property to be disposed.

SSFL is not included in the EA because SSP activities and property usage have been minimal for many years. The infrastructure in place has supported numerous NASA program activities. NASA environmental compliance and restoration activities are ongoing and being conducted by NASA Infrastructure and Administration Office. Consequently, the disposition of assets at SSFL will be addressed outside of the SSP T&R activities. NASA currently is assessing the future needs for SSFL. If NASA decides to excess the property at SSFL, the U.S. General Services Administration (GSA) would be responsible for disposal activities and would prepare the required NEPA documentation. Four other NASA facilities also are not included in this EA because of their limited involvement in the SSP. However, some of these Centers have property that is eligible for listing on the National Register of Historic Places (NRHP). The Ames Research Center has two resources, Buildings N-238 and N-243, that were found eligible for the NRHP for their support of the SSP. These resources, which provided limited support to the SSP, retain their historic integrity. At the Glenn Facilities, the Supersonic Wind Tunnel and the Abe Silverstein Supersonic Wind Tunnel meet the NHRP Criteria A, B, C and exhibit excellent integrity. Wallops Flight Facility is a component of GSFC; it does not have any dedicated Shuttle assets. One structure at Langley Research Center, the Aircraft Landing Dynamics, meets the NRHP criteria for eligibility of the SSP.

Rocketdyne's operations at Canoga Park include the use of the government-owned Pacific Scientific Furnace, which is considered eligible for listing in the NHRP for this association with the SSP (Archaeological Consultants, Inc., 2007a). Every Space Shuttle Main Engine (SSME) flown on the Shuttle was brazed in this furnace. The contractor-owned sites manage the environmental requirements related to their facilities, but coordinate with government property officers to dispose of government-owned property that is operated by the contractor.

1.3 Purpose of the Proposed Action

The purpose of the proposed action is to methodically assess the SSP assets and to provide for their disposition in a manner that fully realizes any remaining value of those assets, and that is compliant with applicable federal, state, and local laws and regulations.

1.3.1 Decisions to be Made

The primary decision to be made by NASA, supported in part by the information contained in this Programmatic EA, is the manner of disposition of the SSP assets.

1.3.2 Public Involvement

The Notice of Availability of the Programmatic EA was announced in the *Federal Register* (FR) on 25 or 26 February 2008. Comments on the Programmatic EA were solicited through notices of availability published in newspapers in Alabama, California, Florida, Louisiana, Mississippi, New Mexico, Texas, and Washington, D.C., as well as in the FR. Appendix B provides a complete list of where these advertisements were published. Public comments were encouraged by offering a variety of means by which to submit comments, including written comments sent through the postal system, electronic mail, and facsimile. NASA received comments from the public as well as Federal and State Agencies. The comments received and the corresponding responses are provided in Appendix B-1.

1.3.3 Issues Considered but Not Carried Forward

NASA applied a systematic and interdisciplinary approach to ensure that the environmental resources at each site were analyzed and potential issues identified for the disposition of Shuttle-related real and personal property. The analyses for the disposition of real property are presented in this Programmatic EA.

Shuttle-related personal property includes hundreds of thousands of items ranging from common parts to complex tooling and flight hardware. The disposition of common parts has no potential for significant impacts to the environment. Consequently, personal properties such as complex tooling and flight hardware may have the potential to adversely affect the environment are analyzed in this Programmatic EA.

Exhibit 1-2 identifies the concerns at each Center that were evaluated and subsequently determined to have no potential for environmental impacts; thus, they were eliminated from further discussion in this document.

1.4 Executive Order 12114

EO 12114 represents the U.S. government's exclusive and complete determination of the procedural and other actions to be taken by federal agencies to further the purpose of the NEPA, with respect to the environment outside the U.S. and its territories and possessions. Although it is based on independent authority, this EO furthers the purpose of NEPA consistent with the foreign policy and national security policy of the U.S. Specifically, EO 12114 defines the environment to mean only the natural and physical environment, but not the social, economic, or other environments.

NASA has various Transoceanic Abort Landing (TAL) sites and Emergency Landing Sites (ELSs) that could be used in an emergency during the Space Shuttle's ascent into orbit. The TAL sites are located in Eastern Europe at Moron Air Force Base (AFB); Spain, Zaragoza AFB, Spain; and Istres-le-Tube AFB, France. The primary role of the personnel at the TAL sites is to remove the astronauts from the Orbiter in the event of an emergency landing. Therefore, the TAL sites are equipped with Shuttle-specific navigational aides, Orbiter grounding equipment, safety equipment, hatch tools, and a crew access vehicle to remove the astronauts from the Orbiter. NASA has a Memorandum of Agreement (MOA) with the respective TAL sites to use these facilities during a launch and contingency landing.

Because of the MOA between NASA and the governments of France and Spain, of the four categories of major federal actions abroad addressed under Section 2-3 of EO 12114, only (c), "Actions significantly affecting the environment of a foreign nation," potentially could apply. However, this category does not apply because the buildings at the TAL sites are not NASA real property and because there would not be any SSP T&R-related activities that potentially could involve radioactive materials. Consequently, neither the Proposed Action nor the No Action Alternative would have potential actions for which EO 12114 would be applicable. Therefore, no further evaluation of the TAL sites under EO 12114 is required.

EXHIBIT 1-2
Issues Considered but Eliminated from Further Analysis

Resource Eliminated from Further Analysis	Rationale
Kennedy Space Center	
Biological Resources – Wetlands	No wetlands will be affected by the disposition of SSP property (NASA, 2003a).
Biological Resources – Floodplains	No floodplains will be affected by the disposition of SSP property, because there are no SSP buildings located in floodplains, according to the KSC 100-year floodplain map.
Cultural Resources – Traditional Cultural Resources	There are no known traditional cultural resources or ethnographic sites at KSC (NASA, 2003a). If any traditional cultural resources are found in the future, KSC must follow all applicable federal regulations.
Cultural Resources – Archaeological Resources	Currently, none of the real property assets owned by the SSP are known to be over archeological sites. Therefore, there would be no impact on known archaeological sites (NASA, 2003a).
Site Infrastructure – Potable Water	Water is supplied to KSC by the City of Cocoa, the Taylor Creek Reservoir, and groundwater wells located in east Orange County. KSC does not provide its own potable water (NASA, 2007u).
Site Infrastructure – Electrical Power	No change to electrical power is anticipated.
Johnson Space Center	
Biological Resources – Wildlife	JSC does not provide high-quality habitat for wildlife because of the high levels of human activity. The small amount of cover and food available, NASA activities, traffic, and a 2.5-m- (8-foot)-high perimeter fence discourage wildlife from inhabiting JSC; therefore, no impacts to wildlife are anticipated as a result of the disposition of SSP real and personal property (NASA, 2004a).
Biological Resources – Protected Species and Habitats	No federal- or state-listed threatened or endangered species are known to inhabit JSC. No critical habitat for protected species exists at JSC (NASA, 2004a).
Cultural Resources – Traditional Cultural Resources	There are no known traditional cultural resources or ethnographic sites at JSC (2004a). If any traditional cultural resources are found in the future, JSC must follow all applicable federal regulations.
Ellington Field	
Biological Resources – Vegetation	No natural plant communities exist at EF because the land at EF is completely developed due to airport operations (NASA, 2005b).
Biological Resource – Wetlands	No wetlands exist at EF (NASA, 2005b).
Biological Resources – Floodplains	No floodplains exist at EF (NASA, 2005b).

EXHIBIT 1-2
Issues Considered but Eliminated from Further Analysis

Resource Eliminated from Further Analysis	Rationale
Biological Resources – Wildlife	EF is located at an airport on completely developed land. Only wildlife associated with human development may be found onsite, including rock dove (*Columba livia*), starling (*Sturnus vulgaris*), sparrows, mockingbird (*Mimus polyglottos*), cardinal (*Cardinalis cardinalis*), and blue jay (*Cyanocitta cristata*). Small mammals such as raccoons (*Procyon lotor*), opossums (*Didelphis virginiana*), and rodents also are found at the airport. A fence at the airport perimeter excludes large wildlife (NASA, 2005b).
Biological Resources – Protected Species and Habitats	No threatened or endangered species exist at EF (NASA, 2005b).
Cultural Resources – Archaeological Resources	No soil disturbance is anticipated to occur due to SSP T&R activities because there are no planned demolition and construction activities (NASA, 2007s).
Cultural Resources – Traditional Cultural Resources	There are no known traditional cultural resources or ethnographic sites at EF (NASA, 2005b). If any traditional cultural resources are found in the future, EF must follow all applicable federal regulations.
Cultural Resources – Historic Resources	There are no known NRHP-eligible historic resources at EF (NASA, 2005b).
Socioeconomics – Population	EF is a satellite facility supporting JSC, located in Houston, only 8 miles northwest of JSC; both facilities are located in Harris County (NASA, 2005b). Therefore, socioeconomic activity associated with EF occurs in the same ROI as JSC, the Houston metropolitan area. NASA expenditures and employment data for EF are included in JSC data. The socioeconomic factors associated with EF are included in the JSC socioeconomic section.
Socioeconomics – Regional Employment and Economic Activity	
Socioeconomics – Community Services	
El Paso Forward Operation Location	
Biological Resources – Vegetation	No natural plant communities exist at EPFOL because the land at EPFOL is completely developed due to airport operations (NASA, 2004c).
Biological Resources – Wetlands	No wetlands exist at EPFOL (NASA, 2004c).
Biological Resources – Floodplains	A 100-year floodplain is located in the northwestern portion of the EPIA. NASA facilities are not within the floodplain and the proposed action and alternatives would not affect this area (NASA, 2004c).
Biological Resources – Wildlife	EPFOL is located at an airport and does not provide quality habitat to wildlife. Only wildlife associated with human development may be found onsite (NASA, 2004c).
Biological Resources – Protected Species and Habitats	Transient protected bird species may occur at areas near the EPFOL, including the bald eagle and arctic peregrine falcon, but these species range widely in the region and are not affected by NASA operations. USFWS consultation indicated that a species of concern, the western burrowing owl, was found in the vicinity of EPIA, but not on the site, due to airport operations (NASA, 2004c).

EXHIBIT 1-2
Issues Considered but Eliminated from Further Analysis

Resource Eliminated from Further Analysis	Rationale
Cultural Resources – Archaeological Resources	There are no known NRHP-eligible archaeological resources at EPFOL (NASA, 2007s). If any archeological resources are found in the future, EPFOL must follow all applicable federal regulations.
Cultural Resources – Traditional Cultural Resources	There are no known NRHP-eligible traditional cultural resources or ethnographic sites at EPFOL (NASA, 2004c). If any traditional cultural resources are found in the future, EPFOL must follow all applicable federal regulations.
Cultural Resources – Historic Resources	There are no known NRHP-eligible historic resources at EPFOL (NASA, 2007s).
Hazardous/Toxic Materials and Waste-contaminated Areas	No RCRA-contaminated sites are located at EPFOL (NASA, 2007s).
Hydrology and Water Quality – Water Quality	There are no jurisdictional surface waters at EPFOL (NASA, 2004c).
Land Use	Land use planning at EPFOL is performed by the Planning Office of the Center Operations Directorate of JSC (NASA, 2004c). EPFOL does not control any property. Real property occupied by EPFOL is leased from EPIA.
Socioeconomics – Population	EPFOL is a satellite facility supporting JSC, located at the EPIA, with only a small workforce. In 2004, the EPFOL employed fewer than 30 NASA and contractor personnel (NASA, 2004c). EPFOL is located within the socioeconomic ROI for WSTF; information about the regional economy is provided in the WSTF socioeconomics section. Effects on the population and the regional economy associated with SSP support activities at EPFOL would be minimal or undetectable, especially in comparison to ongoing economic activity associated with EPIA, WSTF, WSMR, Holoman AFB, and Fort Bliss.
Socioeconomics – Regional Employment and Economic Activity	
Socioeconomics – Community Services	
Stennis Space Center	
Cultural Resources – Traditional Cultural Resources	There are no known traditional cultural resources or ethnographic sites at SSC (NASA, 2005a). If any traditional cultural resources were to be found in the future, SSC would have to follow all applicable federal regulations.
Socioeconomics – Population	The current SSP workforce at SSC represents only approximately 5 percent of the total NASA and non-NASA workforce at SSC (NASA, 2007a). The effects on regional population resulting from SSP economic contributions would be minimal or undetectable, in comparison to all of the other workers and their families associated with SSC. Information about the population of the surrounding region is included in the MAF socioeconomics section.
Socioeconomics – Regional Employment and Economic Activity	For the reasons stated above, regional economic contributions from the SSP at SSC alone are unlikely. Therefore, a detailed analysis of SSC is not necessary. However, because SSC is located within the ROI for MAF, and because that region is still in recovery from the 2005 hurricanes, the combined economic contribution of the SSP at both centers is addressed in the MAF section.

EXHIBIT 1-2
Issues Considered but Eliminated from Further Analysis

Resource Eliminated from Further Analysis	Rationale
Socioeconomics – Community Services	For the reasons stated above, any population-driven effects from the SSP transition on the demand for community services in the communities close to SSC would be minimal or non-existent. Therefore, details about these resources are not required.
Michoud Assembly Facility	
Cultural Resources – Traditional Cultural Resources	There are no known traditional cultural resources or ethnographic sites at MAF (NASA, 2001b). If any traditional cultural resources were to be found in the future, MAF would have to follow all applicable federal regulations.
Marshall Space Flight Center	
Cultural Resources – Traditional Cultural Resources	There are no known traditional cultural resources or ethnographic sites at MSFC (NASA, 2002a). If any traditional cultural resources were to be found in the future, MSFC would have to follow all applicable federal regulations.
White Sands Test Facility	
Biological Resources – Wetlands	No wetlands exist at WSTF (NASA, 2001a).
Biological Resources – Floodplains	No floodplains exist at WSTF (NASA, 2001a).
Cultural Resources – Traditional Cultural Resources	There are no known traditional cultural resources or ethnographic sites at WSTF (NASA, 2001a). If any traditional cultural resources were to be found in the future, WSTF would have to follow all applicable federal regulations.
Hydrology and Water Quality – Groundwater	There are no jurisdictional surface waters at WSTF (NASA, 2001a).
Dryden Flight Research Center	
Biological Resources – Vegetation	There are no biological resources at the Shuttle area (NASA, 2003c).
Biological Resources – Floodplains	Development of floodplains on EAFB has been limited because there are no major stream courses and few courses that are large enough to have developed valleys with floodplains. Floodplains on DFRC are limited to a small portion of the Rogers Dry Lakebed, which is the regional drainage basin (NASA, 2003c). No facilities on DFRC are located in floodplains.
Biological Resources – Wetlands	No wetlands exist at DFRC (NASA, 2003c).
Biological Resources – Wildlife	There are no biological resources at the Shuttle area (NASA, 2003c).
Biological Resources – Protected Species	There are no biological resources at the Shuttle area (NASA, 2003c).
Cultural Resources – Traditional Cultural Resources	There are no known traditional cultural resources or ethnographic sites at DFRC (NASA, 2003c). If any traditional cultural resources were to be found in the future, DFRC would have to follow all applicable federal regulations.

1. PURPOSE AND NEED FOR THE PROPOSED ACTION

EXHIBIT 1-2
Issues Considered but Eliminated from Further Analysis

Resource Eliminated from Further Analysis	Rationale
Hydrology and Water Quality – Groundwater	There are no jurisdictional surface waters at DFRC (NASA, 2003c).
Socioeconomics – Population	DFRC has a small SSP-direct workforce of about 25 workers, primarily contractors, located in leased space at EAFB in California (NASA, 2003c). Effects on the population and the regional economy associated with the SSP support activities at DFRC would be minimal or undetectable in comparison to the ongoing economic activity associated with EAFB. In addition, SSP is only a small portion of overall funding at DFRC (like other NASA research laboratories), so the SSP transition is unlikely to affect DFRC's expenditures and employment substantially (NASA, 2007a).
Socioeconomics – Regional Employment and Economic Activity	
Socioeconomics – Community Services	
Palmdale	
Biological Resources – All	Minimal to no biological resources exist at Palmdale. There is minimal to no natural vegetation onsite. Only human-associated wildlife is found onsite; therefore, no unique habitat exists at Palmdale (NASA, 2007s).
Cultural Resources – Traditional Cultural Resources	There are no known traditional cultural resources or ethnographic sites at Palmdale (NASA, 2002e; California Office of Historic Preservation, February 2007. If any traditional cultural resources were to be found in the future, Palmdale would have to follow all applicable federal regulations.
Hydrology and Water Quality – All	Palmdale is located on property owned by the USAF and operated by the Boeing Company, and is used for other military aircraft operations. It will continue to operate following the cessation of the Shuttle program. A current groundwater remediation effort at Palmdale AFP 42 is being managed and funded by Wright-Patterson AFB (NASA, 2007s) because Palmdale is Wright Patterson's tenant.

No water resources would be affected by the proposed action. No changes in permitted water use or in storm water or wastewater discharges would be expected (NASA, 2007s). |
| Land Use | NASA is a tenant of Wright-Patterson AFB at AFP 42 and does not control real property or land use designations (NASA, 2007s). |

EXHIBIT 1-2
Issues Considered but Eliminated from Further Analysis

Resource Eliminated from Further Analysis	Rationale
Socioeconomics – Population	Palmdale is a GO/CO activity with a small SSP-direct workforce, and is a tenant of Wright-Patterson AFB at AFP 42 (NASA, 2007s). The effects on the regional population, regional economy, or community services would be minimal or undetectable in comparison to the workers and their families associated with the southern California aerospace industry.
Socioeconomics – Regional Employment and Economic Activity	
Socioeconomics – Community Services	

Notes:
AFB = Air Force Base
AFP = Air Force Plant
DFRC = Dryden Flight Research Center
EAFB = Edwards Air Force Base
EF = Ellington Field
EPFOL = El Paso Forward Operation Location
EPIA = El Paso International Airport
GO/CO = Government owned/contractor operated
JSC = Johnson Space Center
KSC = Kennedy Space Center
m = Meter
MAF = Michoud Assembly Facility
MSFC = Marshall Space Flight Center
NASA = National Aeronautics and Space Administration
NRHP = National Register of Historic Places
RCRA = Resource Conservation and Recovery Act
ROI = Region of influence
SSC = Stennis Space Center
SSP = Space Shuttle Program
T&R = Transition and retirement
USFWS = U.S. Fish and Wildlife Service
WSMR = White Sands Missile Range
WSTF = White Sands Test Facility

This page intentionally left blank.

2. Description of Proposed Action and Alternatives

This Programmatic EA for the SSP disposition of real and personal property evaluates two alternatives: the Proposed Action and the No Action Alternative. These two alternatives are described below:

- **Proposed Action**: NASA proposes to implement a centralized process for the disposition of the SSP real and personal property consisting of a coordinated series of actions. SSP real and personal property would be evaluated in accordance with NASA Procedural Requirements (NPR) 8800.15, "Real Estate Management Program Implementation Manual," and NPR 4300.1, "NASA Personal Property Disposal Procedural Requirements," to select the best option for disposition.
- **No Action Alternative:** NASA would not implement the proposed comprehensive and coordinated effort to disposition SSP property under a structured and centralized SSP process. The disposition of SSP property would instead occur on a center-by-center and item-by-item basis in the normal course of NASA's ongoing facility and program management.

The SSP is scheduled for retirement in 2010; NASA is developing this Programmatic EA to fulfill the NEPA requirements. SSP property disposition activities may extend several years beyond 2010. This document provides information about the SSP operations, assets, and environmental activities that are conducted at the major NASA Centers that support SSP. This section of the Programmatic EA describes the Proposed Action and alternatives and summarizes the potential impacts associated with the disposition of assets used in the SSP. Property is defined as follows:

- Real property is defined as land, buildings, and other structures and their associated built-in systems that cannot readily be moved without changing the essential character of the real property.
- Personal property is defined as all assets not classified as real property owned by, leased to, or acquired by the government. Personal property whose disposition may have the potential to significantly affect the environment is analyzed in this Programmatic EA.

This Programmatic EA for the SSP describes the assets related to the SSP activities and evaluates the possible environmental impacts associated with their disposition. Note that the discussions and analyses of impacts are organized by NASA Center (except for Palmdale, which is a USAF-owned, contractor-operated facility). That is, the disposition of assets is linked to their locations and the impacts vary based on the locations.

2.1 Description of the Proposed Action and Preferred Alternative

2.1.1 Disposition of Shuttle Assets

Under presidential direction, NASA will cease operations of its SSP in 2010. A number of assets will be dispositioned during the T&R activities. SSP property disposition activities may extend several years beyond 2010.

NASA proposes to implement a centralized process for the disposition of the SSP real and personal property consisting of a coordinated series of actions. SSP real and personal property would be evaluated in accordance with NPR 8800.15, "Real Estate Management Program Implementation Manual," and NPR 4300.1, "NASA Personal Property Disposal Procedural Requirements," to select the best option for disposition.

2.1.1.1 Real Property

When the SSP disposes of real property, the responsible NASA Center will evaluate whether the property can be used by another NASA program (reutilization), or it may mothball or destroy the property. If NASA decides to convey the property to another federal, state, local, or private individual, NASA relinquishes the property to the GSA. The GSA will convey the property according to federal laws and regulations. The property disposition options that will be evaluated for real property are as follows:

- **Reutilization:** The first option for disposal of government property is reutilization by another NASA program. Property is screened for reutilization by NASA's ongoing programs and for use by future programs.
- **Utilization:** If the property is not required by other NASA programs, it is made available to other federal agencies. The receiving federal agency would be responsible for the applicable NEPA analysis and documentation resulting from the use of the property.
- **Mothball**: Under this option, NASA would mothball particular SSP real property in place. Under this scenario, NASA would maintain these properties at some low level of support in the event that a Center or new program could use them in the future.

- **Destruction**: Under this option, the property would be demolished or otherwise removed from NASA property to an appropriate location, such as a landfill or hazardous waste treatment, storage, or disposal facility (TSDF).
- **Release to GSA**: If the property is no longer needed by NASA, it may be relinquished to the GSA for conveyance to other federal, state, local, or private individuals.

Property Survey. NASA has undertaken a historical survey and evaluation of all NASA-owned facilities and properties (real property assets) to assess their eligibility for listing in the NRHP in the context of the SSP (1969 through 2010). In February 2006, a Shuttle Transition Historic Preservation Working Group (HPWG) was formed that included the Historic Preservation Officers (HPOs) for all NASA Centers.

The HPWG drafted a set of standard criteria for the evaluation of Shuttle program-related properties at all NASA Centers (Appendix C). The SSP estimates that approximately 580 NASA facilities and properties were associated with the SSP. Most of these were existing assets, while others were built specifically for the development and implementation of the SSP. Of these, the HPWG identified more than 300 facilities and properties that were believed to have played significant roles in the SSP. In 2006, NASA surveyed these assets to determine NRHP eligibility. Of these, a total of 223 assets were found to be eligible for listing on the NRHP because of their contributions to the SSP. Of these 223 assets, 205 are real property assets and 18 are considered personal property, aircraft, or unique equipment used by the SSP.

Of the 223 assets, 62 were already NRHP-listed or NRHP-eligible due to a past NASA program or activity. Thus, the HPWG's agency-wide SSP study has identified 161 assets that are considered newly eligible for listing because of their significance to the SSP. Nomination decisions and consultation with the appropriate State Historical Preservation Officer (SHPO) will be made by NASA Centers. NASA HQ is developing a final report of the findings, which will be presented to the Advisory Council on Historic Preservation (ACHP) and National Park Service (NPS) for their information. The results of the surveys are presented by Center in Section 3.

These surveys were completed in accordance with Section 110 of the National Historic Preservation Act (NHPA). They also provide eligibility determinations that will support the Section 106 process for undertakings, because they are planned in support of the development and implementation of future NASA programs or missions such as the Constellation Program. Such future undertakings will not be the SSP's responsibility, but will be led by the NASA projects or programs that plan to use SSP-related assets in the future. The program or project office that proposes to modify listed or eligible assets will be responsible for completing consultation in accordance with the Section 106 process.

2.1.1.2 Personal Property

Shuttle-related personal property includes hundreds of thousands of items ranging from common parts, such as nuts and bolts, to complex tooling and flight hardware. The disposition of common parts has no potential for significant impacts to the environment. Consequently, only personal properties such as complex tooling and flight hardware that may have the potential to adversely affect the environment are analyzed in this Programmatic EA.

When personal property is no longer required by the SSP, it is disposed according to NASA's established procedures for disposal. The disposal procedure progresses through a series of options, as described below:

- **Reutilization:** The first option for the disposal of government property is reutilization by another NASA program. Property is screened for reutilization by NASA's ongoing programs and for use by future programs.
- **Storage:** Under this option, NASA would relocate particular SSP personal property to appropriate storage locations (such as laydown yards or warehouses). At these locations, the property would be maintained at some minimum level of support in the event that a Center or new program could use it in the future. These locations would have an appropriate level of security provided by the location's owner, which either would be NASA or some other federal agency. The storage locations could be located onsite, offsite, or be newly constructed areas or buildings. Because it currently is not known whether any new storage areas would be constructed to store SSP property, the information necessary to analyze the potential environmental impacts for constructing such areas does not exist at this time. Therefore, environmental analyses for the construction of new structures for storage of SSP property are deferred until the construction becomes less speculative, and the information necessary for analyses becomes available. Any additional NEPA analyses will be conducted by the responsible Center.
- **Utilization:** If the property is not required by other NASA programs, it is made available to other federal agencies. The receiving federal agency would be responsible for the applicable NEPA analysis and documentation resulting from the use of the property.
- **Donation:** If the property is not required by another federal agency, it is eligible for donation. Under this option, federal excess property can be provided to the state for screening and then to other eligible applicants, including nonprofit educational and public health activities, nonprofit and public programs (such as museums) for the elderly, educational activities of special interest, public airports, or the homeless.

- **Sales:** Under this option, providing that efforts to reutilize and/or donate have been exhausted, NASA would dispose of the property by means of a competitive bid process such as an auction, sealed bid, or retail sales, in accordance with the guidelines.
- **Destruction:** Under this option, the property would be demolished or otherwise removed from NASA property to an appropriate location, such as a landfill or hazardous waste TSDF.

The evaluation criteria to assess the potential historical significance of personal property and the preservation requirements are being developed by NASA. Once completed, these requirements will be applied to SSP personal property to determine what is historically significant. NASA defines artifacts as unique objects that document the history of the science and technology of aeronautics and astronautics. Their significance and interest stem mainly from their relation to the following: historic flights, programs, activities, or incidents; achievements or improvements in technology; our understanding of the universe; and important or well-known personalities (NASA, 2006e).

Property may be disposed at a landfill or hazardous waste storage facility if no longer needed, or may be engineered for re-use by NASA, or put on display by NASA or a museum. Some of the property will contain hazardous substances such as lead paint, asbestos, chromium coatings, hypergols, oxidizers, heavy metals, and other materials. NASA currently is planning to address "end-state" requirements for those assets that contain hazardous substances. The end-state requirements for each asset will include the tasks of decontamination and safing each item to meet the requirements for its end-use (final disposition) and to be in compliance with applicable state, federal, and local laws. For example, an asset that will be on public display at a museum will require a higher level of decontamination and safing than will an asset that will be reutilized by future space programs.

2.1.1.3 Property Disposition Schedule

In 2007 NASA had approximately 600,000 property line items planned to be excessed between 2008 and 2015 (Exhibit 2-1) and approximately 350,000 property line items to be transferred during the same timeframe (Exhibit 2-2). Bar graphs depicting the planned property to be excessed and transferred by location are shown in Exhibits 2-3 and 2-4, respectively. These property totals are based on 2007 data and will likely increase, based on the trends depicted in the bar graphs in Exhibits 2-3 and 2-4.

2. DESCRIPTION OF PROPOSED ACTION AND ALTERNATIVES

EXHIBIT 2-1
Property Excess Planned Burndown

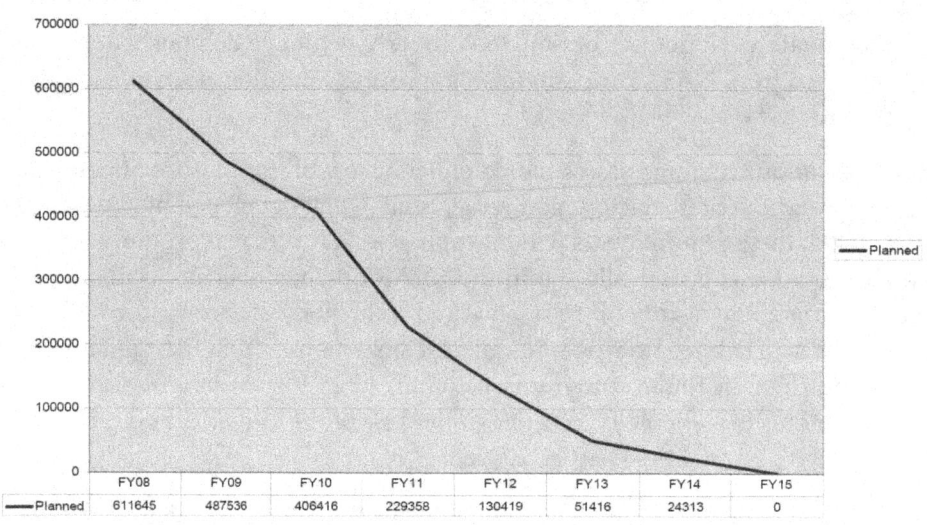

Total Excess Planned Burndown FY 08 - FY15

	FY08	FY09	FY10	FY11	FY12	FY13	FY14	FY15
Planned	611645	487536	406416	229358	130419	51416	24313	0

EXHIBIT 2-2
Property Transfer Planned Burndown

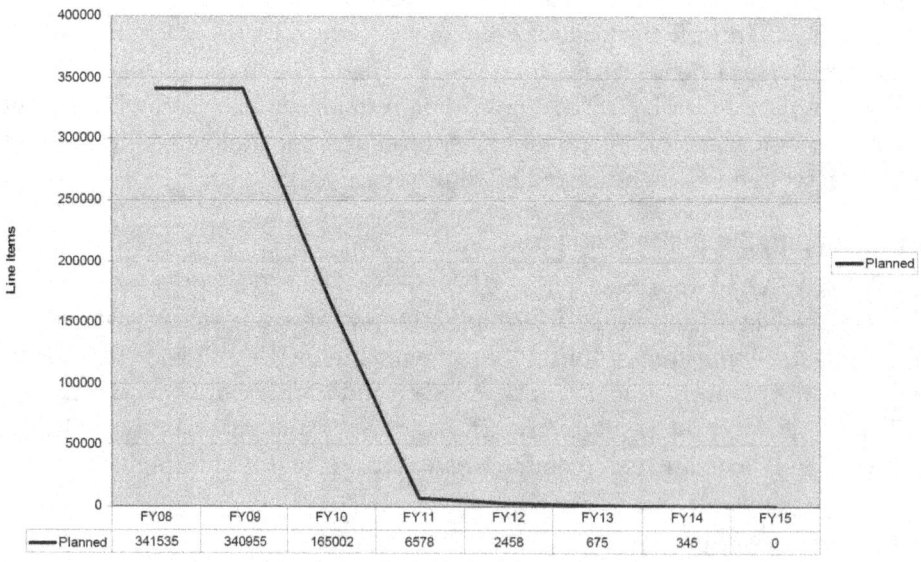

Planned Property Transfer Burndown

	FY08	FY09	FY10	FY11	FY12	FY13	FY14	FY15
Planned	341535	340955	165002	6578	2458	675	345	0

EXHIBIT 2-3
Property Excess Planned Burndown by Location

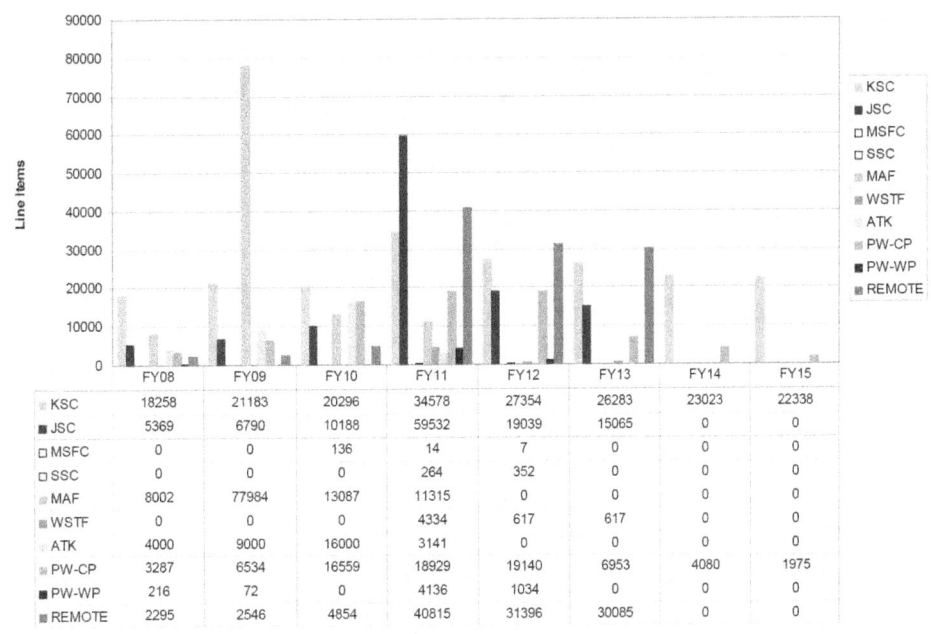

	FY08	FY09	FY10	FY11	FY12	FY13	FY14	FY15
KSC	18258	21183	20296	34578	27354	26283	23023	22338
JSC	5369	6790	10188	59532	19039	15065	0	0
MSFC	0	0	136	14	7	0	0	0
SSC	0	0	0	264	352	0	0	0
MAF	8002	77984	13087	11315	0	0	0	0
WSTF	0	0	0	4334	617	617	0	0
ATK	4000	9000	16000	3141	0	0	0	0
PW-CP	3287	6534	16559	18929	19140	6953	4080	1975
PW-WP	216	72	0	4136	1034	0	0	0
REMOTE	2295	2546	4854	40815	31396	30085	0	0

EXHIBIT 2-4
Property Transfer Planned Burndown by Location

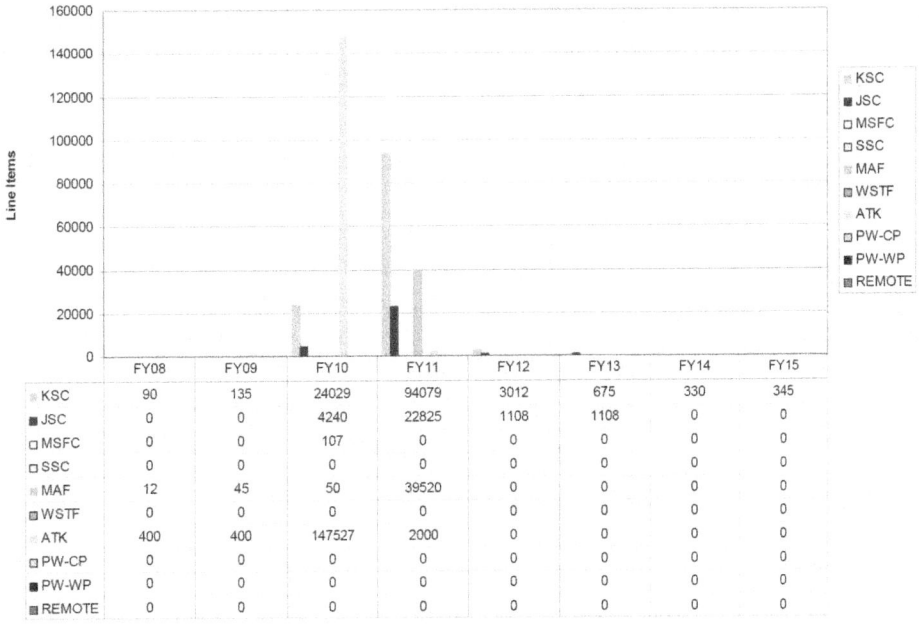

	FY08	FY09	FY10	FY11	FY12	FY13	FY14	FY15
KSC	90	135	24029	94079	3012	675	330	345
JSC	0	0	4240	22825	1108	1108	0	0
MSFC	0	0	107	0	0	0	0	0
SSC	0	0	0	0	0	0	0	0
MAF	12	45	50	39520	0	0	0	0
WSTF	0	0	0	0	0	0	0	0
ATK	400	400	147527	2000	0	0	0	0
PW-CP	0	0	0	0	0	0	0	0
PW-WP	0	0	0	0	0	0	0	0
REMOTE	0	0	0	0	0	0	0	0

2.1.2 Space Shuttle Operations and Elements

2.1.2.1 Space Shuttle Operations

SSP-related operations are conducted at numerous sites nationwide. The locations of the major SSP-related sites are shown in Exhibit 2-5. Exhibits 2-6 and 2-7 illustrate the SSP hardware flow and associated facilities. Additional SSP-related operations such as testing and training are conducted at these and other sites. The major Centers and their roles in supporting the SSP are described below:

- KSC – Space Shuttle assembly, launch, and landing
- JSC – SSP management, astronaut training, and mission control
- EF – Astronaut flight training
- EPFOL – Astronaut flight training
- SSC – SSME testing
- MAF – SSP ET manufacturing
- MSFC – Space Shuttle propulsion management
- WSTF – Hypergol testing and astronaut Shuttle landing training facility (White Sands Space Harbor [WSSH])
- DFRC – Space Shuttle back-up landing facility
- Palmdale – Thermal Control System (TCS) development, cold plates, ET disconnects, and logistics manufacturing

The prime contractor facilities associated with SSP operations include ATK (Promontory, Utah), Boeing (Huntington Beach, California), Lockheed Martin (at MAF), Pratt Whitney Rocketdyne (West Palm Beach, Florida; and Canoga Park, California), and USA (primarily KSC and JSC locations). These facilities were not included (except for MAF's NASA Operations) because they are responsible for the disposition of their own properties. However, government-owned property at contractor sites is included in this EA. Exhibit 2-6 outlines the flow of SSP hardware between the prime contractor facilities and the NASA Centers.

Facilities at which SSP operations are conducted, including government owned/government-operated (GO/GO) and government owned/contractor-operated (GO/CO), are assessed for potential environmental impacts. The design, manufacture, testing, and operation of numerous SSP components are accomplished at several contractor facilities around the U.S. These facilities are covered by existing environmental permits and state regulations and are not assessed for potential environmental impacts in this Programmatic EA.

EXHIBIT 2-5
SSP Facilities

This page intentionally left blank.

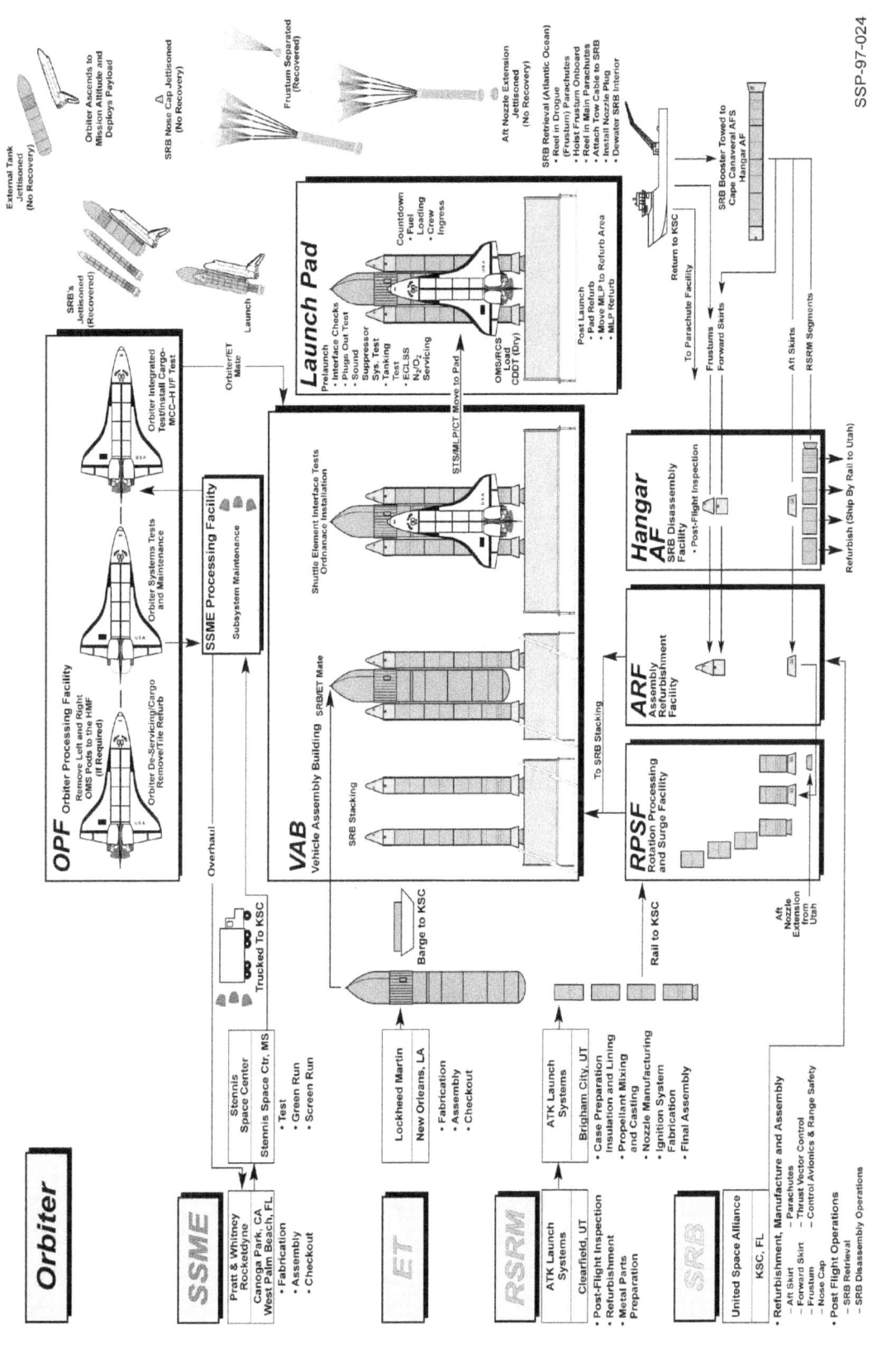

EXHIBIT 2-6
SSP Hardware Flow

This page intentionally left blank.

EXHIBIT 2.7
Space Shuttle Elements Flow at SSP Related NASA Centers

Space Shuttle Elements	Orbiter	Space Shuttle Main Engines	External Tank	Solid Rocket Booster	Reusable Solid Rocket Motor (RSRM)
Dryden Flight Research Center (DFRC)	Alternate landing site for the Orbiter if conditions are not favorable at KSC. Maintains GSE and a Shuttle hangar in case of a Shuttle landing at DFRC.	Not applicable.	Not applicable.	Not applicable.	Not applicable.
Ellington Field (EF)	Maintains aircraft including the STA for training the astronauts by simulating the flight controls of the Orbiter. In the past, the Shuttle, transported on a Boeing 747 carrier, has stopped at EF for transport to KSC.	Not applicable.	Not applicable.	Not applicable.	Not applicable.
El Paso Forward Operating Location (EPFOL)	Astronauts fly T-38 aircraft from EF to EPFOL to prepare for flights in the STA. The astronauts are briefed at EPFOL for their training mission in the STA.	Not applicable.	Not applicable.	Not applicable.	Not applicable.
Johnson Space Center (JSC)	Manages the Orbiter Project. The office also manages program engineering support activities for operation elements and flight crew equipment hardware and flight preparation activities. The USA Flight Crew Equipment Facility is located offsite, but supports numerous requirements associated with Orbiter-owned hardware.	Not applicable.	Not applicable.	Not applicable.	Not applicable.
Kennedy Space Center (KSC)	After completing a space mission, the Orbiter is returned to KSC to undergo preparations for its next flight. In the OPF, the vehicle is safed, residual propellants and other fluids are drained, and returning horizontal and middeck payloads are removed. Any problems that may have occurred with Orbiter systems and equipment on the previous mission are checked out and corrected. Equipment is repaired or replaced and extensively tested. Any modifications to the Orbiter that are required for the next mission also are made in the OPF. Following extensive testing and verification of all electrical and mechanical interfaces, the Orbiter is transferred to the nearby VAB, where it is mated to the ET with attached SRBs. The MLPs provide GSE for Shuttle checkout, servicing, and launch. They are a two-story transportable launch base for the Shuttle stack. The exterior of the MLPs provide for SRB hold-down posts, Orbiter tail service masts, and sound suppression water nozzles for deluge water. The MLPs are transported from the VAB to the launch pad by a large tracked vehicle called the Crawler-Transporter. At the launch pad, final preflight and interface checks of the Orbiter, its payloads, and the associated GSE are conducted. After a positive Flight Readiness Review, the decision to launch is made and the final countdown begins.	The SSMEs arrive at KSC via truck from SSC. Three SSMEs are readied for installation on the Orbiter at the SSME Processing Facility. The SSME Processing Facility also performs maintenance on the SSMEs. The SSMEs are moved to the OPF for installation on the Orbiter.	The ET is sent to KSC from MAF for installation in VAB via barge.	SRBs are built at KSC. SRBs are manufactured, assembled, and refurbished at the ARF. The SRBs are sent through Post Flight Operations at Hangar AF. These operations entail recovering and towing the SRBs, disassembly, safing, and surface coating removal. SRBs are then sent to the RPSF and then to the VAB for final assembly.	RSRMs are constructed at a contractor's facility in Utah and shipped by rail to KSC. The RSRM is run through the RPSF and is then sent to the VAB for final assembly.
Marshall Space Flight Center (MSFC)	Not applicable.	Manages the SSME Project.	Manages the ET Project.	Manages the RSRB (combined motor and booster project)	Manages the RSRM Project.

2. DESCRIPTION OF PROPOSED ACTION AND ALTERNATIVES

EXHIBIT 2-7
Space Shuttle Elements Flow at SSP Related NASA Centers

Space Shuttle Elements	Orbiter	Space Shuttle Main Engines	External Tank	Solid Rocket Booster	Reusable Solid Rocket Motor (RSRM)
Michoud Assembly Facility (MAF)	Not applicable.	Not applicable.	ET is manufactured, assembled, and tested at MAF.	Not applicable.	Not applicable.
Palmdale	TPS manufacturing and testing, cold plate manufacturing, and logistic manufacturing are conducted at Palmdale.	Not applicable.	ET umbilical manufacturing and assembly are conducted at Palmdale.	Not applicable.	Not applicable.
Stennis Space Center (SSC)	Not applicable.	SSME testing is conducted at SSC. NASA operates nine barges at SSC to transport liquid hydrogen (three barges) and liquid oxygen (six barges). The SSME is tested to meet an SSP requirement, whether it is to test an engine component or to prepare an entire engine for flight. After testing, the engine remains on the test stand for further testing or is removed and sent to Building 9101 for storage or to be rebuilt. If the engine is being tested for flight, the flight testing profile is completed through a series of tests. The engine is removed and then shipped via truck to KSC for installation on an Orbiter.	Not applicable.	Not applicable.	Not applicable.
White Sands Test Facility (WSTF)	NASA evaluates materials and components at WSTF for use in propulsion, power generation, and life-support systems, crew cabin equipment, payloads, and experiments carried aboard the Shuttle Orbiter and the ISS. The WSSH is the Orbiter approach and landing training facility. It also is a contingent landing site for the Orbiter if the conditions at KSC or EAFB are not favorable.	Not applicable.	Not applicable.	Not applicable.	Not applicable.

Notes:
ARF = Assembly and Refurbishment Facility
DFRC = Dryden Flight Research Center
EF = Ellington Field
EPFOL = El Paso Forward Operating Location
ET = External Tank
GSE = Ground support equipment
ISS = International Space Station
JSC = Johnson Space Center
KSC = Kennedy Space Center
MAF = Michoud Assembly Facility
MLP = Mobile Launch Platform
MSFC = Marshall Space Flight Center
NASA = National Aeronautics and Space Administration
NBL = Neutral Buoyancy Laboratory
OPF = Orbiter Processing Facility
RPSF = Rotation, Processing and Surge Facility
RSRM = Reusable Solid Rocket Motor
SRB = Solid Rocket Booster
SSC = Stennis Space Center
SSME = Space Shuttle Main Engine
SSP = Space Shuttle Program
STA = Shuttle Training Aircraft
TPS = Thermal Protection System
USA = United Space Alliance
VAB = Vehicle Assembly Building
WSSH = White Sands Space Harbor
WSTF = White Sands Test Facility
WSTF = White Sands Test Facility

2.1.2.2 Space Shuttle Space Flight Hardware Elements

The primary Space Shuttle elements are a piloted, reusable orbiting vehicle called the Orbiter, three SSMEs, an ET, two Reusable Solid Rocket Motors (RSRMs), and two Solid Rocket Boosters (SRBs). The configuration of the vehicle's elements is shown in Exhibit 2-8. Ground support equipment (GSE), logistics support, and flight crew equipment also are critical components of the SSP. These groups work together with the Systems Engineering and Integration Office to support the assembly, launch, flight, landing, and refurbishment of the Space Shuttle.

EXHIBIT 2-8
Space Shuttle Configuration

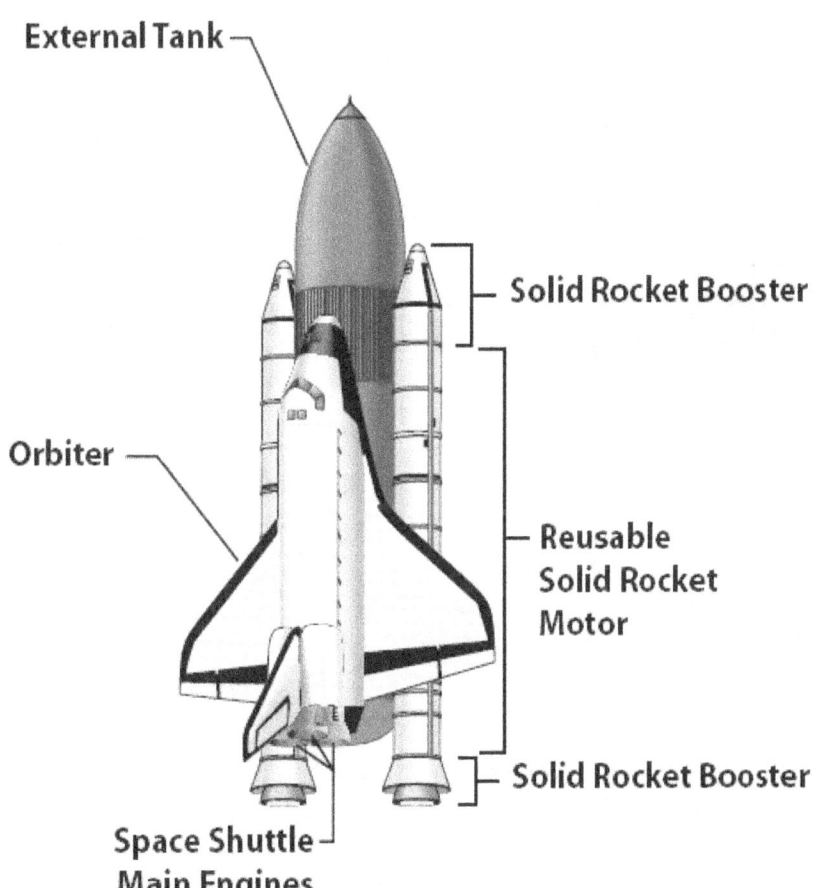

Orbiter. The Orbiter, shown in Exhibit 2-9, is about the same size and weight as a DC-9 aircraft. The Orbiter contains a pressurized crew compartment that normally can carry up to 7 crew members, and has a payload bay to carry cargo that is 18 meters (m) (60 feet [ft]) long and 4.5 m (15 ft) wide, and 3 main engines mounted on its aft end. To protect its aluminum structure during ascent and descent into Earth's atmosphere, the Orbiter is covered with heat-resistant tiles and reinforced carbon panels (NASA, 2004e).

EXHIBIT 2-9
Space Shuttle Orbiter

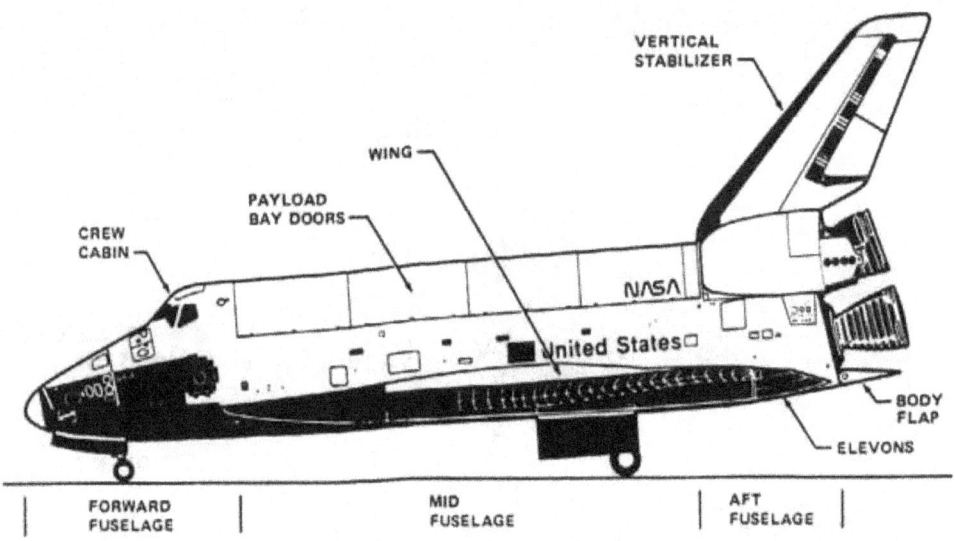

After completing a space mission, the Orbiter is returned to KSC to undergo preparations for its next flight in a sophisticated aircraft-like hangar called the Orbiter Processing Facility (OPF). In the OPF, the vehicle is safed, residual propellants and other fluids are drained, and returning horizontal and middeck payloads are removed. The Orbiter is refurbished and processed by USA at KSC.

Any problems that may have occurred with Orbiter systems and equipment on the previous mission are checked out and corrected. Equipment is repaired or replaced and extensively tested. Modifications to the Orbiter that are required for the next mission also are made in the OPF.

Orbiter refurbishment operations and processing for the next mission also begin in the OPF. Large horizontal payloads are installed in the Orbiter cargo bay. Vertical payloads are installed at the launch pad.

Following extensive testing and verification of the electrical and mechanical interfaces, the Orbiter is transferred to the nearby Vehicle Assembly Building (VAB), where it is mated to the ET with attached SRBs. Then, the assembled Space Shuttle vehicle is carried to the launch pad by a large tracked vehicle called the Crawler-Transporter.

At the launch pad, final preflight and interface checks of the Orbiter, its payloads, and associated GSE are conducted. After a positive Flight Readiness Review, the decision to launch is made and the final countdown begins (NASA, 1992).

Space Shuttle Main Engine. The three main engines on the Orbiter are the SSMEs, as shown in Exhibit 2-10. With a maximum thrust at sea level of more than 418,000 pounds each, they work in tandem with the SRBs from liftoff until the SRBs separate, about 2 minutes after launch, after which they are the sole means of propelling the Orbiter into space. They use liquid hydrogen (LH2) for fuel and cooling and liquid oxygen (LOX) as an oxidizer. The propellant is carried in separate tanks in the ET and supplied to the main engines under pressure. Each SSME is 4 m (14 ft) long and 2.3 m (7.5 ft) in diameter at the nozzle exit, and weighs approximately 3,175 kilograms (kg) (7,000 pounds). The SSME's major components are the fuel and oxidizer turbopumps, preburners, hot gas manifold, main combustion chamber, nozzle, oxidizer heat exchanger, and propellant valves.

SSME components are manufactured by Pratt-Whitney/Rocketdyne in Canoga Park, California, and shipped to SSC for assembly and testing. SSMEs are hot-fired tested and prepared for flight at SSC. SSC tests new engine components as well as entire engines for flight. After an SSME successfully completes a test series that determines its flight readiness, it is transported via truck to KSC. The SSME arrives at the SSME Processing Facility, where it is readied for installation on the Orbiter. The SSME Processing Facility also performs maintenance on the SSME. The SSME is moved to the OPF for installation on an Orbiter.

External Tank. The ET contains the propellants used by the SSMEs, as shown in Exhibit 2-11. The ET also provides structural support for the Shuttle stack during the launch at the attachment points for the SRBs and Orbiter.

The ET, which is the only major component of the Space Shuttle that is not reusable, is 47 m (154 ft) long and 8.7 m (28.6 ft) in diameter, and weighs slightly more than 71,000 pounds without fuel. The largest and heaviest (when loaded) element of the space shuttle, the ET has three major components: the forward LOX tank, an unpressurized intertank that contains most of the electrical components, and the aft LH2 tank. To meet the need for flights to the ISS, a new super lightweight tank was developed that incorporates aluminum-lithium in its internal structures, thus reducing the overall tank weight by 7,500 pounds.

2. DESCRIPTION OF PROPOSED ACTION AND ALTERNATIVES

EXHIBIT 2-10
Space Shuttle Main Engine

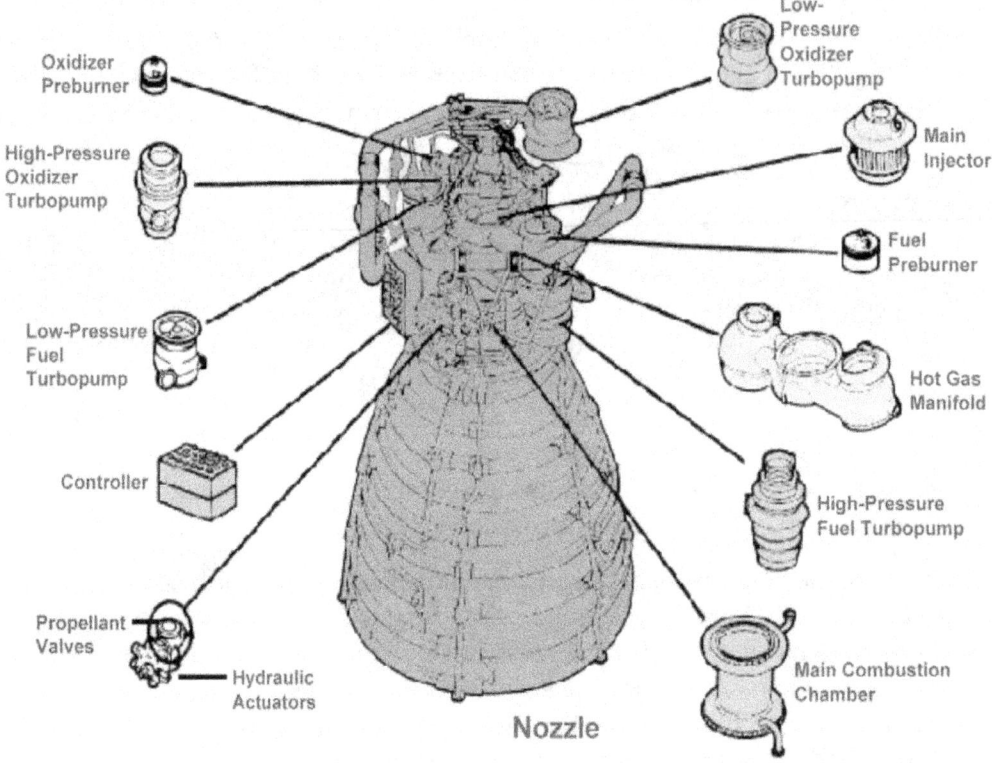

EXHIBIT 2-11
Space Shuttle External Tank

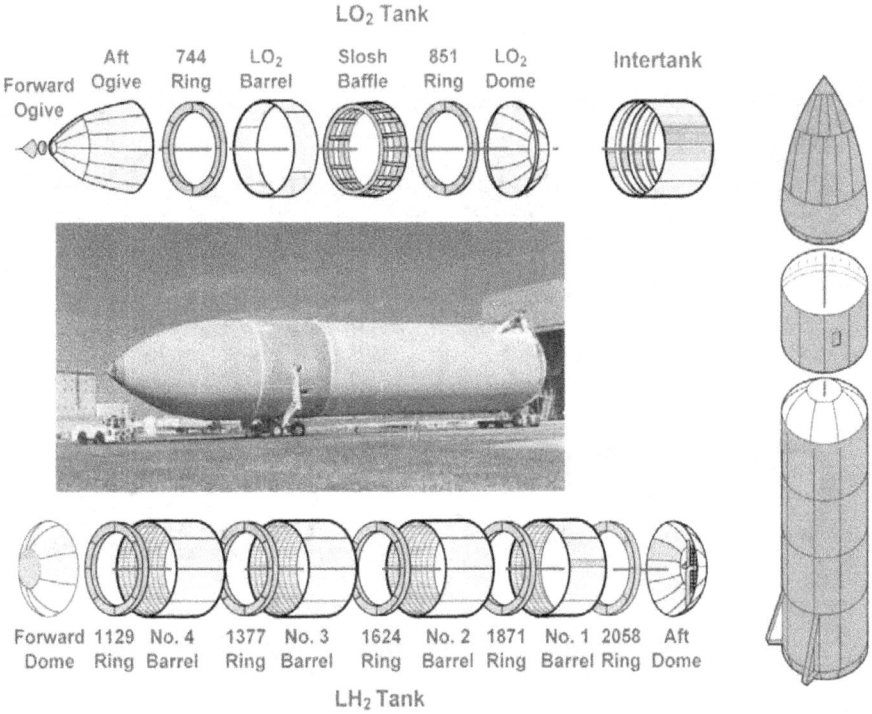

The skin of the ET is covered with a thermal protection system (TPS) coating of spray-on polyisocyanurate foam. The purpose of the TPS is to maintain the propellants at an acceptable temperature, to protect the skin surface from aerodynamic heat, and to minimize ice formation.

The ET includes a propellant feed system to duct the propellants to the Orbiter engines, a pressurization and vent system to regulate the tank pressure, an environmental conditioning system to regulate the temperature and render the atmosphere in the intertank area inert, and an electrical system to distribute power and instrumentation signals and provide lightning protection. The tank's propellants are fed to the Orbiter through a 43-centimeter (cm) (17-inch)-diameter connection that branches inside the Orbiter to feed each main engine (NASA, 2007q).

The ET is manufactured by Lockheed Martin at MAF in New Orleans, Louisiana. Upon completion, the tanks are shipped via barge to KSC, where they are mated to the Shuttle in the VAB.

Reusable Solid Rocket Motor. The Space Shuttle RSRM is the largest Solid Rocket Motor (SRM) ever to fly and the only SRM rated for human flight (Exhibit 2-12).

EXHIBIT 2-12
Space Shuttle Reusable Solid Rocket Motor

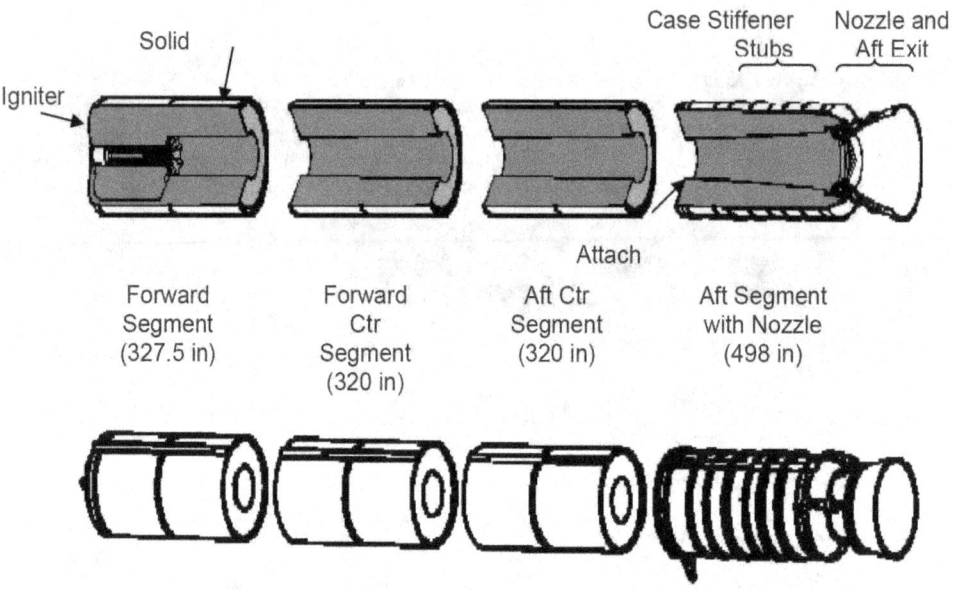

Each RSRM consists of four rocket motor segments, a nozzle, and an aft exit cone assembly. Each motor is just slightly more than 38 m (126 ft) long and 3.7 m (12 ft) in diameter. The propellant mixture in each motor consists of aluminum powder (fuel), polymer (binder), iron oxide (a catalyst), and a curing agent.

Each Space Shuttle launch requires the boost of two RSRMs to lift the 4.5-million-pound shuttle vehicle. From ignition to end of burn, each RSRM generates an average thrust of 2.6 million pounds and burns for approximately 123 seconds. By the time the twin RSRMs have expended their fuel, the Space Shuttle Orbiter has reached an altitude of 39 kilometers (km) (24 nautical miles) and is traveling at a speed in excess of 4,828 km per hour (km/h) (3,000 miles per hour [mph]). Hardware for each RSRM can be used as many as 20 times.

ATK manufactures and assembles the RSRM segments and nozzles at Promontory, Utah, and then ships them by rail to KSC. At KSC, they are stacked with additional assemblies to become SRBs, as described below.

After flight, the RSRMs are retrieved and towed by boat to the CCAFS Hangar AF, where they are disassembled, rinsed, and placed on railcars for shipment back to ATK. ATK refurbishes the RSRM hardware, prepares the case segments, mixes and casts the propellant, and assembles the segments in preparation for shipment back to KSC.

Solid Rocket Booster. The SRBs include forward skirt and aft skirt assemblies stacked fore and aft with the RSRM segments (Exhibit 2-13). The SRB is manufactured and assembled by USA at KSC. The SRB forward and aft skirts are assembled and refurbished in the SRB Assembly and Refurbishment Facility (ARF). The RSRM aft segment is attached to the SRB aft skirt in the Rotation, Processing, and Surge Facility. In the VAB, the additional RSRM segments and the SRB forward skirt are stacked on top of the aft assembly. The aft skirt is assembled in the RSRM stack in the Rotation, Processing, and Surge Facility.

EXHIBIT 2-13
Space Shuttle Solid Rocket Booster

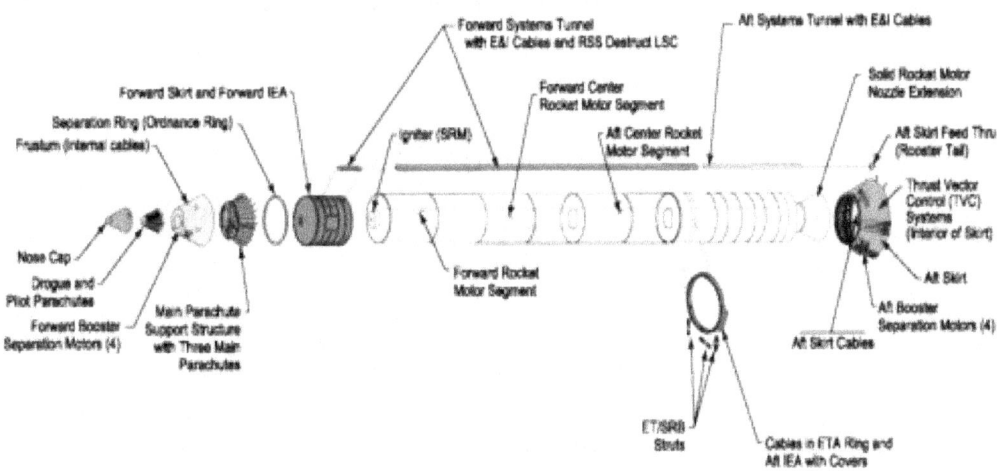

The forward skirt is assembled to the RSRM stack in the VAB. The aft skirt assembly consists of the aft skirt, which houses the steering system called the thrust vector control system, cables, and four separation motors. The forward skirt assembly consists of the nose cap (houses pilot and drogue parachutes), four booster separation motors, frustum (houses three main parachutes and cables), and the forward aft skirt (houses guidance gyros).

Two minutes after SSP launch, at an altitude of about 39 km (24 miles), the two SRB and RSRM assemblies separate from the ET and descend by parachute into the ocean, where they are collected by recovery ships for refurbishment and reuse. Post-flight inspection is conducted in Hangar AF. After inspection, the motor segments are shipped back to ATK in Utah to be reloaded with solid propellant.

Shuttle Processing. The Shuttle Processing operations include all of the integration, maintenance, processing, and repairs to the Space Shuttle vehicle upon landing until launch. Therefore, Shuttle Processing uses most of the facilities located at KSC to perform the operations, including the Launch Pad Complexes, VAB, OPFs, and Shuttle Landing Facility (SLF). During the course of a Shuttle ground operations flow, the Orbiter is processed and integrated with the SSMEs, and eventually mated to the ET and SRBs atop the Mobile Launch Platform. Propellant operations take place at the Launch Pad before a launch.

2.1.3 Proposed Action Schedule

The SSP is scheduled for retirement in 2010. Under the Proposed Action, once an asset is determined to no longer be needed by the SSP, it would become slated for disposition. Disposition could occur for some assets before SSP retirement in 2010. However, many assets will be needed until the final Space Shuttle mission is completed. Furthermore, the evaluation of the potential usefulness of some assets for other NASA programs may not be possible until those programs reach a certain level of maturity. Therefore, so that NASA may best use its SSP-related assets, final disposition of SSP-related assets under the Proposed Action would continue for several years past 2010.

2.2 Description of the No Action Alternative

Under the No Action alternative, NASA would not implement the proposed comprehensive and coordinated effort to disposition SSP property under a structured and centralized SSP process. The disposition of SSP property would instead occur on a center-by-center and item-by-item basis in the normal course of NASA's ongoing facility and program management.

2.3 Alternatives Considered but Not Carried Forward

There were no other alternatives considered. The Vision for Space Exploration issued by the President directed NASA to use the Space Shuttle to fulfill its obligation to complete assembly of the ISS and then to retire the Shuttle in 2010; therefore, no other alternatives were considered.

2.4 Summary Comparison of Alternatives

Exhibit 2-14 summarizes the potential environmental impacts, which are presented in detail in Section 4. Potential impacts to resources resulting from the implementation of the two alternatives were identified and placed into one of the following pre-determined classifications (NASA, 2007h):

- No Impact–no impacts are expected
- Minimal–Impacts are not expected to be measurable, or are measurable but are too small to cause any change in the environment

- Minor–Impacts that are measurable but are within the capacity of the affected system to absorb the change, or the impacts can be compensated for with little effort and few resources so that the impact is not substantial
- Moderate–Impacts that are measurable but are within the capacity of the affected system to absorb the change, or the impacts can be compensated for with effort and resources so that the impact is not substantial
- Major–Environmental impacts that, individually or cumulatively, could be substantial

EXHIBIT 2-14
Summary Comparison of Alternatives

Resource Area	Potential Impact of Proposed Action	Potential Impact of No Action Alternative
Kennedy Space Center		
Air Quality	minimal to no impact	minimal to no impact
Biological Resources	minimal impact	minimal impact
Cultural Resources	moderate impact	moderate impact
Hazardous/Toxic Materials and Waste	minimal impact	minimal impact
Health and Safety	minimal impact	minimal impact
Hydrology and Water Quality	minimal impact	minimal impact
Land Use	minimal impact	minimal impact
Noise	minimal impact	minimal impact
Site Infrastructure	minimal impact	minimal impact
Socioeconomics	minimal to no impact	minimal to no impact
Solid Waste	minimal impact	minimal impact
Traffic and Transportation	minimal impact	minimal impact
Johnson Space Center		
Air Quality	minimal to no impact	minimal to no impact
Biological Resources	no impact	no impact
Cultural Resources	moderate impact	moderate impact
Hazardous/Toxic Materials and Waste	minimal impact	minimal impact
Health and Safety	minimal impact	minimal impact
Hydrology and Water Quality	minimal impact	minimal impact
Land Use	minimal impact	minimal impact
Noise	minimal impact	minimal impact
Site Infrastructure	minimal impact	minimal impact

EXHIBIT 2-14
Summary Comparison of Alternatives

Resource Area	Potential Impact of Proposed Action	Potential Impact of No Action Alternative
Socioeconomics	minimal to no impact	minimal to no impact
Solid Waste	minimal impact	minimal impact
Traffic and Transportation	minimal impact	minimal impact
Ellington Field		
Air Quality	minimal to no impact	minimal to no impact
Hazardous/Toxic Materials and Waste	minimal to no impact	minimal impact
Health and Safety	minimal impact	minimal impact
Hydrology and Water Quality	minimal impact	minimal impact
Land Use	minimal impact	minimal impact
Noise	minimal impact	minimal impact
Site Infrastructure	minimal impact	minimal impact
Solid Waste	minimal impact	minimal impact
Traffic and Transportation	minimal impact	minimal impact
El Paso Forward Operating Location		
Air Quality	minimal to no impact	minimal to no impact
Hazardous/Toxic Materials and Waste	minimal to no impact	minimal impact
Health and Safety	minimal impact	minimal impact
Hydrology and Water Quality	minimal impact	minimal impact
Noise	minimal impact	minimal impact
Site Infrastructure	minimal impact	minimal impact
Solid Waste	minimal impact	minimal impact
Traffic and Transportation	minimal impact	minimal impact
Stennis Space Center		
Air Quality	minimal to no impact	minimal to no impact
Biological Resources	minimal impact	minimal impact
Cultural Resources	moderate impact	moderate impact
Hazardous/Toxic Materials and Waste	minimal impact	minimal impact
Health and Safety	minimal impact	minimal impact
Hydrology and Water Quality	minimal impact	minimal impact

EXHIBIT 2-14
Summary Comparison of Alternatives

Resource Area	Potential Impact of Proposed Action	Potential Impact of No Action Alternative
Land Use	minimal impact	minimal impact
Noise	minimal impact	minimal impact
Site Infrastructure	minimal impact	minimal impact
Solid Waste	minimal impact	minimal impact
Traffic and Transportation	minimal impact	minimal impact
Michoud Assembly Facility		
Air Quality	minimal to no impact	minimal to no impact
Biological Resources	minimal impact	minimal impact
Cultural Resources	moderate impact	moderate impact
Hazardous/Toxic Materials and Waste	minimal impact	minimal impact
Health and Safety	minimal impact	minimal impact
Hydrology and Water Quality	minimal impact	minimal impact
Land Use	minimal impact	minimal impact
Noise	minimal impact	minimal impact
Site Infrastructure	minimal impact	minimal impact
Socioeconomics	minimal to no impact	minimal to no impact
Solid Waste	minimal impact	minimal impact
Traffic and Transportation	minimal impact	minimal impact
Marshall Space Flight Center		
Air Quality	minimal to no impact	minimal to no impact
Biological Resources	minimal impact	minimal impact
Cultural Resources	moderate impact	moderate impact
Hazardous/Toxic Materials and Waste	minimal impact	minimal impact
Health and Safety	minimal impact	minimal impact
Hydrology and Water Quality	minimal impact	minimal impact
Land Use	minimal impact	minimal impact
Noise	minimal impact	minimal impact
Site Infrastructure	minimal impact	minimal impact
Socioeconomics	minimal to no impact	minimal to no impact

EXHIBIT 2-14
Summary Comparison of Alternatives

Resource Area	Potential Impact of Proposed Action	Potential Impact of No Action Alternative
Solid Waste	minimal impact	minimal impact
Traffic and Transportation	minimal impact	minimal impact
White Sands Test Facility		
Air Quality	minimal to no impact	minimal to no impact
Biological Resources	minimal impact	minimal impact
Cultural Resources	moderate impact	moderate impact
Hazardous/Toxic Materials and Waste	minimal impact	minimal impact
Health and Safety	minimal impact	minimal impact
Hydrology and Water Quality	minimal impact	minimal impact
Land Use	minimal impact	minimal impact
Noise	minimal impact	minimal impact
Site Infrastructure	minimal impact	minimal impact
Socioeconomics	minimal to no impact	minimal to no impact
Solid Waste	minimal impact	minimal impact
Traffic and Transportation	minimal impact	minimal impact
Dryden Flight Research Center		
Air Quality	minimal to no impact	minimal to no impact
Cultural Resources	moderate impact	moderate impact
Hazardous/Toxic Materials and Waste	minimal impact	minimal impact
Health and Safety	minimal impact	minimal impact
Hydrology and Water Quality	minimal impact	minimal impact
Land Use	minimal impact	minimal impact
Noise	minimal impact	minimal impact
Site Infrastructure	minimal impact	minimal impact
Solid Waste	minimal impact	minimal impact
Traffic and Transportation	minimal impact	minimal impact
Palmdale		
Air Quality	minimal to no impact	minimal to no impact
Cultural Resources	moderate impact	moderate impact

EXHIBIT 2-14
Summary Comparison of Alternatives

Resource Area	Potential Impact of Proposed Action	Potential Impact of No Action Alternative
Hazardous/Toxic Materials and Waste	minimal impact	minimal impact
Health and Safety	minimal impact	minimal impact
Noise	minimal impact	minimal impact
Site Infrastructure	minimal impact	minimal impact
Solid Waste	minimal impact	minimal impact
Traffic and Transportation	minimal impact	minimal impact

Notes:
No Impact–No impacts expected
Minimal–Impacts are not expected to be measurable, or are measurable but are too small to cause any change in the environment
Minor–Impacts that are measurable but are within the capacity of the affected system to absorb the change, or the impacts can be compensated for with little effort and few resources so that the impact is not substantial
Moderate–Impacts that are measurable but are within the capacity of the affected system to absorb the change, or the impacts can be compensated for with effort and resources so that the impact is not substantial
Major–Environmental impacts that, individually or cumulatively, could be substantial

This page intentionally left blank.

3. Affected Environment

This section describes the environmental characteristics of each resource area that may be affected by implementing the Proposed Action and the No Action Alternatives. For each of the resource areas, a region of influence (ROI) is established and the existing conditions are described. The affected environment is described succinctly to provide a context for understanding the potential impacts. Those components of the affected environment that are of greater concern relevant to the potential impacts are described in greater detail.

3.1 Overview of Resource Areas

Twelve resource areas are considered to provide a context for understanding the potential effects of the Proposed Action and to provide a basis for assessing the severity of the potential impacts. Section 4 contains the impact analysis. Several of these environmental components are regulated by federal and/or state environmental statutes, many of which set specific guidelines, regulations, and standards. These standards provide a benchmark that assists in determining the significance of the environmental impacts under the NEPA evaluation process. The compliance status of each project area or installation with respect to the environmental requirements was included in the information collected about the affected environment. The 12 areas of environmental consideration are as follows:

- Air Quality
- Biological Resources
- Cultural Resources
- Hazardous and Toxic Materials and Wastes
- Health and Safety
- Hydrology and Water Quality
- Land Use
- Noise
- Site Infrastructure
- Socioeconomics
- Solid Waste
- Traffic and Transportation

3. AFFECTED ENVIRONMENT

SSP-related operations are conducted at numerous sites nationwide. The locations of the major SSP-related sites are shown in Exhibit 3-1. The major Centers and their roles in supporting the SSP are described below:

- KSC – Space Shuttle assembly, launch, and landing
- JSC – SSP management, astronaut training, and mission control
- SSC – SSME testing
- EF – Astronaut flight training
- EPFOL – Astronaut flight training
- MAF – SSP ET manufacturing
- MSFC – Space Shuttle propulsion management
- WSTF – Hypergol testing and astronaut Shuttle landing training facility (WSSH)
- DFRC – Space Shuttle back-up landing facility
- Palmdale – TCS development

EXHIBIT 3-1
SSP Facilities

Other SSP-related facilities include the TAL sites. Because these sites are outside of the U.S. and its territories and possessions, NEPA does not apply at these locations. However, EO 12114, "Environmental Effects Abroad of Major Federal Actions," does apply at these locations. EO 12114 requires an environmental analysis, similar to NEPA, at these locations. The EO 12114 analysis for these sites is provided in Section 4.

The ROI for each resource area is as follows, unless otherwise explicitly stated within each section:

Air Quality: The airshed surrounding the Center.

Biological Resources: The boundaries of the Center.

Cultural Resources: The boundaries of the Center.

Hazardous and Toxic Materials and Waste: The operational areas within the Center.

Health and Safety: The health and safety of employees directly involved with SSP operations at the Center.

Hydrology and Water Quality: SSP activities and operations that have the potential to affect hydrology and water quality in and around the Center.

Land Use: All land within the Center boundaries.

Noise: The area in and around the Center that could be exposed to 8-hour, time-weighted average (TWA), sound pressure levels (SPLs) equal to or greater than 85 decibels A-rated (dBA).

Site Infrastructure: The SSP-related facilities at the Center and the potable water and energy sources that supply the SSP activities.

Socioeconomics: The counties surrounding the Center where approximately 90 percent of the workforce resides.

Solid Waste: The NASA operational areas at the Center associated with the SSP and the landfill receives the SSP solid waste from that Center.

The regulatory settings for each resource area are provided in Appendix D. All federal, state, and local regulations were evaluated for their potential applicability, in addition to NASA HQ and Center policy and procedural requirements. Appendix D provides a list of the applicable regulations and policies.

3.2 Kennedy Space Center

KSC is located on the eastern coast of Florida. The Center itself is situated approximately 242 km (150 miles) south of Jacksonville and 64 km (40 miles) due east of Orlando, on the northern end of Merritt Island adjacent to Cape Canaveral (Exhibit 3-2).

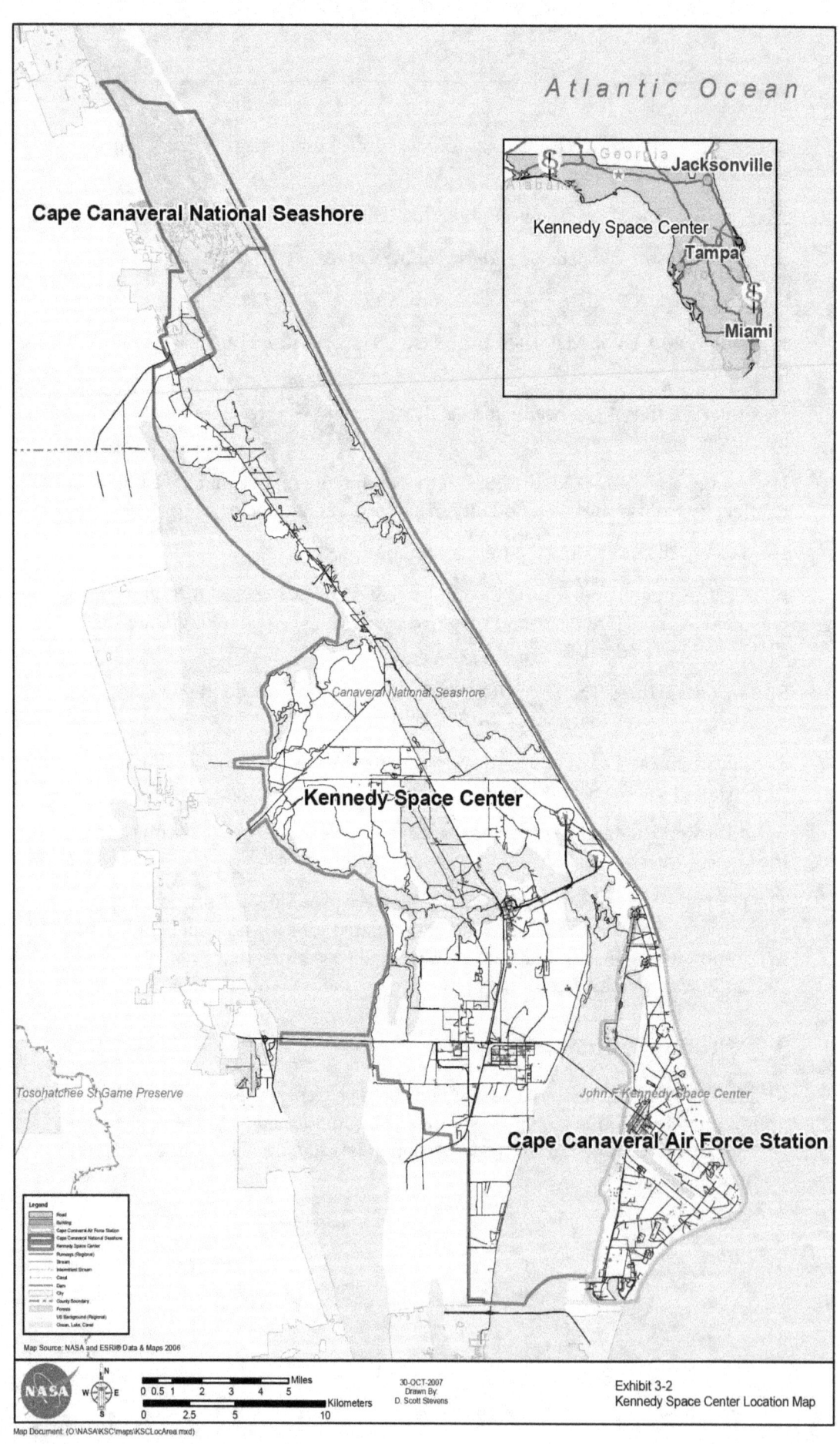

Exhibit 3-2
Kennedy Space Center Location Map

KSC is relatively long and narrow, being approximately 56 km (35 miles) in length and varying from 8 to 16 km (5 to 10 miles) in width. The Center is bordered on the west by the Indian River and on the east by the Atlantic Ocean and the CCAFS. The northernmost end of the Banana River lies between Merritt Island and the CCAFS and is included as part of KSC's submerged lands. The southern boundary of KSC runs east-west along the Merritt Island Barge Canal, which connects the Indian River with the Banana River and Port Canaveral at the southern tip of Cape Canaveral. The northern border lies in Volusia County near Oak Hill across Mosquito Lagoon. The Indian River, Banana River, and Mosquito Lagoon, collectively, make up the Indian River Lagoon System (IRLS). Merritt Island consists of prime habitat for unique and endangered wildlife; therefore, NASA entered into an agreement with the U.S. Fish and Wildlife Service (USFWS) to establish a wildlife preserve, known as the Merritt Island National Wildlife Refuge (MINWR), within the boundaries of KSC. Public Law 93-626 created the Canaveral National Seashore (CNS); thereby, an agreement with the U.S. Department of the Interior (USDI) also was formed because of the location of CNS within KSC's boundaries.

Part of KSC's mission is to process, launch, and recover Space Shuttle vehicles. Payloads are processed in the Space Shuttle and in the Expendable Launch Vehicles (ELVs) that are launched from KSC and CCAFS.

The following activities are conducted at KSC in support of NASA's mission:

- Assembling, integrating, and validating the Space Shuttle elements, along with associated payloads, including ISS elements and upper stage boosters
- Conducting launch, recovery, and landing operations
- Designing, developing, constructing, operating, and maintaining each launch and landing facility and the associated support facilities
- Maintaining the GSE required to process launch vehicle systems and their associated payloads
- Serving as the NASA point-of-contact for DoD launch activities and providing logistics support to NASA's activities at KSC, CCAFS, Patrick Air Force Base (PAFB), Vandenberg Air Force Base (VAFB), and various contingency and secondary landing sites around the world
- Managing Shuttle flight hardware logistics
- Researching and developing new technologies to support space launch and ground-processing activities
- Providing government oversight and approval authority for commercial ELV operations

The buildings at KSC that primarily are used by the SSP are shown in Exhibit 3-3. USA is the primary contractor for the SSP and operates most of the facilities at KSC. Implementing the requirements for the Shuttle are the support contractor's responsibility. The contractor manages the requirements for all of the Shuttle

This page intentionally left blank.

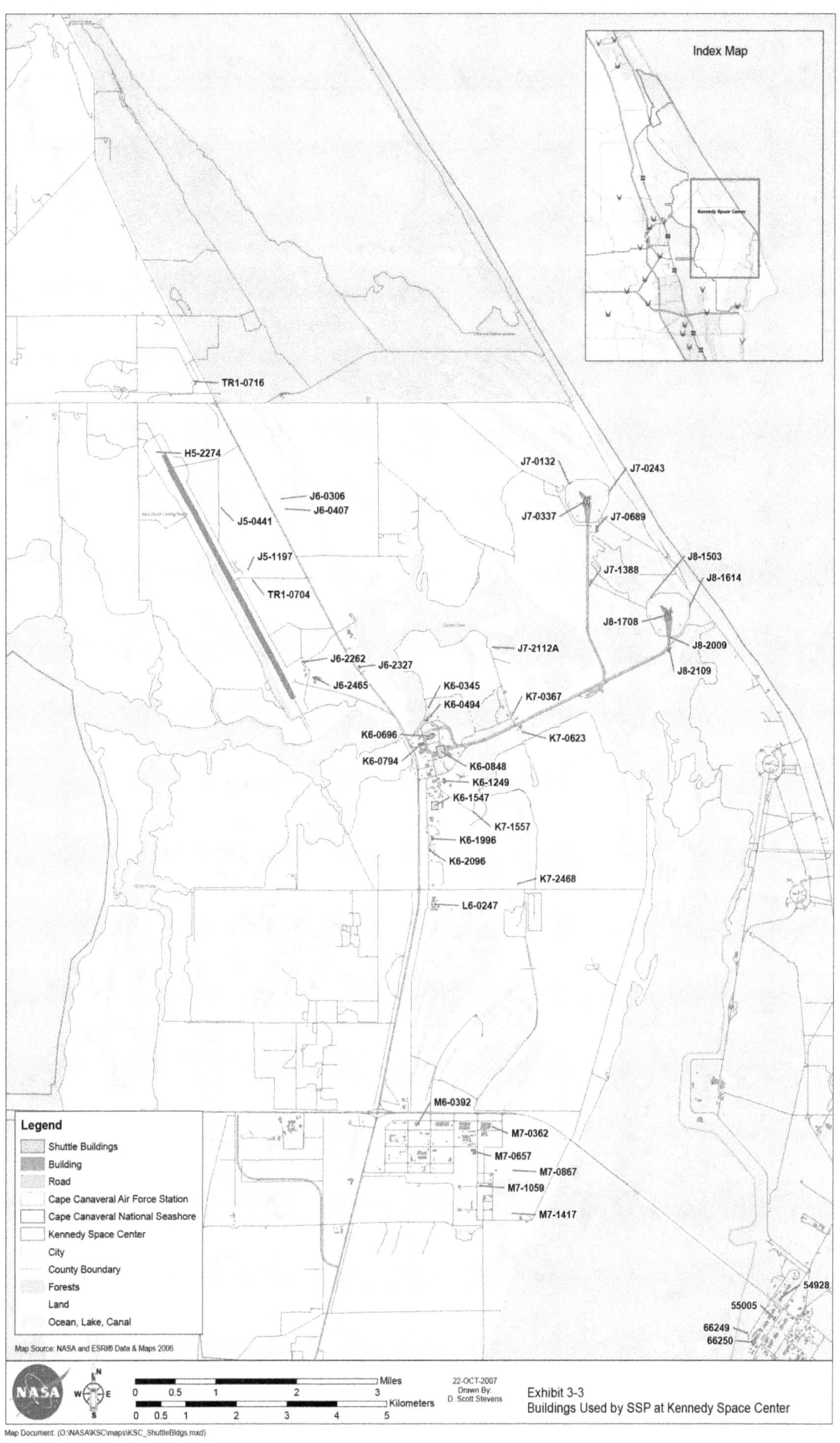

Exhibit 3-3
Buildings Used by SSP at Kennedy Space Center

This page intentionally left blank.

element activities at KSC, including processing of the Shuttle elements and the environmental requirements for construction of facilities to support Shuttle activities. The contractor also supports the KSC Emergency Management Office (EMO) with SSP environmental requirements.

3.2.1 Air Quality

3.2.1.1 Affected Environment

Regional Climate. KSC has a subtropical climate with short, mild winters and hot, humid summers. There are no recognizable spring or fall seasons. Summer weather prevails for about 9 months of the year. Rainfall ranges from an average of 2.48 inches in January to an average of 7.27 inches in August (The Weather Channel [TWC], 2007b).

The dominant weather pattern from May to October at KSC is characterized by southeast winds, which travel clockwise around the Bermuda High. The southeast winds bring moisture and warm air, which help produce almost daily thundershowers, thus creating a wet season. Approximately 70 percent of the average annual rainfall occurs during this period. In contrast to localized, heavy thundershowers in the wet season, rains are light and tend to be uniform in distribution during the dry season (NASA, 2003a).

Emission Sources. Brevard County is classified as being in attainment for all National Ambient Air Quality Standards (NAAQS) and for Florida's state-specific Ambient Air Quality Standards. Therefore, KSC follows the Prevention of Significant Deterioration (PSD) Program, instead of the more stringent New Source Review (NSR) Program (Florida Department of Environmental Protection [FDEP], 2007a).

KSC is a major source of hazardous air pollutants (HAPs), and therefore, falls under the applicable National Emission Standards for Hazardous Air Pollutants (NESHAP). The amount of regulated asbestos-containing materials (ACMs) on KSC meets the threshold of the Asbestos NESHAP, which requires that asbestos surveys and remediation must take place before remediation and demolition projects can occur (NASA, 2003a). The facility currently is operating under Title V Permit Number 0090051-012-AV. This permit was revised on September 9, 2004, and expires on August 30, 2008. It includes emission units at several buildings that are dedicated to the SSP (FDEP, 2007b).

The KSC Title V permit lists 15 hot water generators, 6 surface coating operations, 3 diesel- and gasoline-fired engines (including the emergency power plant), 19 Hypergol servicing operations and activities, and miscellaneous insignificant and unregulated emission units and/or activities (FDEP, 2007b). Halon is the only effective fire suppressant for the fuels used in the Orbiter. A "Halon Bank" has been established at KSC to fulfill the needs of the SSP. Class I and Class II ozone depleting substances (ODSs) either are recycled or disposed in a manner consistent with Title VI of the Clean Air Act (CAA) (NASA, 2007s).

3.2.2 Biological Resources

3.2.2.1 Affected Environment

Vegetation. Because of its geologic history and physical location, KSC is composed of many diverse plant communities (Exhibit 3-4). Three major vegetation types exist on KSC, including upland, anthropogenic, and wetland communities.

Approximately 14,182 hectares (ha) (35,044 acres) of KSC are considered upland vegetation, with natural communities on sites that are not flooded for extended periods. Upland vegetation consists of scrub, flatwoods, and hardwoods and mixed hardwood-coniferous forests.

Saw palmetto, wax myrtle, fetterbrush, and other shrubs and herbs comprise the understory. The hardwood and mixed forest vegetation types on KSC primarily are closed-canopy hardwood forests on upland sites (NASA, 2003a).

Anthropogenic communities are found in areas affected by development, agriculture, or other human alteration. Approximately 3,982 ha (9,840 acres) on KSC are included in this vegetation type, which consists of Australian pines (*Casuarina* spp.), citrus groves, disturbed herb shrub brush, barren land, and urban and developed areas (NASA, 2003a).

Wildlife. The proximity of uplands and wetlands and the mixing of temperate and subtropical flora provide habitat for a large number of wildlife species on KSC (NASA, 2003a).

MINWR is considered one of the top 10 birding spots in the U.S. A total of 267 bird species have been identified on KSC.

The IRLS has the highest fish species diversity of any estuary in North America. Nearly 150 species of fish have been identified in the lagoon surrounding KSC, with the highest diversity generally near inlets and toward the southern end of the lagoon.

Protected Species and Habitats. More federally protected species are found at MINWR than at any other national wildlife refuge in the continental U.S. Exhibit 3-5 lists the state and federally protected species at KSC.

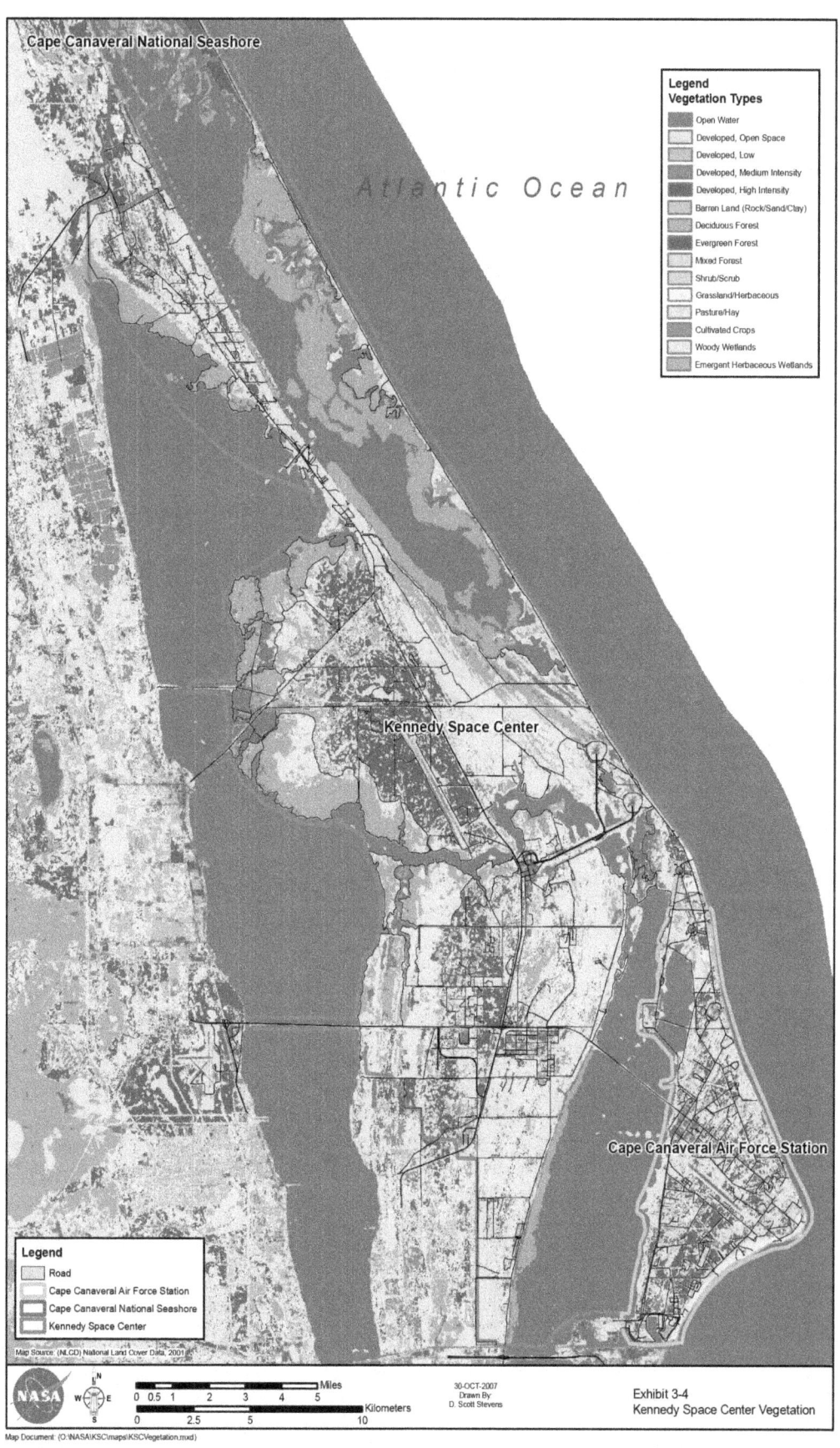

EXHIBIT 3-5
State and Federally Protected Species at KSC

Scientific Name	Common Name	Level of Protection	
Amphibians and Reptiles		State	Federal
Rana capito aesopus	Florida gopher frog	SSC	
Alligator mississippiensis	American alligator	SSC	T(S/A)
Caretta caretta	Loggerhead	T	T
Chelonia mydas	Atlantic green turtle	E	E
Dermochelys coriacea	Leatherback sea turtle	E	E
Gopherus polyphemus	Gopher tortoise	SSC	
Drymarchon couperi	Eastern indigo snake	T	T
Pituophis melanoleucus mugitus	Florida pinesnake	SSC	
Birds			
Pelecanus occidentalis carolinensis	Eastern brown pelican	SSC	
Egretta thula	Snowy egret	SSC	
Egretta caerulea	Little blue heron	SSC	
Egretta tricolor	Tricolored heron	SSC	
Egretta rufescens	Reddish egret	SSC	
Eudocimus albus	White ibis	SSC	
Ajaia ajaja	Roseate spoonbill	SSC	
Mycteria americana	Wood stork	E	E
Falco peregrinus tundrius	Arctic peregrine falcon	E	
Falco sparverius paulus	Southeastern American kestrel	T	
Sterna antillarum	Least tern	T	
Rynchops niger	Black skimmer	SSC	
Aphelocoma coerulescens	Florida scrub-jay	T	T

EXHIBIT 3-5
State and Federally Protected Species at KSC

Scientific Name	Common Name	Level of Protection	
Mammals			
Peromyscus polionotus niveiventris	Southeastern beach mouse	T	T
Podomys floridanus	Florida mouse	SSC	
Trichechus manatus	West India manatee	E	E

Notes:
E = Endangered
SSC = Species of special concern
T(S/A) = Threatened because of similarity of appearance to another protected species
T = Threatened

Turtles and manatees are discussed further below, because these species rely heavily on the habitats at KSC.

Turtles. Three of the top 10 turtle nesting beaches in the U.S. are located within the KSC, CNS, and CCAFS property. The KSC and MINWR ocean beaches, as well as those of CNS and CCAFS, provide excellent nesting habitat for marine turtles and are used by the loggerhead (*Caretta caretta*), the green turtle, and the leatherback sea turtle (*Dermochelys coriacea*) (NASA, 2003a).

Gopher tortoises (*Gopherus polyphemus*) are terrestrial turtles that also nest on KSC. The gopher tortoise is considered a "keystone species" because its burrows provide habitat for hundreds of invertebrates and vertebrate species. Several wildlife species that use tortoise burrows also are federally or state protected, and having healthy, reproductive gopher tortoise colonies is essential for these other species' survival (NASA, 2003a).

Manatees. As much as 15 percent of the total manatee population of the U.S. is located in the waters immediately surrounding KCS. In 1990, the USFWS created a sanctuary for manatees that covers the majority of the KSC section of the Banana River. The USFWS also designated the following areas at KSC as critical habitat: 1) the entire inland section of water known as the Indian River, from its northernmost point immediately south of the intersection of U.S. Highway 1 and Florida State Route (SR) 3; 2) the entire inland section of water known as the Banana River north of the Kennedy Athletic, Recreational, and Social Organization (KARS) Park; and 3) all waterways between the Indian and Banana Rivers (exclusive of existing manmade structures or settlements that are not necessary to the normal needs of survival of the species) (NASA, 2003a).

3.2.3 Cultural Resources

3.2.3.1 Affected Environment

Historic Resources. The following buildings and structures at KSC are listed individually on the NRHP:

- VAB (K6-0848)
- Launch Control Center (LCC) (K6-0900)
- Crawlerway (UK-0008)
- Two Missile Crawler Transporter Facilities
- Press Site: Clock and Flag Pole
- The Launch Complex (LC) 39: Pad A Historic District is a multiple property NRHP listing originally listed as significant to the Apollo program; it has since been determined to be significant to the SSP. The district includes the following buildings:
  - Pad A (J8-1708)
  - High Pressure Gaseous Hydrogen (GH2) Facility (J8-1462)
  - LOX Facility (J8-1502)
  - Operations Support Building (OSB) A-1 (J8-1503)
  - Camera Pad A No. 1 (J8-1512)
  - LH2 Facility (J8-1513)
  - Electrical Equipment Building No. 2 (J8-1553)
  - Camera Pad No. 6 (J8-1554)
  - Electrical Equipment Building No. 1 (J8-1563)
  - Operations Support Building A-2 (J8-1614)
  - Slidewire Termination Facility (J8-1703)
  - Water Chiller Building (J8-1707)
  - Camera Pad A No. 2 (J8-1714)
  - Camera Pad A No. 4 (J8-1956)
  - Camera Pad A No. 3 (J8-1961)

Resources newly determined eligible at Pad A as contributing to the SSP include the following:

- Water Tank (J8-1610)
- Flare Stack (J8-1611)
- Electrical Equipment Building No. 3 (J8-1811)
- Electrical Equipment Building No. 4 (J8-1856)
- Hypergol Oxidizer Facility (J8-1862)
- Hypergol Fuel Facility (J8-1906)

LC-39 Pad B Historic District is a separate multiple property NRHP listing for resources associated with the Apollo context and now with the SSP. The buildings in this district include the following:

- Pad B (J7-0337)
- OSB B-1 (J7-0132)
- High Pressure GH2 Facility (J7-0140)
- LOX Facility (J7-0182)
- Camera Pad B No. 6 (J7-0183)
- Camera Pad B No. 1 (J7-0191)
- LH2 Facility (J7-0192)
- Electrical Equipment Building No. 2 (J7-0231)
- Electrical Equipment Building No. 1 (J7-0241)
- OSB B-2 (J7-0243)
- Slidewire Termination Facility (J7-0331)
- Camera Pad B No. 2 (J7-0342)
- Water Chiller Building (J7-0385)
- Camera Pad B No. 4 (J7-0584)
- Camera Pad B No. 3 (J7-0589)

The following resources were newly determined eligible for the Pad B Historic District: J7-0240, J7-0288, J7-0490, J7-0491, J7-0534, and J7-0535.

The following properties at KSC have been recommended as being eligible individually for listing on the NRHP for their association with the SSP, per the recent SSP survey:

- Parachute Refurbishment Facility (M7-657)
- Canister Rotation Facility (M7-777)
- Payload Canister (2)
- Retrieval Ships (Liberty Star and Freedom Star)
- SRB ARF Manufacturing Building
- Three Mobile Launcher Platforms
- Rotation/Processing Building

Four new districts were identified in the recent SSP survey–the SLF Historic District (HD); the Orbiter Processing HD; the SRB Assembly, Disassembly and Refurbishment HD; and the Hypergol Maintenance and Checkout Area (HMCA) HD. The SLF HD includes the following individually significant properties–the Shuttle Runway, Landing Aids Control Building (J6-2313), and Mate-Demate Device (J6-2262). The Orbiter Processing HD includes the following individually significant properties–OPF (K6-894), OPF High Bay (HB) 3 (includes the SSME Processing Facility) (K6-0696), and TPS Facility (K6-794). The Hypergol Module Processing North (M7-961) is an individually significant property that contributes to the HMCA

HD; the Hypergol Support Building (M7-1061) is a contributing resource to the HMCA HD (Archaeological Consultants, Inc., 2007b).

The SRB Assembly and Refurbishment Complex HD includes the following facilities that are not individually eligible, but that are contributing to the newly established HDs at KSC:

- Hangar AF (66260)
- High Pressure Gas Facility (66251)
- High Pressure Wash Facility (66240)
- First Wash Building (66242)
- SRB Recovery Slip (66244)
- SRB Paint Building (66310)
- Robot Wash Building (66320)
- Thrust Vector Control Deservicing Building (66249)
- Multi-Media Blast Facility (66340)

Two buildings–M6-0399 and M7-0355–were listed but are not eligible under the SSP context.

Exhibit 3-6 shows the properties listed and eligible for listing on the NRHP at KSC.

3.2.4 Hazardous and Toxic Materials and Waste

3.2.4.1 Affected Environment

Storage and Handling. Operations at KSC involve numerous types of hazardous materials to support the SSP. Each contractor procures its own materials, which are received at several points around KSC. The purchase, transport, and temporary storage of propellants are controlled by the Joint Propellants Contractor (JPC). Releases that occur at KSC are reported to the Environmental Program Branch (EPB). The determination of whether the release is a reportable quantity of a reportable substance is done by the EPB. Notification and correspondence with offsite authorities regarding releases that have occurred are coordinated by the EPB (NASA, 2007aa). Compliance with the Resource Conservation and Recovery Act (RCRA) is achieved by the directives listed in applicable permits issued to KSC (Kennedy NASA Procedural Requirements [KNPR] 8500.1 "KSC Environmental Requirements") (NASA, 2002b).

Waste Management. KSC has an FDEP operating permit for the storage, treatment, and disposal of hazardous waste. The main facility that operates under this permit is the Hazardous Waste Storage Facility (K7-165) in the LC-39 area, which handles liquid and solid hazardous wastes. There are four cells at the facility, each of which

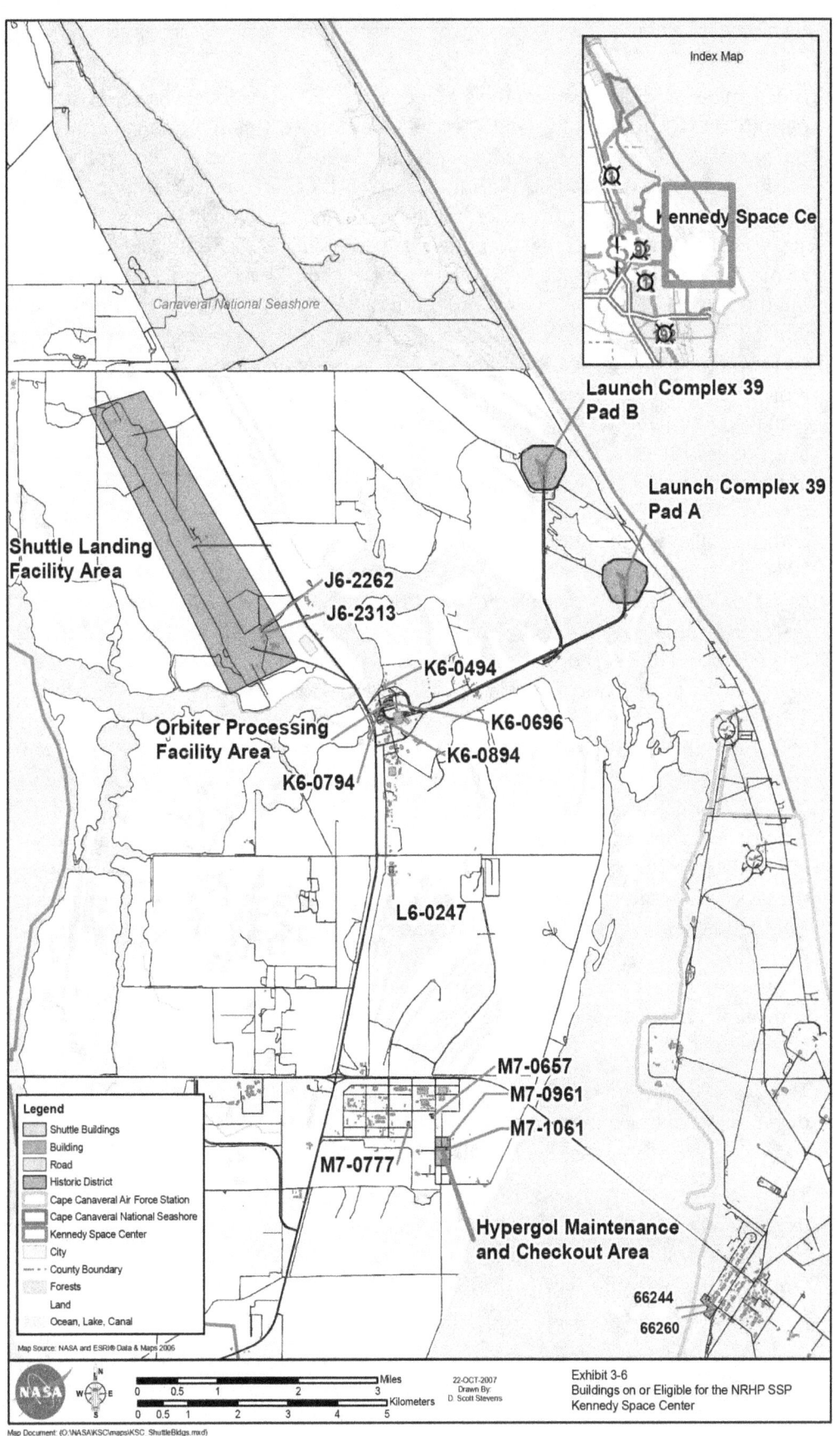

is designated and designed for the storage of specific hazardous wastes. Wastes permitted to be stored at the facility include flammable, organic, and toxic waste; caustic toxic reactive wastes; acidic waste; and solid hazardous and controlled wastes. Joint Base Operations Support Contract (JBOSC) Waste Operations operates the facility and maintains the records and reports associated with the waste activities at the facility to ensure Center compliance (NASA, 2003a). KSC maintains a comprehensive inventory of RCRA-defined hazardous wastes and controlled wastes not regulated by RCRA. This inventory is maintained by a manifest records system, which tracks the generation, onsite storage, treatment, and reclamation of hazardous and controlled wastes. Various types of wastes being managed include used oil, which is recycled; used antifreeze, which is recycled; and fluorescent lamps that are managed as universal waste and also are recycled. The manifest records system is integrated with an automated data processing system, which provides the capability to generate current waste status reports, as well as quarterly and annual summary reports. The JBOSC contractor is responsible for the maintenance of the hazardous and controlled waste database inventory, including the KSC Biennial Hazardous Waste Disposal Report (NASA, 2003a). The amount of hazardous waste generated at KSC classifies the Center as a large-quantity generator (LQG) of hazardous waste. NASA is regulated by RCRA Permit #FL68000014585 for KSC, for the storage, treatment, and disposal of hazardous waste. As noted above, the main facility regulated by this permit is the Hazardous Waste Storage Facility (K7-165) in the LC-39 area (NASA, 2003a). NASA has developed a program of managing and handling hazardous and controlled wastes at KSC in compliance with the provisions of its permit, RCRA, and the implementing regulations adopted by the State of Florida (62-730, Florida Administrative Code [F.A.C.]). The organizational and procedural requirements of the KSC hazardous waste management program are contained in KNPR 8500.1, "KSC Environmental Requirements." This report clearly delineates the procedures and methods for obtaining and providing hazardous waste support; establishing and approving operations and maintenance (O&M) instructions; and providing instructions to maximize resource recovery and minimize costs. Additionally, the Center uses the JBOSC to provide contractor support for the management and storage of wastes to be disposed offsite from the Center's permitted TSDF.

The Center's hazardous and non-RCRA-regulated waste generation activities are dependent on launch processing, construction, and associated activities (NASA, 2003a).

The number of hazardous waste collection sites maintained at the Center is dynamic. The contractors continually review the processes to reduce the amount of hazardous waste being generated, which in turn reduces the number of sites required to manage the wastes (NASA, 2003a).

Contaminated Areas. The U.S. Environmental Protection Agency (EPA) has conducted an RCRA Facility Assessment (RFA) at KSC to identify solid waste management units (SWMUs) for RCRA corrective action (NASA, 2007s).

Asbestos and Lead-based Paint. There are three instances in which KSC is required to report to the FDEP the abatement and/or demolition of regulated ACMs (RACMs) (NASA, 2007z) as follows:

- Individual abatement projects
- Annual abatement projects
- Demolition of facility projects

Lead-based paint (LBP) is a potential concern for the demolition of buildings and structures built before 1978. Demolition debris potentially containing LBP is subject to landfilling restrictions. Maintenance activities could have created the potential for localized lead contamination in soils in areas around those older buildings or structures.

PCB-contaminated Paint or Coating. The potential for the presence of polychlorinated biphenyls (PCBs) in paints and coatings on KSC structures has been documented. Exterior and structural surface paints and coatings either must be sampled and analyzed, or must be considered to contain greater than 50 parts per million (ppm) PCBs. Materials that have PCB concentrations greater than or equal to 50 ppm are regulated by and must be managed in accordance with the requirements specified in 40 CFR 761 (NASA, 2007aa).

If PCBs are detected in the sampling analyses of the paints and coatings to be removed, the paint or coating waste must be stored and managed according to the Toxic Substances Control Act (TSCA) regulations for PCB wastes. The storage and management procedures vary according to the PCB concentrations in the paints or coatings. After the PCB paints or coatings are removed from metal and/or concrete, the remaining material can be handled and recycled as non-PCB material. However, if the removal of paints or coatings and material segregation and recycling are not possible, the PCB bulk product waste from construction and demolition debris must be transported to the KSC Landfill on Schwartz Road, in accordance with the site's operating permit and associated procedures, or to an approved landfill or incinerator under 40 CFR 761 (NASA, 2007z).

It has been demonstrated that paint chips containing PCBs have caused or contributed to environmental contamination at KSC, thus resulting in site cleanup.

3.2.5 Health and Safety

The discussion of human health and safety includes both workers (NASA and other government personnel, and contractor personnel) and the general public. Safety issues include injuries that may result from one-time accidents. Health issues result from activities wherein people may be affected over a long period of time rather than immediately. The affected environment for health and safety will include those areas that have the potential to be affected by the SSP T&R. This discussion will include existing hazards such as emergency preparedness and response, explosion and fire hazards, and other Center-specific hazards. In addition, existing safety procedures will be described. Issues related to the use of hazardous materials and the generation of hazardous wastes will be addressed in detail under the hazardous materials and hazardous waste sections of this EA.

3.2.5.1 Affected Environment

The potential impacts are outlined in the following subsections.

Hazardous Materials Exposure. Hazardous materials are used in the production and processing of the SSP. The hazardous materials used and hazardous wastes generated are discussed in detail in Section 3.2.4. The degree of exposure to hazardous materials is minimized by the implementation of work practices and control technologies. The risks associated with hazardous materials are managed under NASA Policy Directive (NPD) 1820.1B.

Hazardous Materials Transportation Safety. Hazardous materials such as propellant and chemicals are transported in accordance with U.S. Department of Transportation (DOT) regulations for the interstate shipment of hazardous substances (49 CFR 100 through 199). Hazardous materials such as liquid rocket propellant are transported in specially designed containers to reduce the potential for spills or accidents.

Explosions and Fire Hazards. The storage and use of certain hazardous materials, including propellants, in SSP production presents a risk of explosions and fire hazards. To minimize these risks, NASA has implemented several physical and procedure controls.

Although unlikely, explosions of propellants or other hazardous materials could result in damage to structures and personnel thousands of feet from the ignition site. Additionally, KSC has implemented the use of quantity-separation distances (QD arcs), or the minimum safe distances required to separate two given sites or buildings where at least one of the sites has a potential for an explosion or fire. The implementation of control technologies and QD arcs has minimized the risk of explosions and fire hazards associated with the SSP operations at KSC (NASA, 2003a).

3.2.6 Hydrology and Water Quality

3.2.6.1 Affected Environment

Surface Waters. Surface waters at KSC are associated with the IRLS and the Atlantic Ocean. The IRLS consists of the Mosquito Lagoon and the Banana and Indian Rivers (Exhibit 3-7). These waters are shallow, aeolian lagoons with depths averaging

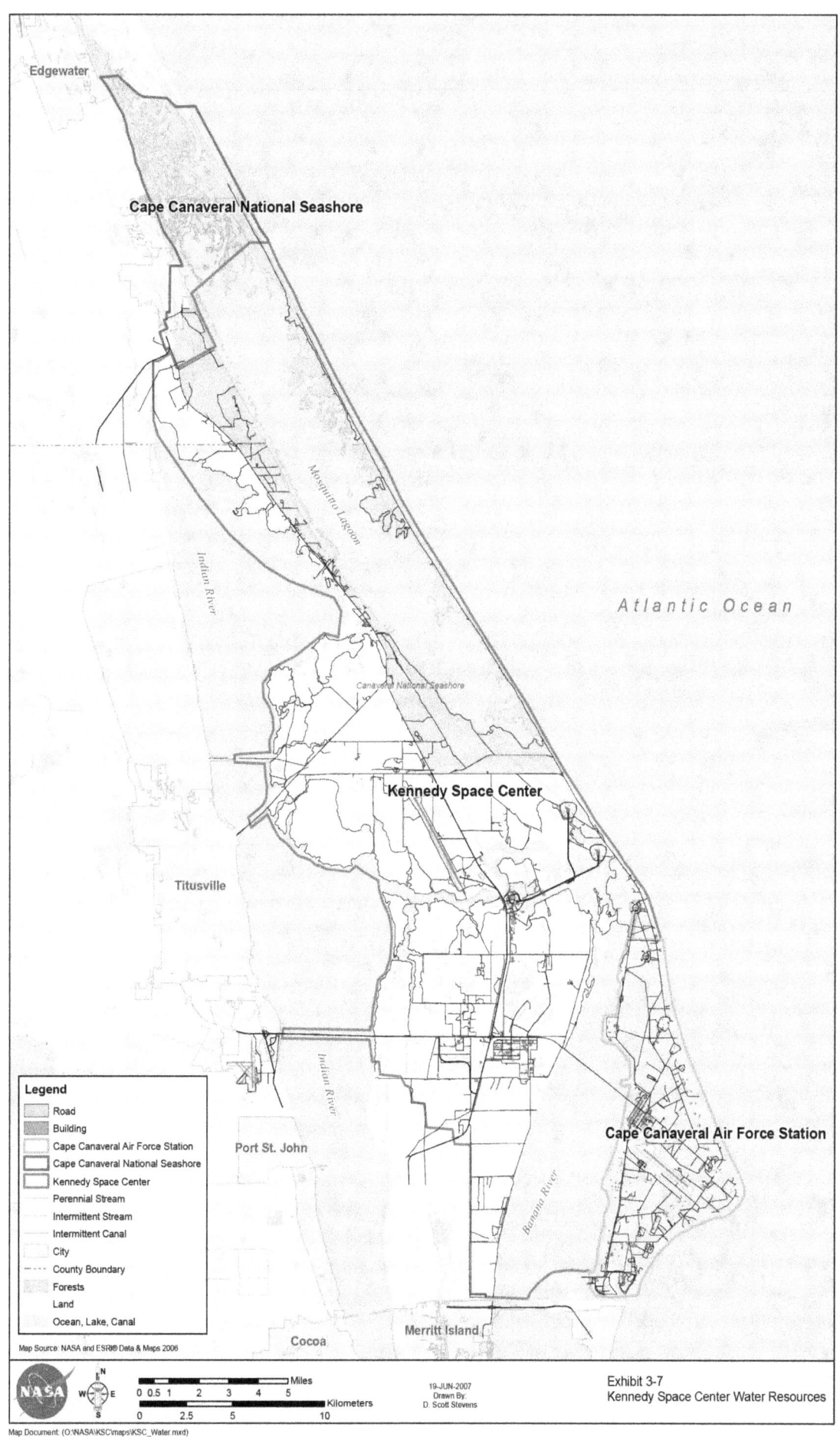

Exhibit 3-7
Kennedy Space Center Water Resources

This page intentionally left blank.

1.8 m (5.9 ft) and maximum depths of 9 m (29 ft), generally restricted to dredged basins and channels.

Mosquito Lagoon and the Indian River are connected by Haulover Canal and the Intercoastal Waterway along the western edge of KSC. Water flow between these two systems is primarily wind-driven. No circulation occurs between Mosquito Lagoon and the Banana River. The Indian and Banana Rivers connect in the region near Eau Gallie and through a canal located just south of KSC.

Mosquito Lagoon connects to the Atlantic Ocean through the Ponce de Leon Inlet approximately 49 km (31 miles) north of KSC. Port Canaveral provides an oceanic connection to the Banana River approximately 12 km (7.5 miles) south of KSC.

However, navigation locks in Port Canaveral eliminate significant oceanic influence on the Banana River. The Sebastian Inlet, located 80 km (50 miles) south of KSC, is the nearest southerly oceanic connection to the Indian River. The remoteness of the estuarine waters from oceanic influence and the restrictions imposed by constructed causeways minimize water circulation within the lagoon basins. Surface water movement and flushing are primarily functions of wind-driven forces. Salinity regimes are controlled mostly by precipitation, upland runoff, evaporation, and groundwater seepage.

The primary freshwater body in KSC is Banana Creek, which drains the estuaries adjacent to the Space Shuttle launch pads via a canal located northwest of the VAB. Salinity usually increases in a westward direction, but depending on the wind direction, the Indian River system can have a greater or lesser effect on salinity in Banana Creek. Other freshwater inputs to the estuarine system surrounding KSC include direct precipitation, storm water runoff, discharges from impoundments, and groundwater seepage (NASA, 2003a). These input sources are generally of high quality and do not adversely affect the water quality of the receiving waters.

Groundwater. KSC is underlain by two aquifer systems. The largest is the Floridan aquifer, one of the highest-producing aquifers in the world. This aquifer system is composed of a sequence of limestone and dolomite, which thickens from about 250 ft in Georgia to about 3,000 ft in south Florida. The Floridan aquifer system has been divided into an upper and lower aquifer, separated by a unit of lower permeability. The upper Floridan aquifer is the principal source of water supply in most of north and central Florida. In the southern and coastal portions of the aquifer, it contains brackish water and is non-potable. Groundwater flow is generally from near the center of the state toward the coast.

The surficial aquifer system, the smaller aquifer system in Florida, encompasses KSC and includes undefined aquifers that are present at the land surface. The surficial aquifer system is generally under unconfined, or water-table, conditions and is made up mostly of unconsolidated sand, shelly sand, and shell. The aquifer thickness is typically less than 50 ft. Groundwater in the surficial aquifer generally flows from areas of higher elevation toward the coast or streams, where it can discharge as baseflow. Water enters the aquifer from rainfall and exits as baseflow to streams, discharge to the coast, evapotranspiration, and downward recharge to deeper aquifers. Because of its lower yield, the surficial aquifer mainly is used for domestic, commercial, or small municipal supplies (FDEP, 2006).

Water Quality. The FDEP has established minimum water quality standards for five classifications of surface waters based on their potential use. In addition to the use designations listed in Exhibit 3-8, waters also may be assigned additional protection though designations such as Outstanding Florida Waters or Aquatic Preserves.

EXHIBIT 3-8
Designated Uses for Surface Waters on KSC

Water	Florida Surface Water Classification	Description of Classification
Banana River Banana Creek Majority of Indian River	Class 3	Standards are established to ensure safe recreation and fish and wildlife propagation.
Northernmost portion of Indian River Mosquito Lagoon	Class 2	Standards are established to protect shellfish propagation and harvesting. This designation carries more stringent limits on bacterial and fluoride concentrations and prohibits discharges of treated wastewater.
All waters within Merritt Island National Wildlife Refuge (MINWR)	Outstanding Florida Waters	Water quality may not be degraded below ambient water quality conditions.
Mosquito Lagoon	Designated Aquatic Preserve	A management plan has been prepared for the system.

Notes:
Class 2: Shellfish Propagation and Harvesting
Class 3: Recreation and Fish and Wildlife Propagation
Source: NASA, 2003a

Water quality is monitored by several different monitoring programs. NASA, the St. John's River Watershed Management Division (SJRWMD), and Brevard County maintain water quality monitoring stations around and within KSC's boundaries. Surface water quality at KSC is considered to be generally good (NASA, 2003a). However, some segments of the Banana River, Mosquito Lagoon, and Indian River near KSC are considered impaired and are included on Florida's draft 2006 303d list. The Indian and Banana Rivers above the 520 Causeway are listed as impaired because of mercury contamination in fish tissue and low dissolved oxygen (DO) concentrations. The Indian River above the NASA Causeway is listed as impaired because of mercury in fish tissue and elevated concentrations of nutrients. The Atlantic Coast is listed as impaired because of mercury in fish tissue (FDEP, 2006).

Regulated Water Discharges and Withdrawals. KSC has more than 100 surface water management systems to control storm water runoff. One National Pollutant Discharge Elimination System (NPDES) permit has been issued (FLR05F574) for a storm water system. A Storm Water Pollution Prevention Plan (SWP3) was implemented to meet the permit requirements. Implementation of the SWP3 includes conducting analytical and visual monitoring of storm water runoff (NASA, 2007s).

Raw wastewater is pumped from KSC to the permitted CCAFS Regional Wastewater Treatment Facility (FLA010292), located on CCAFS, for treatment (NASA, 2007u). Three facilities have operating permits to treat industrial wastewater. The three facilities and their discharges are as follows:

- SRB Refurbishment Area–Water from hydrolase cleaning of SRBs is filtered, treated onsite, and reused.
- Visitors Center Bus Wash–Water generated from vehicle cleaning is treated in a 100-percent closed-loop, recycled washwater plant and reused.
- LICON Recycling System (Component Refurbishment and Cleaning Area)– Waste streams from component cleaning, an analytical laboratory, and a compressor discharge storage tank are treated and reused in the testing laboratory. A wet concentrated residual is obtained, which is tested for hazardous characteristics and disposed offsite (NASA, 2003a).

There are a number of septic tank systems throughout KSC that typically support small offices or temporary facilities. Only a small percentage of the existing septic tanks is permitted by the State of Florida (Chapter 64E-6, F.A.C). The remaining septic tanks were constructed before the permitting regulations were implemented, and therefore, are not subject to these rules (NASA, 2007s).

KSC has a consumption use permit (#50054) for water for household, industrial, aesthetic, and agricultural and landscaping uses. The permit allows for withdrawals of up to 353.27 million gallons per year (mgy). Most of that amount is provided by the cities of Cocoa and Titusville, although up to 13.23 mgy are pumped from the Floridan and surficial aquifers beneath KSC (NASA, 2007u).

3.2.7 Land Use

3.2.7.1 Affected Environment

NASA has developed two plans that guide current and future land use planning at KSC. Current land uses were established in accordance with the KSC Master Plan and the Cape Canaveral Spaceport Master Plan (CCSMP). The KSC Master Plan, produced in 1984, focused on NASA operational areas (NASA, 2003a). In 2000, NASA formed financial partnerships at KSC with the Air Force 45th Space Wing at CCAFS and the Space Florida to perform joint planning for KSC and CCAFS, which collectively are known as the Cape Canaveral Spaceport (CCS) (NASA, 2003a). CCS land use is managed in accordance with the CCSMP.

The USFWS is developing the MINWR Comprehensive Conservation Plan, which will guide USFWS operations in the MINWR for a 15-year period (USDI, 2006).

The NPS is developing a general management plan for CNS that will identify methods to protect and manage the seashore for the next 15 to 20 years. The NPS has developed a Resource Management Plan that summarizes its immediate and long-term resource management objectives (NASA, 2003a).

NASA oversees the 56,449 ha (139,490 acres) that make up KSC. The overall land management objectives of NASA and KSC are to maintain the nation's space mission operations, while supporting alternative land uses that are in the nation's best interests and maximizing environmental protection. All zoning and land use planning falls under NASA's directive for implementation of the nation's Space Program. Essential safety zones, clearance areas, lines-of-sight, and similar restrictions were developed as guides to master planning and, where applicable, as mandatory operational requirements. All facility sitings and projects are reviewed extensively, with special attention given to the requirements described in this subsection. For areas not directly used for NASA operations, land planning and management responsibilities have been delegated to the NPS and the USFWS (Exhibit 3-9). These agencies exercise management control over agricultural, recreational, and environmental programs at KSC (NASA, 2003a).

Undeveloped lands dominate KSC. Undisturbed areas include uplands, wetlands, mosquito control impoundments, and open water areas, comprising approximately 95 percent of the total KSC area. Nearly 40 percent of KSC consists of open water areas, including portions of the Indian and Banana Rivers, Mosquito Lagoon, and all of Banana Creek (NASA, 2003a).

NASA Operational Areas. NASA has devised the following 11 land use categories to describe the areas within KSC in which various types of operational or support activities are conducted (NASA, 2003a):

- **Launch.** The Launch land use classification includes all facilities directly related to vehicle launch operations and is subdivided into horizontal and vertical launch subcategories.
- **Launch Support.** The Launch Support land use classification includes all facilities and operations not classified as Launch that are essential to processing and launching a vehicle from the Spaceport, recovering and processing a vehicle returning to the Spaceport, and supporting a mission during flight.
- **Airfield Operations.** The Airfield Operations land use classification includes runways and helipads.
- **Spaceport Management.** The Spaceport Management land use classification includes all administrative functions that provide for management and oversight of Spaceport operations, plus the services administered by those managing entities for the benefit of the overall Spaceport complex, including O&M, service and utilities, and infrastructure.

EXHIBIT 3-9
KSC Administrative Areas

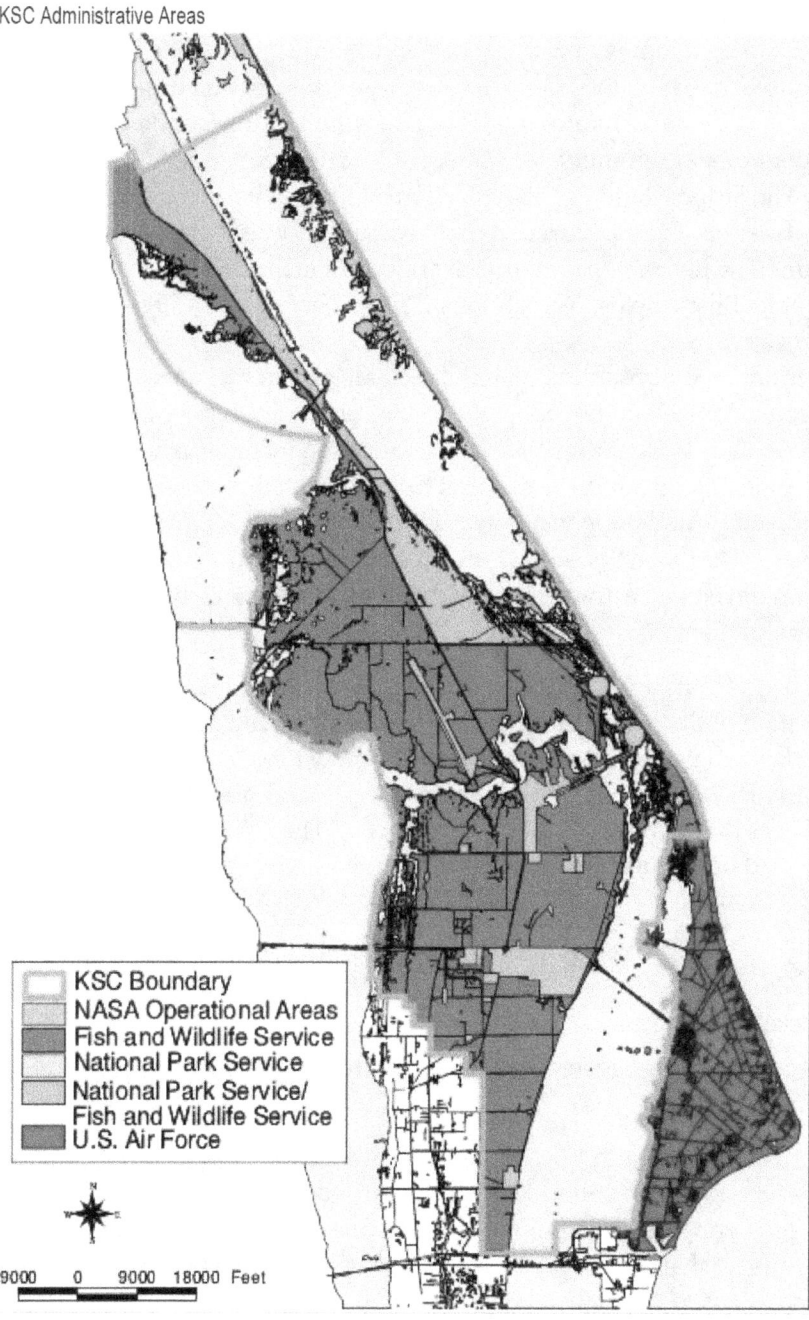

- **Research and Development (R&D).** The R&D land use classification includes laboratories and related facilities that perform testing and experimentation for the purpose of developing new programs and technologies at the Spaceport.
- **Public Outreach.** The Public Outreach land use classification designates facilities that provide an informational or educational connection between the Spaceport and the community.
- **Seaport.** The Seaport land use classification includes wharves used for the docking of vessels and facilities that directly support wharf operations.
- **Recreation.** The Recreation land use classification includes parks, outdoor fitness areas, athletic fields, recreation buildings, centers, and clubs in the Spaceport complex.
- **Conservation.** The Conservation land use classification includes all natural areas and all undeveloped land not assigned to another land use classification.
- **Agriculture.** The Agriculture land use classification includes land areas used for the cultivation of crops or plant material for commercial purposes or for Spaceport facility landscape maintenance.
- **Open Space.** The Open Space land use classification includes undeveloped open land within developed activity centers identified as being likely locations for future development.

Special land use permits are considered during the review of facility siting requests. Both the duration of the permit and which department within NASA assigns the permit vary. Special permits are for activities that take place at KSC and can cover a variety of activities. One example of a current special land use permit is the U.S. Army Corps of Engineers' (USACE's) spoil site. The USACE has a permit for a spoil area located on the northern bank of the Barge Canal at the southern boundary of KSC (NASA, 2003a).

3.2.8 Noise

3.2.8.1 Affected Environment

The two categories of noise generated at KSC are from industrial activities and other man-made noises.

The closest residential areas to KSC are to the south, in the cities of Cape Canaveral and Cocoa Beach. Expected sound levels in these areas are normally low, with higher levels occurring in industrial areas (Port Canaveral) and along transportation corridors. Residential areas and resorts along the beach would be expected to have low overall noise levels, normally about 45 to 55 dBA (NASA, 2003a).

A number of permanent and temporary measures are taken to reduce noise levels at KSC and to protect employee noise exposures at KSC. Noise abatement measures for any facility or operation include the following:

- Property acquisition for use as a buffer zone
- Landscaping with high, dense vegetation or earthen berm
- Noise insulation of buildings
- Permanent noise barriers erected
- Proper scheduling (day or night) of a specified activity might eliminate or alleviate noise impacts during critical periods

For construction projects, portable sound screens and the strategic placement of stationary machinery to avoid noise impacts are used when possible to minimize noise levels (NASA, 2003a).

The typical noise levels associated with the activities at KSC are listed in Exhibit 3-10 and discussed further in the following subsections.

EXHIBIT 3-10
Noise Generated at KSC

Noise Type	Noise Range (dBA)
Aircraft Noise	87-158[1]
Industrial Operations	45-199[2]
Construction	54-111
Traffic Noise	51-110

Notes:
[1] Calculated from ground zero. Clearance zones are established to preclude significant adverse impacts to humans.
[2] Noise at upper range is generated by the operation of hydraulic pumps within enclosed spaces.
dBA = Decibel A-rated
Source: NASA, KSC ERD (August 2003a).

Industrial Noise. Industrial operations are associated with the assembly and preparation of the Shuttle for launches and maintenance of GSE, which generate noise. Hydraulic pumps operating within the confines of their enclosures produce the loudest noise generated by industrial activities at KSC. Operators of these pumps and of other industrial operations are required by Occupational Safety and Health Administration (OSHA) regulations to be equipped with ear protection devices when exposed to noise levels above 90 decibels (dB) for an 8-hour work day. KSC maintains an occupational hearing program to ensure that employees are protected from industrial noises. Other intermittent raised levels of noise occur during the operation of the following:

- Lifting equipment
- Diesel-powered generators and locomotives
- Heavy-duty service vehicles
- The Crawler-Transporter
- Sheet metal forming and cutting processes
- Aqualaser removal of residual thermal protection materials from recovered SRBs (NASA, 2003a)

Other Man-made Noises. General sources of noise at KSC include traffic and construction. Average ambient noise levels at KSC over a 24-hour period are appreciably lower than 70 dBA and have no impact outside the KSC boundaries. The intermittent noise of arriving and departing vehicles, including visitors, is no greater than that experienced in a major shopping center parking lot (NASA, 2003a).

A number of aircraft are used at KSC for payload delivery, ferry support, NASA executives, security, and astronaut training. Typically, noise levels are no greater than those experienced by a small commercial airport (NASA, 1997).

3.2.9 Site Infrastructure

3.2.9.1 Affected Environment

Wastewater System. See the Hydrology and Water Quality section for KSC (Section 3.2.6).

Storm Water System. See the Hydrology and Water Quality section for KSC (Section 3.2.6).

3.2.10 Socioeconomics

3.2.10.1 Region of Influence

The economic ROI for KSC is defined as the Florida counties of Brevard, Orange, Seminole, and Volusia (Exhibit 3-11), where approximately 96 percent of KSC employees live. The majority (78.5 percent) of KSC employees live in Brevard County, many of those in the City of Titusville, which is considered the "gateway" to KSC (Personal Communication, 2007b).

The four counties of the KSC ROI include the "Space Coast" area and are part of the larger central Florida Region, which is composed of Brevard, Flagler, Lake, Orange, Osceola, Seminole, and Volusia counties. Brevard County is designated as the Palm

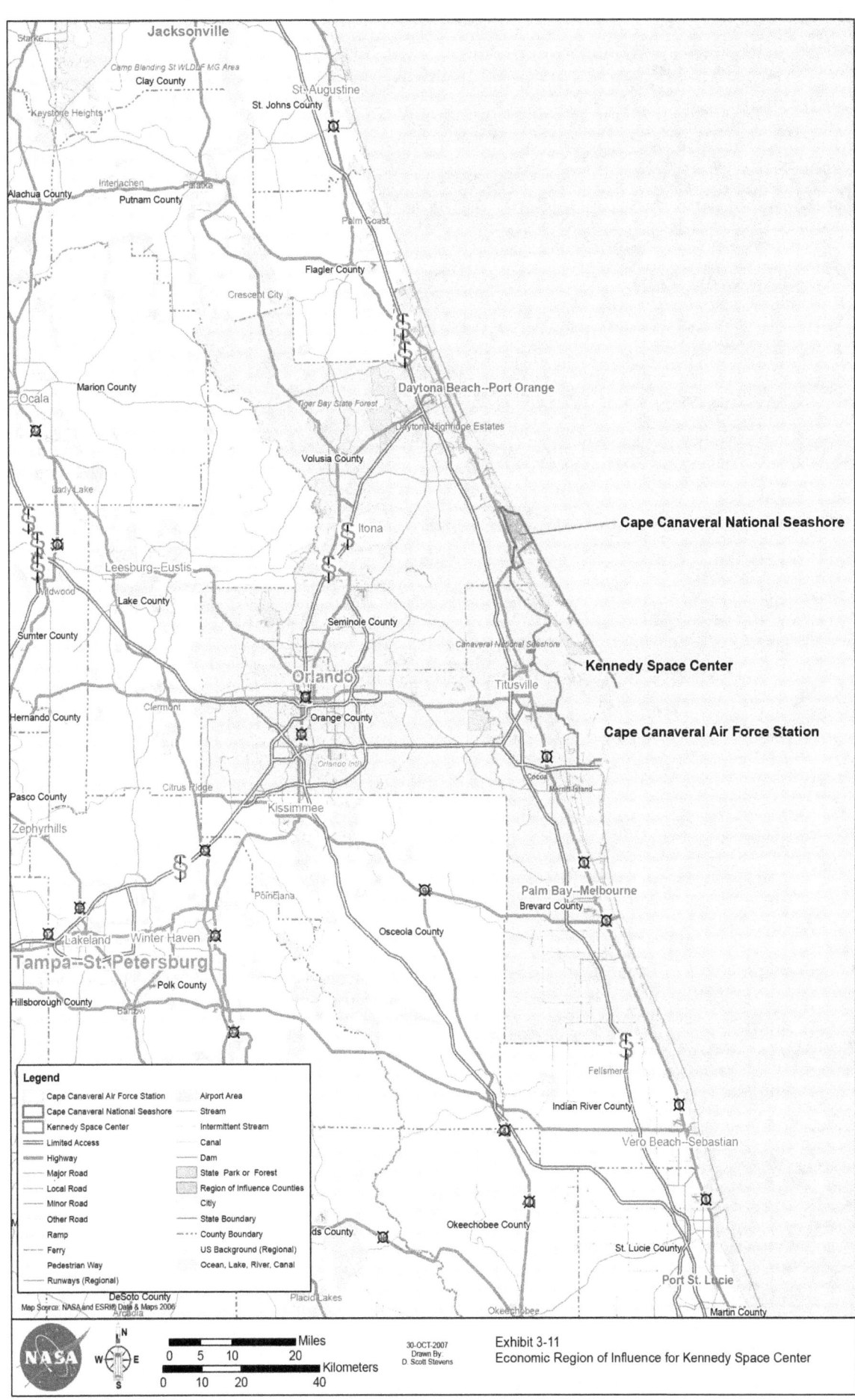

Exhibit 3-11
Economic Region of Influence for Kennedy Space Center

This page intentionally left blank.

Bay-Melbourne-Titusville Metropolitan Statistical Area[1] (MSA) by the U.S. Office of Management and Budget (OMB) (2006). NASA historically has maintained, and will continue to maintain, a close reciprocal relationship with Brevard County (NASA, 2003a).

3.2.10.2 Affected Environment

Population. In 2006, more than 2.5 million people lived in the four counties of the ROI, an estimated increase of 17 percent from the 2000 Census. By 2010, the population in the ROI is projected to grow another 9 percent, with Orange County showing the greatest rate of growth (U.S. Census Bureau, 2007b; Florida Office of Economic and Demographic Research [EDR], 2007).

Regional Employment and Economic Activity. The economic base of the region is tourism (attracting more than 20 million visitors annually) and manufacturing. Tourist attractions include resorts such as Disney World, Universal Orlando, Sea World, and the KSC Visitor Complex (along with the MINWR and seashore areas on the KSC property). In FY 2005, more than 800,000 out-of-state visitors spent more than $48 million on goods and services at the KSC Visitor Complex (NASA, 2003a; NASA, 2006g).

The total labor force in the 4-county ROI was 1.3 million persons in 2006, with an unemployment rate of 2.9 percent, which is similar to that of the state (3 percent) and below the national rate (4.6 percent). By 2014, employment is projected to increase by a modest 2.2 percent in the 4-county ROI (Florida Agency for Workforce Innovation, 2007).

KSC is the heart of the Space Coast and a key part of the central Florida technology corridor. NASA provides an important source of revenue for local firms through the procurement of goods and services, including contracts in Florida funded by other NASA centers besides KSC.

KSC is also Brevard County's largest single place of employment, with approximately 13,500 onsite and near-site workers in 2005 and more than 15,540 in 2006. In the other counties of the ROI, the only employers larger than KSC are Walt Disney World (53,500 employees), Orange County Public Schools (22,807 employees), and Universal Orlando (14,500 employees) (NASA, 2006g; NASA, 2007t; Enterprise Florida, Inc. [eFlorida], 2006).

Historically, the highest recorded employment level at KSC (nearly 26,000 people) was under the Apollo program in 1968, and the lowest (close to 8,500) was in 1976 after the Apollo program ended. Employment rose again in 1979 when KSC was designated as the Launch and Operations Support Center for the Space Shuttle. The

---

[1] An MSA is an area, defined by the U.S. Office of Management and Budget (OMB) for federal statistical purposes, consisting of a core urban area with 50,000 or more population and adjacent counties that have a high degree of social and economic integration (as measured by commuting patterns) with that urban core (OMB, 2006).

loss of the shuttle Challenger caused an employment drop of 2,400 people in 1986 (NASA, 2003a).

NASA economic studies have estimated that each job in the space industry generates an additional 1.93 jobs in central Florida and that each direct job at KSC generates 1.5 total jobs in the state of Florida. In FY 2005, KSC and all other NASA operations created a total economic impact in central Florida of approximately $3.7 billion in economic output, $1.8 billion in income, and 35,000 jobs (NASA, 2003a; NASA, 2006g).

Economic Contribution of the Space Shuttle Program. In 2006, more than 7,200 full-time equivalent personnel (FTEs) (approximately 470 civil service and 6,800 prime contractor), or about half of the total KSC employees, worked directly on the SSP at KSC. This estimate includes only direct charges to the SSP budget; it excludes other functions such as ground and base support, financial management, and administrative, as well as unmanned launch, R&D, and other programs at KSC (NASA, 2007a).

In FY 2006, the SSP put nearly $0.96 billion into the regional economy, including civil service and prime contractor salaries and non-payroll procurements to subcontractors and suppliers. Those expenditures generate additional economic output, jobs, and income into supporting industries within the 4-county ROI. The total (direct plus indirect and induced[2]) effect of the SSP on economic output was approximately $2.9 billion (which represents less than 3 percent of the nearly $120 billion[3] in overall economic activity in the region), $1.1 billion in earnings, and more than 20,000 jobs (NASA, 2007aa).

3.2.11 Solid Waste

3.2.11.1 Affected Environment

KSC has two unlined landfills that are permitted by the FDEP. The permits cover the Class III and the Closed Class III Landfills on Schwartz Road. At KSC, the Center Operations Directorate, EPB, oversees the requirements associated with the landfills' management. The EPB is responsible for implementing an inspection program to monitor the landfills for compliance with F.A.C. 62-701 and specific conditions of the permits. The EPB coordinates permit-required groundwater, surface water, and gas monitoring at the landfills. All samples, laboratory analyses, and records are maintained as required by F.A.C. 62-701 and permit-specific conditions, and are inspected routinely. Records of daily operations, maintenance, load checking, and training are maintained by the Center's contractor responsible

---

[2] Based on the "multiplier effect" using economic multipliers for the 10-county region from the U.S. Bureau of Economic Analysis.

[3] U.S. Bureau of the Census, 2002 Economic Census—Total sales, shipments, receipts, or revenue for all establishments (2-digit North American Industry Classification System [NAICS] codes) in the 4-county-ROI for KSC.

for operating the landfill and are provided to the EPB for transmittal to the FDEP in accordance with the permit conditions (NASA, 2007m).

The Schwartz Road Closed Landfill was the primary land disposal site at KSC until December 1995. The landfill was placed in operation in 1968 and operated initially as a Class II facility until 1982. After 1982, the landfill accepted only Class III waste material, which included trash and paper products; plastic; glass; and debris from land clearing, construction, or demolition activities. The landfill site encompasses approximately 25 ha (64 acres), with about 20 ha (51 acres) being used for waste disposal. The renewal of the facility operations permit in March 1993 resulted in the completion of a site-specific hydrogeologic investigation and the construction of a new network of groundwater monitoring wells (NASA, 2003a).

Waste was disposed in excavated cells at depths of 0.9 to 1.8 m (3 to 6 ft) below original grade, with cell dimensions being roughly 15 m (50 ft) wide and 106 m (350 ft) long. Trenching began along the eastern side of the site and progressed westward, with trenches generally oriented in the east and west directions. The closed trenches have been covered with approximately 0.6 m (2 ft) of sandy soil. The final closure of the Schwartz Road Landfill was in January 1996. Long-term, post-closure monitoring of the site will continue for 30 years from the date of closure (NASA, 2003a).

3.2.12 Traffic and Transportation

3.2.12.1 Region of Influence

The transportation ROI for KSC is defined as the counties of Brevard, Orange, Seminole, and Volusia in Florida, where approximately 96 percent of all KSC civil service and prime contractor employees live, based on zip code data for civil servants. The majority of KSC employees currently live in Brevard County, which is also the Palm Bay-Melbourne-Titusville MSA. MSAs are defined on the basis of commuting patterns found in the U.S. Census journey-to-work data (NASA, 2007s; OMB, 2006).

3.2.12.2 Affected Environment

Transportation Routes. The geography of the KSC area, with the center located on Merritt Island between the ocean and inland waterways bordering the mainland, creates a distinctive transportation pattern. The result is a strong north-south transportation system orientated parallel to the coast, with relatively few east-west connections from Merritt Island to the mainland communities.

Interstate (I)-95 is the largest traffic artery serving the area, running north-south along the inland (western) edge of Titusville, Cocoa, Melbourne, and other communities located on the Indian River. Highway 1 (U.S. 1, also designated as Florida Highway 5 in this area) parallels I-95 to the east, passing directly through these communities. SR 3 enters KSC from the north via U.S. 1 near Oak Hill and continues southward (as Courtenay Parkway south of KSC) to Indian Harbour Beach. Part of this road through KSC is designated as Kennedy Parkway and is closed to the public (Exhibit 3-12).

Access Roads to KSC. There are four access roads into KSC. NASA Parkway West serves as the primary access road for cargo, tourists, and personnel. This four-lane road originates in Titusville as SR 405 and crosses the Indian River Lagoon onto KSC. After passing through the Industrial area, the road narrows to two lanes. It then crosses the Banana River and enters the CCAFS. The second point of entry onto KSC is from the south via South Kennedy Parkway, which originates on north Merritt Island as SR 3. This road, the major north-south artery for KSC, is a four-lane highway. The third entry point is accessible from Titusville along Beach Road, which connects to North Kennedy Parkway. The final access point is south of Oak Hill at the intersection of U.S. 1 and North Kennedy Parkway. All of the roads into KSC have controlled access points that are manned 24 hours per day, 7 days per week.

Railroads. A railroad spur runs from the Florida East Coast rail line to KSC. The spur spans the Indian River and Intracoastal Waterway via a causeway and bascule bridge from Wilson, on the mainland, to Merritt Island. Approximately 65 km (40 miles) of rail track provide heavy freight transportation to KSC.

Airports. The region has three major airports–Orlando International, Daytona Beach International, and Melbourne International. KSC contains an SLF for government aircraft, astronaut training, and delivery of launch vehicle components.

Transit. There is currently no public transit service to KSC. Space Coast Area Transit operates fixed route and paratransit service throughout Brevard County, excluding KSC (Brevard County, 1988).

3.3 Johnson Space Center

JSC is located in Harris County, Texas, on 656 ha (1,620 acres), approximately 40 km (25 miles) southeast of central Houston, and controls manned space missions and provides training to astronauts (Exhibit 3-13). Mission control at JSC requires continuous fully functional communications links, computers, and simulation equipment. Space research also is conducted at JSC, including the following:

- Development of communications devices
- Materials testing
- Lunar sample chemistry
- Physiological adaptation to microgravity
- Remote sensing and space simulation

NASA owns the property at JSC. However, JSC also is responsible for the operations conducted at EF and EPFOL. JSC also has Memorandums of

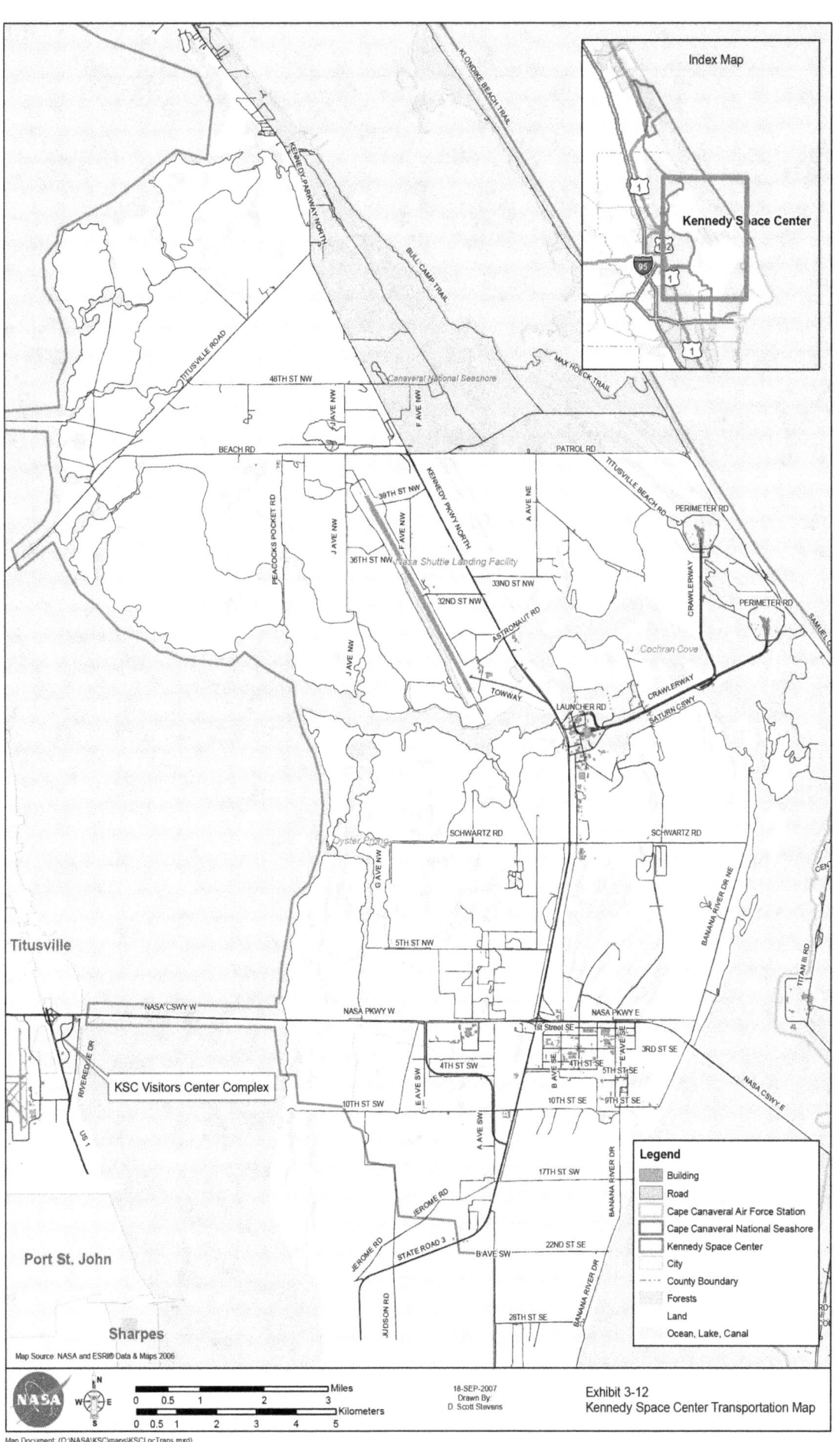

Exhibit 3-12
Kennedy Space Center Transportation Map

This page intentionally left blank.

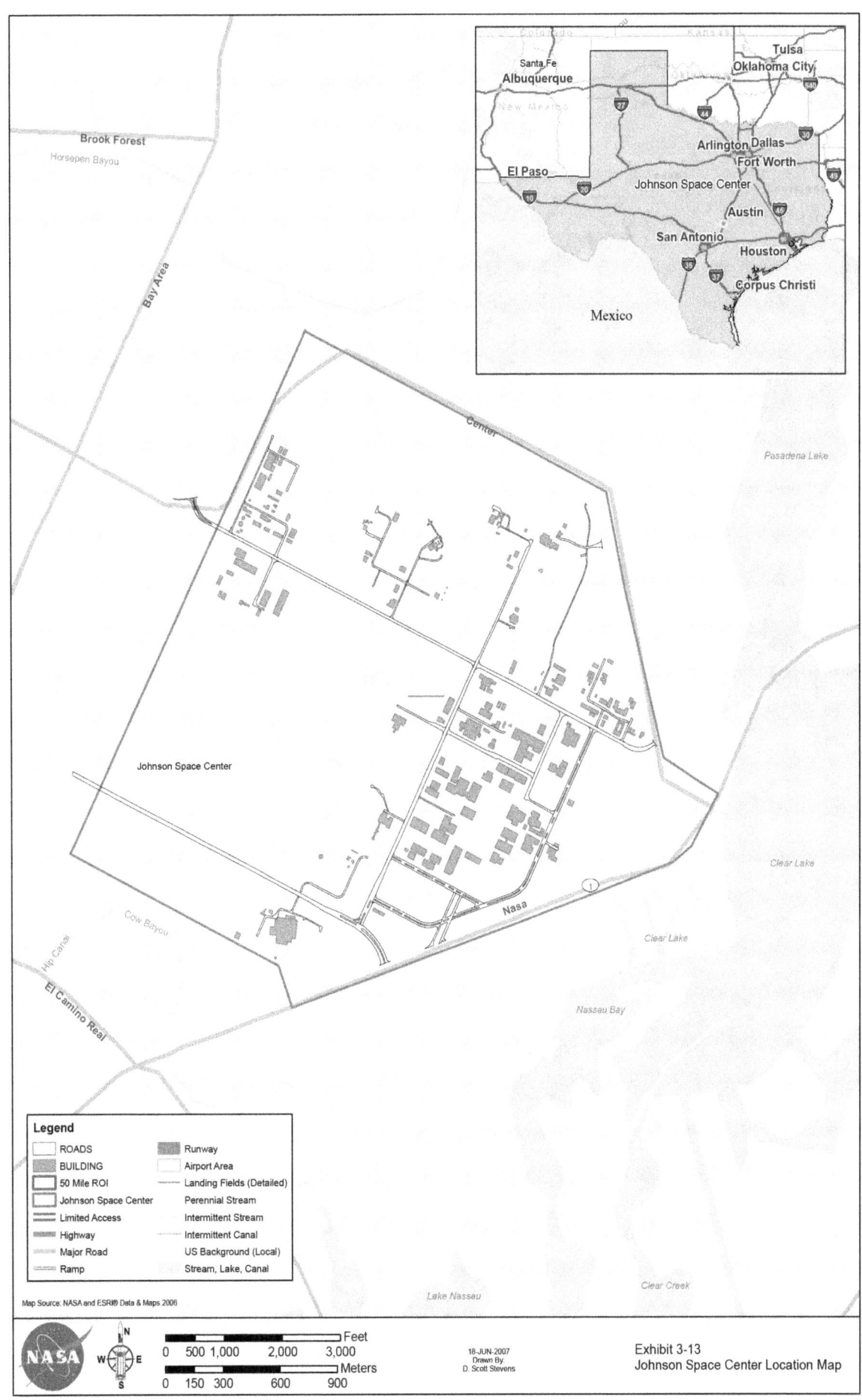

Exhibit 3-13
Johnson Space Center Location Map

This page intentionally left blank.

Understanding (MOUs) with WSTF and WSSH. The affected environments at these facilities are discussed in later subsections.

3.3.1 Air Quality

3.3.1.1 Affected Environment

Regional Climate. Houston has a warm subtropical climate. Warm tropical winds from the Gulf of Mexico control the climate during the spring, summer, and fall. Summers are hot and winters are mild, and the relative humidity is more than 50 percent for most of the year (NASA, 2004a).

Average annual rainfall is about 47 inches. From June to November, the Gulf Coast may be struck by hurricanes and tropical storms, with sustained heavy rain and strong winds. Flooding may occur in coastal areas such as JSC due to storm surges (extremely high tides caused by wind action). Winds at JSC are predominantly from the south and southeast (NASA, 2004a).

Emission Sources. JSC is categorized as a major source of criteria air emissions, with a Title V Federal Operating Permit. It also is categorized as a minor source of HAP emissions, with a synthetic minor limit. JSC is located in a "moderate" ozone non-attainment area. Therefore, JSC follows the NSR program and also must evaluate all new projects under the General Conformity rule. The area is listed as being in attainment for carbon monoxide (CO), nitrogen dioxide, sulfur oxides, particulate matter (PM) (2.5 and 10 microns), and lead (NASA, 2007s; EPA, 2007a). Texas does not have any state-specific air quality standards; however, it does have a "Watch List" of HAPs. JSC is on the watch list for benzene, styrene, and 1,3-butadiene (Texas Council on Environmental Quality [TCEQ]a, 2007).

NASA is regulated at JSC by two construction air permits for boilers and a Title V Federal Operating Permit for 4 boilers, 1 groundwater stripper, 1 classified waste incinerator, 7 solvent cleaners, 12 stationary diesel back-up generators, and 1 paint booth. The Title V permit also incorporates by reference 11 registered Permits-by-Rule (PBRs) that are minor sources of air emissions, and dozens more unregistered PBRs (NASA, 2007s).

The Shuttle air lock in Building 7 uses hydrochlorofluorocarbon (HCFC) 21 in its cooling system. The used HCFC 21 is shipped to KSC, where it is purified and then returned to JSC for reuse (NASA, 2007s; EPA, 1993).

3.3.2 Biological Resources

3.3.2.1 Affected Environment

Wetlands. Five palustrine emergent wetlands, one palustrine forested wetland, and four palustrine unconsolidated bottom wetlands have been indicated at the JSC facility through the USFWS National Wetland Inventory (NWI) maps (NASA, 2004a). Several site-specific wetland surveys also have identified an additional

11 wetland areas not depicted on the USFWS NWI maps. Because comprehensive wetland delineations have not been conducted on JSC, there may be other wetlands onsite that have not been described previously (NASA, 2004a). The wetlands for the area around JSC are shown in Exhibit 3-14.

Floodplains. The Federal Emergency Management Act (FEMA) publishes Floodplain Insurance Rate Maps (FIRMs) for insurance ratings; the 1996 and 2000 maps for JSC show the majority of JSC lying outside the 500-year floodplain (Exhibit 3-15). However, the eastern corner of JSC near the intersection of NASA Parkway and Space Center Boulevard and a section located along a tributary to Mud Lake in the northeastern portion of JSC are designated as lying within the 100-year and 500-year floodplains (NASA, 2004a).

3.3.3 Cultural Resources

3.3.3.1 Affected Environment

Archaeological Resources. There are no identified archaeological sites within JSC's boundaries, according to the Environmental Resources Document (ERD), although there are records of prehistoric occupation in the area (NASA, 2004a:94).

Historic Resources. Two properties have been designated as National Historic Landmarks (NHLs):

- The Space Environment Simulation Laboratory (SESL), Chambers A and B (Building 32)
- The Apollo Mission Control Center (Building 30) (NASA, 2004a:93; NPS, 2007d).

All NHL properties automatically are listed in the NRHP. The NHL designation recognizes properties that exemplify important trends in U.S. history (NPS, 2007d). The Mission Control Center and the SESL also are eligible for their association with the SSP.

The following structures are individually eligible for listing in the NRHP for their association with the SSP: *Discovery, Orbiter Vehicle (OV)-103, Atlantis, OV-104,* and *Endeavour OV-105.*

The survey of properties associated with the SSP included two new historic districts: the Astronaut Training Facilities HD and the R&D HD. The Astronaut Training Facilities HD includes the Jake Garn Mission Simulator and Training Facility (Building 5), Systems Integration Facility (Building 9), and the SCTF/Neutral Buoyancy Laboratory (NBL) that have been determined to be individually eligible and contributing to the historic district. The Mission Simulation Development Facility (Building 35) is a contributing property to the district.

The R&D HD includes the Crew Systems Laboratory (Building 7), Avionics Systems Laboratory (SAIL) (Building 16), the Communications and Tracking Development Lab, and Atmospheric Reentry Materials and Structures Evaluation Building

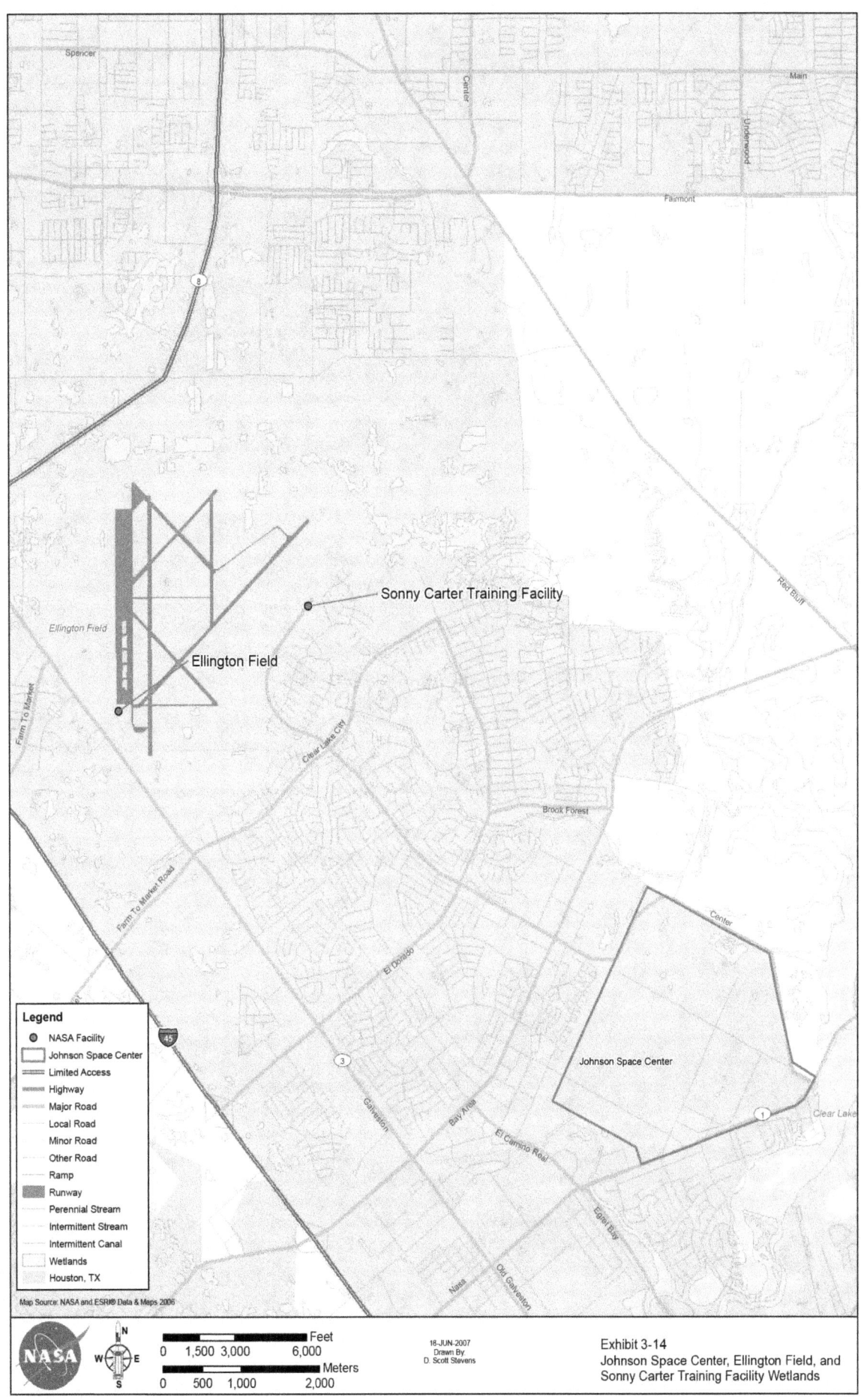

Exhibit 3-14
Johnson Space Center, Ellington Field, and Sonny Carter Training Facility Wetlands

This page intentionally left blank.

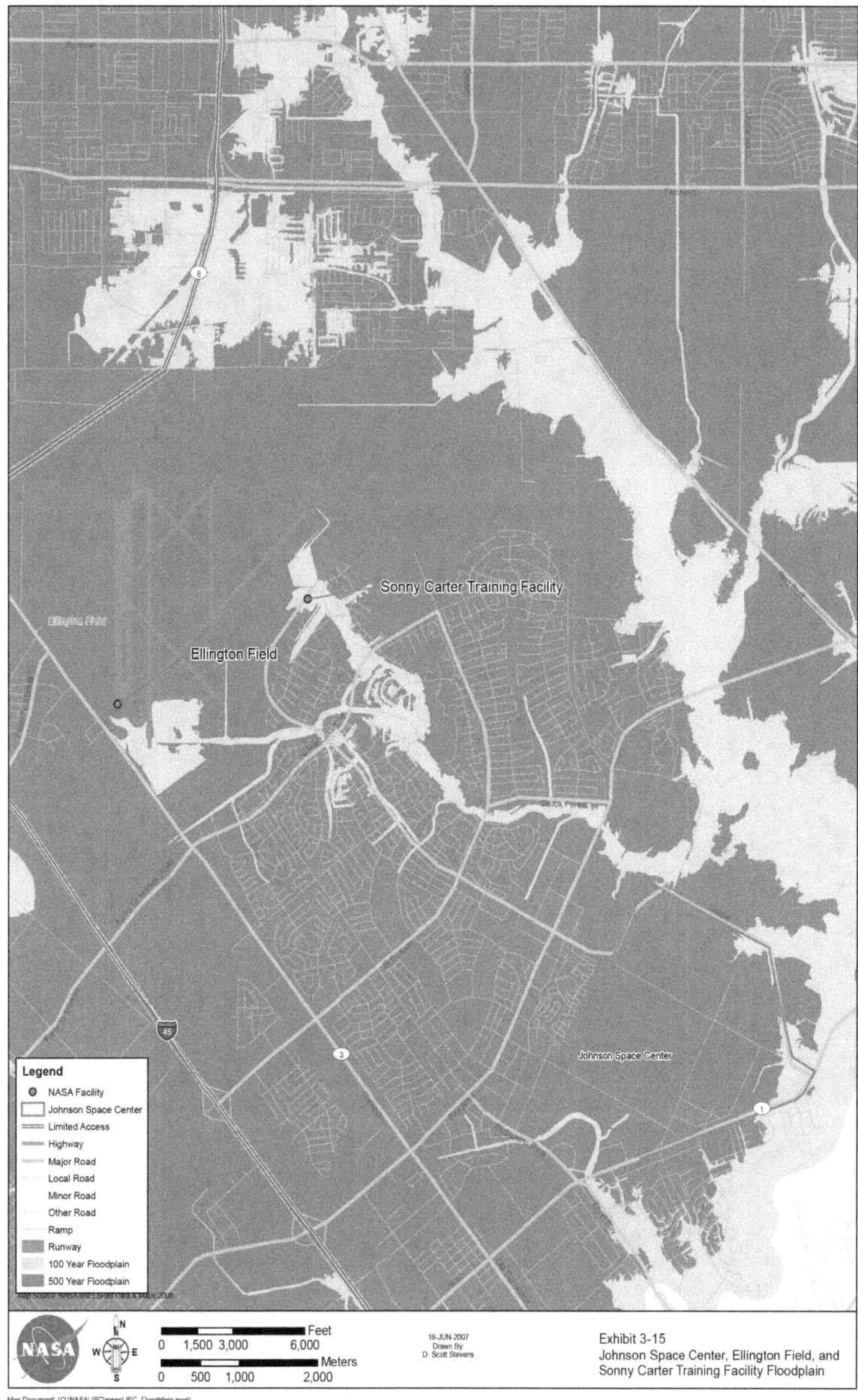

This page intentionally left blank.

(Building 222). These properties are all individually eligible and contributing to the R&D HD (Archaeological Consultants, Inc., 2007c). The locations of these properties are shown in Exhibit 3-16.

3.3.4 Hazardous and Toxic Materials and Waste

3.3.4.1 Affected Environment

Storage and Handling. JSC is registered by the TCEQ and generates and stores large quantities of solid and hazardous wastes. Hazardous wastes are held at the site for less than 90 days (NASA, 2004a).

For each hazardous waste generated at JSC, there is a notice of registration on file with the TCEQ (NASA, 2007s).

Waste Management. The Hazardous Waste 90-day Storage Facility (Building 358) is the central storage site for hazardous waste. Waste is generated at various points around the Center and transferred to this building to be prepared for shipment to disposal sites. Transport vehicles take the wastes to private hazardous waste disposal operations (NASA, 2004a).

Contaminated Areas. JSC is a not a federal Comprehensive Environmental Response, Compensation, and Liability Act of 1980 (CERCLA) National Priorities List (NPL) site (NASA, 2007s). Past contamination at JSC occurred at the sandblasting area near the Surplus Equipment Staging Warehouse (Building 338), the Fire Prevention Training Facility (Building 384), and the Energy Systems Test Area, where contaminated groundwater is being treated to remove Freon 113. The plume of Freon 113 was caused by a leaking process sewer and extends through about 10 ha (25 acres). Remediation of the groundwater using a pump–and-treat system began about 1990. The pump-and-treat system was later replaced with a potassium permanganate ($KMnO_4$) chemical oxidation technology system (NASA, 2004a).

Toxic Substances. Asbestos is present in buildings at JSC and is removed as buildings are renovated. NASA has procedures for handling asbestos while performing maintenance and while renovating or demolishing buildings at JSC (NASA, 2007s). Electrical equipment that contains PCBs is disposed as equipment is replaced due to attrition. NASA has had an aggressive program to eliminate PCB-containing equipment at the site; however, the Center does still have a small inventory of PCB-containing equipment.

This page intentionally left blank.

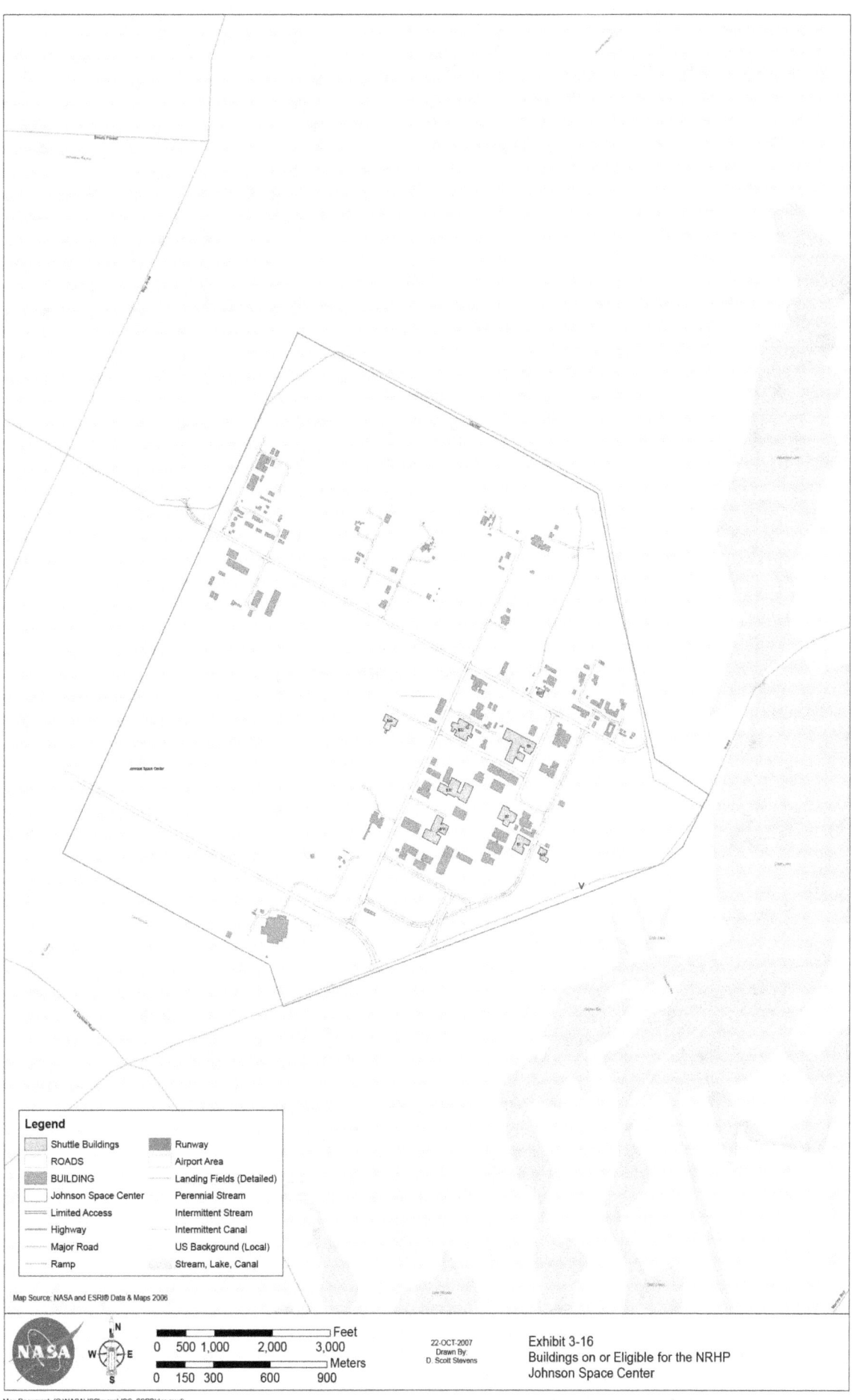

This page intentionally left blank.

3.3.5 Health and Safety

3.3.5.1 Affected Environment

JSC operates a variety of test facilities, research laboratories, simulators, and mock-up facilities in support of SSP. The following subsections outline JSC's programs for protecting the health and safety of JSC employees, as well as the public. Noise hazards at JSC are outlined in Section 3.3.8.

JSC's health and safety program must meet or exceed NASA, federal, and OSHA Voluntary Protection Program (VPP) requirements. JSC is a VPP Star site and must continue to improve its program beyond the minimum requirements. JSC's program is organized around the following four major elements:

- Management leadership and employee involvement
- Worksite analysis
- Hazard prevention and control
- Health and safety training

Hazardous Materials. Hazardous materials are used to conduct SSP operations at JSC. The hazardous materials used and hazardous wastes generated are discussed in detail in Section 3.3.4.

The implementation of work practices and control technologies minimizes employee exposure to hazardous materials. The JSC Safety and Mission Assurance (S&MA) Directorate is responsible for training employees to handle the hazardous chemicals kept at their worksites (hazard communications) and for implementing appropriate spill response protocols in case of emergencies. Hazardous material spills or releases that are too large to be handled by the shop employees where they occur are handled by a NASA spill response team and/or by the JSC fire department, depending on the location and severity of the incident (NASA, 2004a).

Buildings at JSC contain asbestos in the form of insulation and other building materials. JSC complies with the 29 CFR 1910.1001 standard for the protection of employees from asbestos exposure. Buildings at JSC also contain LBP. The JSC Safety and Health Handbook Policy, Requirements and Instructions, Chapter 9.4, (NASA, 2002c), outlines the requirements for protecting employees from exposure to lead, including activities that may disturb surfaces coated with LBP.

Hazardous Materials.
Transportation Safety. Hazardous materials such as fuels, chemicals, and hazardous wastes are transported in accordance with DOT regulations for interstate shipment of hazardous substances (49 CFR 100 through 199).

Explosions and Fire Hazards. The use and storage of certain hazardous materials, including fuels, in R&D operations presents a risk of explosions and fire hazards. The JSC Safety and Health Handbook, Chapter 3.8 (NASA, 2002c), outlines the requirements at JSC for emergency responses, including fire prevention and response. Each building at JSC has a fire warden to oversee building fire safety. Fire protection at JSC is contracted with the City of Houston.

3.3.6 Hydrology and Water Quality

3.3.6.1 Affected Environment

Surface Waters. There is no natural aquatic habitat on JSC. However, JSC is bordered by several water bodies, as shown in Exhibit 3-17. Waters on and near JSC are tidal streams and estuaries associated with Galveston Bay. Clear Lake is located along the southeastern corner; Mud Lake and Armand Bayou are northeast of JSC. Cow Bayou is located to the southwest and Horsepen Bayou is to the north of JSC. Horsepen Bayou flows east to its confluence with Armand Bayou. Armand Bayou and its tributaries drain about 140 square kilometers [km^2] (54 square miles) of southeastern Harris County. Armand Bayou then flows into the northern end of Mud Lake, part of the Clear Lake estuary, which is connected to western Galveston Bay. Cow Bayou flows into Clear Creek, which drains to Clear Lake.

Artificial water bodies on JSC include a canal that carries cooling water from the former Houston Lighting & Power Company's Webster Power Station. The canal traverses the southern side of JSC and drains 2 km (1 mile) to the south, into Clear Lake. Three connected artificial concrete ponds are located in the central mall. The storm water system includes a series of underground conduits and ditches. Most storm water collects in four main ditches; two ditches discharge to Mud Lake and the other two discharge to Cow Bayou and Horsepen Bayou. Clear Lake, and ultimately Galveston Bay, receives all of the drainage from JSC (NASA, 2004a).

Groundwater. Groundwater is found in soil strata under JSC, usually beginning about 2 to 3 m (8 to 11 ft) below the ground surface. The water table fluctuates with the weather and may reach the ground surface during wet periods. Several strata of soil contain silty and sandy zones; these zones may contain perched groundwater.

The most shallow confined groundwater aquifer under JSC is a sand layer 18 m (60 ft) below the surface. This aquifer is contained between clay layers at a depth of approximately 26 m (85 ft). The aquifer dips to the southeast by 4 m per km (20 ft per mile). Its thickness ranges from 6 to 10 m (21 to 32 ft), with the thickest part toward the east.

Two important fresh water aquifers are located under JSC and the Houston area-the Chicot and the Evangeline. Both aquifers are comprised of discontinuous sand, silt, and clay. In the southern and eastern parts of the region, the aquifers are artesian. At JSC, the base of the Chicot aquifer is between 180 and 210 m (600 and 700 ft) below the surface, and the base of the Evangeline aquifer is between 790 to 910 m (2,600 to 3,000 ft) below the surface.

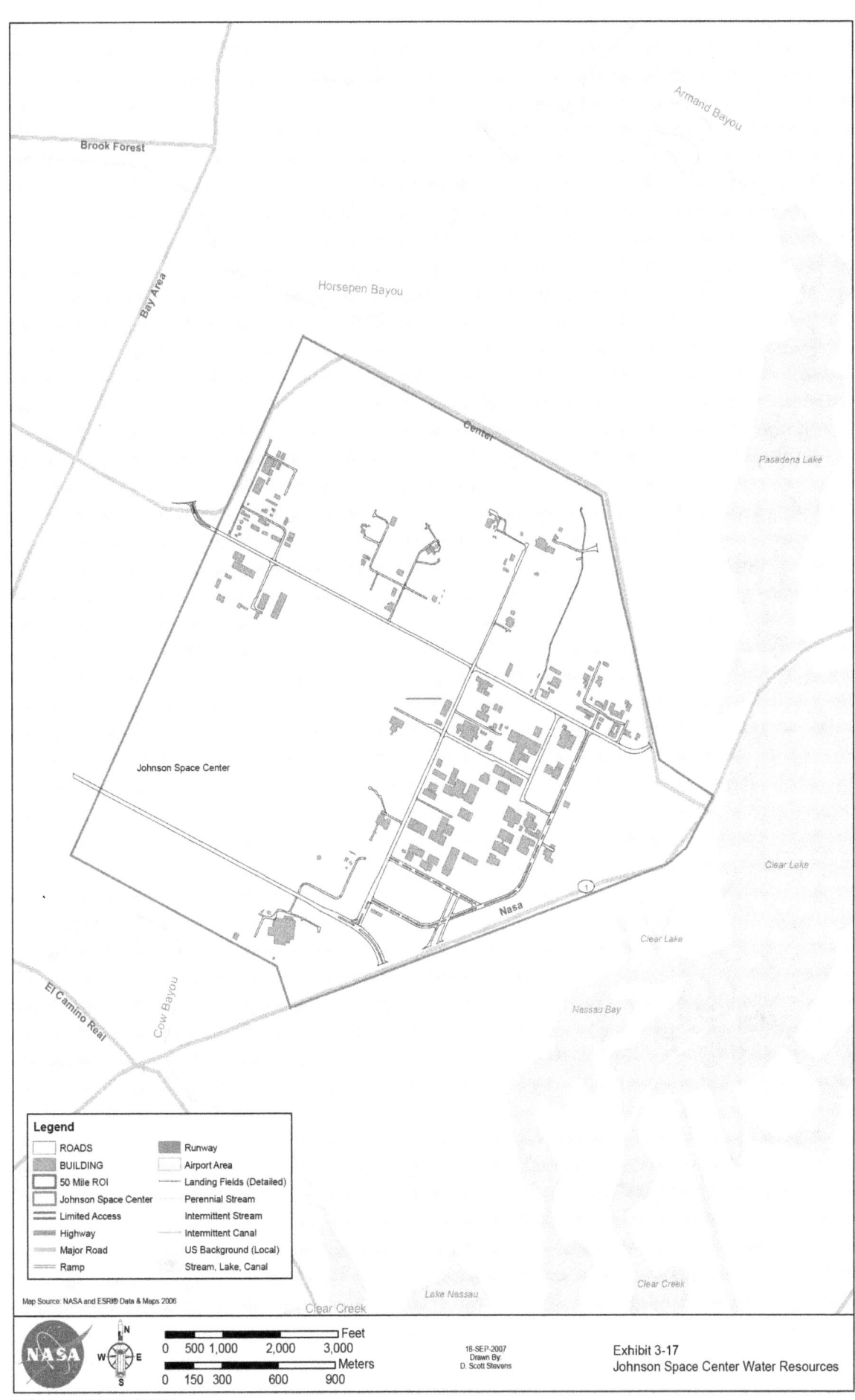

This page intentionally left blank.

Water Quality. Texas waters are classified according to one or more of the following use designations:

- Recreation–contact or non-contact
- Drinking water supply–domestic supply or aquifer protection
- Aquatic life–limited, intermediate, high, exceptional, or oyster waters

Numeric criteria have been established to ensure that these uses are maintained. Armand Bayou, Mud Lake, Clear Lake, and Clear Creek are designated as contact recreation and high aquatic life waters (30 Texas Administrative Code [TAC] 307). Those water bodies are included on the 2004 303d list of impaired waters in Texas (TCEQ, 2005).

Clear Lake and Clear Creek are listed for elevated levels of bacteria. Armand Bayou and Mud Lake are listed for low DO and elevated levels of bacteria.

The TCEQ ranks the water quality of Texas estuaries. An estuary's rank is determined by its levels of nitrogen, DO, degree of eutrophication, and concentration of fecal coliform bacteria. Out of 80 Texas estuaries, the tidal reach of Clear Creek ranks fourth worst, Cow Bayou ranks sixth worst, Armand Bayou ranks tenth worst, and Clear Lake ranks sixteenth worst. The pollutants that most affect these estuaries' ranks are fecal coliform bacteria and nutrients.

Armand Bayou is classified by the TCEQ as "water quality limited." The water body is designated for contact recreation and high-quality aquatic habitat; however, the stream has low levels of DO and high levels of ortho and total phosphorus, chlorophyll *a*, nitrite, and nitrate-nitrogen. High levels of fecal coliform bacteria cause restrictions on the recreational use of the bayou.

Clear Lake also is classified as "water quality limited" by the TCEQ. The lake is designated for contact recreation and high-quality aquatic habitat. High phosphorus levels and high fecal coliform bacterial counts occur in Clear Lake. Chlorophyll *a* levels in the western end of the lake are high, which indicates eutrophication (NASA, 2004a).

Galveston Bay is part of EPA's National Estuary Program (NEP). Armand Bayou is a coastal preserve in the Galveston Bay NEP. As part of the NEP, a Comprehensive Conservation and Management Plan (CCMP) has been developed to address all aspects of environmental protection for the estuary, including water quality (EPA, 2007g).

Regulated Water Discharges and Withdrawals. JSC purchases approximately 3,780,000 liters (L) (1 million gallons) of water per day from the Clear Lake City Water Authority. This water is conveyed by pipeline from the Clear Lake City Water Authority plant. The sources are the San Jacinto and Trinity Rivers. The Center uses about 1.02 billion L (272 million gallons) of water per year.

JSC does not use groundwater routinely. However, two water wells (Well Nos. 2 and 4–TCEQ identifier 1010250) are maintained for contingency and emergency use only. Groundwater is pumped from these wells only for preventive maintenance.

Approximately 3.21 million L per day (850,000 gallons per day [gpd]) of sanitary sewage from buildings and wastewaters from the NASA operations flow in underground sewer pipes through a series of lift stations and force mains to a wastewater treatment plant (WWTP) operated by the Clear Lake City Water Authority. This WWTP is located offsite, to the northeast of the Center. The plant discharges treated effluent to Horsepen Bayou. Discharges to the Clear Lake City Water Authority are treated to meet the pre-treatment requirements (NASA, 2004a).

Storm water discharges from industrial sources also require discharge permits. The TCEQ has developed general permits that cover those discharges, as well as construction activities, as long as the facility complies with the permits' conditions, including preparing a Pollution Prevention (P2) Plan, monitoring effluent quality, and keeping records. NASA has implemented an SWP3 and is covered under a general permit (Permit ID TXR05K587) (TCEQ, 2007b).

3.3.6.2 Affected Environment

JSC adjoins homes and offices in the Clear Lake City development to the north and west. To the south are shops, offices, and homes in the City of Nassau Bay. Armand Bayou Nature Center is northeast of JSC. To the east are the West Mansion and Clear Lake. The West Mansion once housed the Lunar and Planetary Institute of Rice University (NASA, 2004a).

JSC is almost entirely within the limits of the City of Houston. Space Center Houston, the new visitor center, is in the extra-territorial jurisdiction of the City of Houston (NASA, 2004a).

Land Use Planning.
Land Use at JSC. A map showing the land use at JSC is provided in Exhibit 3-18. JSC's Master Plan divides JSC into four areas by major activities to guide future development (NASA, 2004a).

Area I, the southeastern section, includes the main complex of permanent buildings in the primary architectural style of JSC. These buildings house administration, training, operations, major testing, engineering, development sciences, and management associated with manned space missions and tourism. The Space Center Houston visitor center is in this area. The southern part is allocated to administration, management, and engineering development. The northern part has mission operations, training, major testing, and science laboratories (NASA, 2004a).

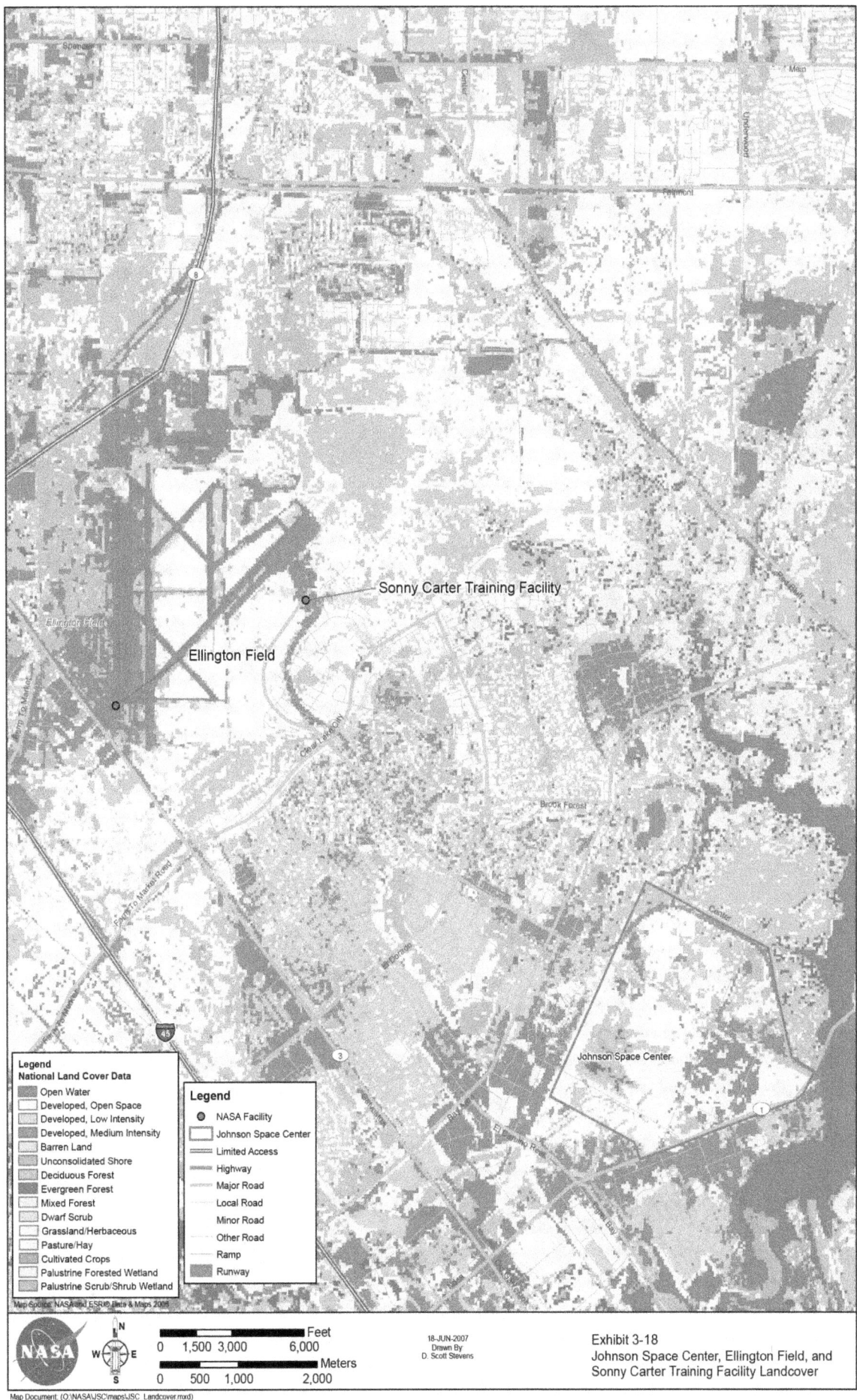

This page intentionally left blank.

Area II, the northeastern section, includes the electrical substation and various support facilities. The southeastern part of this area is restricted for development because it is vulnerable to tidal surges from hurricanes. The far northern part of the area is for recreation (NASA, 2004a).

Area III, to the northwest, is used for hazardous activities. This area contains the Energy Systems Test Area and includes storage areas for hazardous materials, explosives, and until recently, a training area for fire control. It also includes industrial-type support for JSC such as maintenance operations, central waste collection, service contractor construction activities, and warehouses (NASA, 2004a).

Area IV is the southwestern quadrant of JSC and is reserved for activities that require large open areas. The northwestern part of the area is used for warehouses, shipping and receiving, motor pool, logistic support, and other housekeeping functions (NASA, 2004a).

The remainder of JSC's land is zoned as follows:

- Restricted Use: specific development controls, building restrictions, limits on physical characteristics or features, activity limitations, etc.
- Semi-restricted Use: continued development for an established specific purpose or activity; or general restrictions not as stringent as those for restricted use
- General Use: unrestricted, multipurpose development

Easements and Rights-of-Way. Easements for non-NASA entities cover 200 ha (500 acres) of JSC. These easements include rights-of-way (ROWs) for storm sewers, cooling water canals, electrical transmission lines, gas pipelines, and telecommunications cables. NASA grants easements to entities when they do not interfere with JSC's functions. NASA has dedicated easements for Space Center Boulevard on the northeast and for widening NASA Parkway to the southeast. Exxon Oil Company has an easement for oil drilling on 8 ha (19.8 acres) in the northwestern part of the site (NASA, 2004a).

3.3.7 Noise

Sensitive receptors to noise generated at JSC include the Child Care Facility (Building 210); the Gilruth Recreation Facility (Building 207); the Visitor Center; and homes, stores, and offices outside JSC. Sensitive receptors are those locations where low noise levels serve a public need and where the preservation of those qualities is important if the area is to continue to serve its intended purpose. These areas may include picnic areas, recreation areas, playgrounds, active sports areas, parks, residences, motels, hotels, schools, churches, libraries, and hospitals.

3.3.7.1 Affected Environment

JSC's noise sources do not exceed the typical conversation level of 60 dBA at receptors outside the Center. The Center evaluates and controls noise in work areas so that it will not cause hearing loss or physical impairment (NASA, 2004a). The two main sources of noise at JSC–utility-related noise and noise from research and testing activities–are discussed in the following subsections.

Utility Noise. The Central Heating and Cooling Plant (Building 24), Auxiliary Chiller Facility (Building 28), and Emergency Power Building (Building 48) are the primary sources of utility-related noise from the operation of boilers, compressors, and chillers. Employees working in these facilities in support of utilities at JSC are required to wear hearing protection when the generators are operating.

Research and Testing. Many of the facilities at JSC are designed to help evaluate whether spacecraft systems and materials can be used on space vehicles.

The Vibration and Acoustic Test Facility (Building 49) houses an acoustical chamber that subjects flight hardware to noise levels up to 165 dBA for 1- to 2-minute intervals.

The Atmospheric Re-entry Materials and Structures Evaluation Facility (Building 222), known as the arcjet, is used for testing materials and components under aero-thermodynamic heating conditions similar to those encountered during space flight and reentry.

The Propulsion Test Facility (Building 353) is in the northern part of the Energy Systems Test Area. It is equipped with a steam ejection system to produce a vacuum during routine test procedures. Employees at these test facilities are required to wear hearing protection while tests are being conducted.

3.3.8 Site Infrastructure

3.3.8.1 Affected Environment

Potable Water Supply. NASA receives drinking water from the Clear Lake City Water Authority, but the two water wells at JSC qualify the facility as having its own water supply.

JSC purchases water from the Clear Lake City Water Authority. This water comes from the San Jacinto and Trinity Rivers and flows through the Coastal Water Authority canal system to the City of Houston's Southeast Water Plant. There it is purified and then conveyed by pipeline to the Clear Lake City Water Authority plant, just southwest of the Center. JSC operates and maintains two potable water wells and a potable water treatment system that would be used if City of Houston and Clear Lake City Water Authority water service were interrupted. The water system most recently was audited by the TCEQ and determined to be in compliance on April 27, 2007 (NASA, 2004a).

Water from the Clear Lake City Water Authority (and the water wells, when used) flows under pressure to JSC's two aboveground potable water storage tanks (Buildings 339 and 341). These tanks store 1 million gallons and 600,000 gallons, respectively. The tanks are connected in series; water comes into Tank No. 2 (Building 341) and then moves to Tank No. 1 (Building 339). If necessary, water is chlorinated at the Water Treatment Plant (WTP) Building (Building 322) before distribution. Four booster pumps bring water to an elevated storage tank (Building 40) for distribution; the tank holds up to 950,000 L (250,000 gallons) (NASA, 2004a).

Wastewater System. See the Hydrology and Water Quality, Section 3.3.6.

Storm Water System. See the Hydrology and Water Quality, Section 3.3.6.

Energy Sources. JSC does not generate electricity using natural gas. JSC does have five diesel generators that burn fuel oil to provide electricity for mission support as back-up systems and primarily are used as contingency generators for weather support or could be used if electrical service were to be interrupted to the Building 30 Mission Control Center. These generators are authorized by the TCEQ to operate under JSC's Title V permit. JSC does not provide electricity into the grid. JSC purchases natural gas from Centerpoint Energy. JSC is under a blanket GSA contract. Through modifications to the GSA contract, the facility also purchases short-term blocks from the market to match its current load. This approach saves approximately $500,000 each year (NASA, 2007y).

JSC purchases electricity from Constellation New Energy. Contracts last for 1 to 2 years because of the volatility of the market. JSC is currently under a 1-year block and index contract. The Center buys specific base loads and also purchases additional blocks for its daily consumption (NASA, 2007y).

Both the electricity and natural gas distribution systems are owned by NASA.

3.3.9 Socioeconomics

3.3.9.1 Region of Influence

The economic ROI for JSC is defined as the 10 counties of the Houston-Baytown-Sugarland MSA[4] (Exhibit 3-19). Approximately 87 percent (or about 7,500 employees) of civil service and contractor employees live in 3 of the 10 MSA counties (Harris, Galveston, and Brazoria) and the remainder live elsewhere in the Houston metropolitan area (NASA, 2007a).

The Clear Lake area, which includes Harris and Galveston counties and parts of the cities of Houston and Pasadena, is the center of Houston's aerospace industry and the part of the MSA that is most closely associated with JSC.

---

[4] An MSA is an area, defined by the OMB for federal statistical purposes, consisting of a core area with 50,000 or more population and adjacent counties that have a high degree of social and economic integration (measured by commuting patterns) with that urban core (OMB, 2006).

This page intentionally left blank.

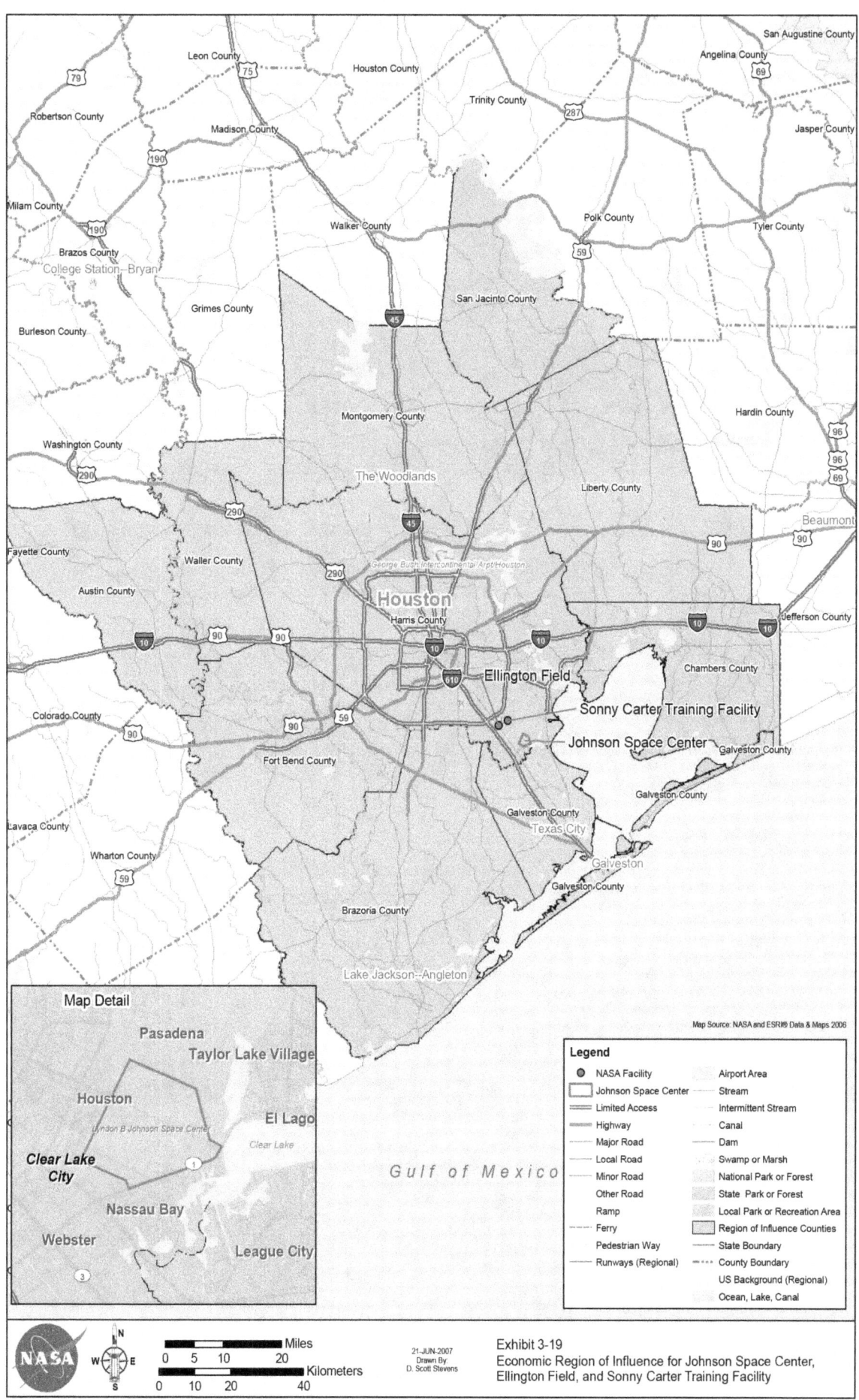

Exhibit 3-19
Economic Region of Influence for Johnson Space Center, Ellington Field, and Sonny Carter Training Facility

This page intentionally left blank.

3.3.9.2 Affected Environment

Population. In 2005, more than 5.2 million people lived in the Houston-Baytown-Sugarland MSA, an increase of 12 percent from the 2000 Census. By 2010, the population of the MSA is projected to grow by 5 percent (U.S. Census Bureau, 2006c; Texas State Data Center, 2006).

Regional Employment and Economic Activity. Houston is one of the world's largest manufacturing centers for petrochemicals. Although historically reliant on the fortunes of the oil industry, Houston's economy has diversified strongly into the technology and service industries. The Port of Houston is a major transportation hub, moving nearly 250,000 tons of cargo in 2006. Tourism, to which JSC contributes, is the fastest growing industry in the Clear Lake area. An estimated 1 million tourists each year visit JSC and its visitor center, Space Center Houston, which was designed by Disney (NASA, 2004a; Greater Houston Partnership, 2007a).

In 2006, the total labor force of the ROI was slightly more than 2.7 million people, with an overall unemployment rate of 4.9 percent, the same as the state and only slightly higher than the national unemployment rate of 4.6 percent (U.S. Bureau of Labor Statistics [BLS], 2006). The oil and energy industry employs nearly half of the labor force. The Texas Medical Center, one of the world's largest, employs more than 65,000 health care professionals. Major NASA contractors with offices in the area employ an estimated 16 percent of the local labor force. The Bayport Industrial Complex near JSC employed nearly 8,000 workers in 2004 (NASA, 2004a).

JSC, as NASA's largest R&D facility, employs approximately 16,000 people–3,000 civil service and 13,000 contractor personnel. Prior NASA studies have estimated that each job in the aerospace industry generates an additional 2.2 jobs in the Houston region (Personal Communication, 2007h; NASA, 2004a; NASA, 2007a).

JSC generates billions of dollars in contracts annually. The total economic impact on the City of Houston and Texas includes more than 26,435 jobs with personal incomes of more than $2.5 billion and total spending that exceeds $3.5 billion (Bay Area Houston Economic Partnership [BAHEP], 2007).

Economic Contribution of the Space Shuttle Program. In 2006, approximately 4,700 FTE employees (770 civil service and 3,900 prime contractor) worked directly on the SSP at JSC, or less than one third of total JSC employment. This estimate includes only direct charges to the SSP budget; it excludes base operations and administrative personnel, R&D, time spent supporting other programs at JSC, and jobs at offsite suppliers and subcontractors within and outside of the region (NASA, 2007a).

In FY 2006, the SSP put nearly $0.6 billion into the regional economy, including civil service and prime contractor salaries and non-payroll procurements to subcontractors and suppliers. Those expenditures generate additional economic output, jobs, and income in supporting industries within the ROI. The total (direct plus indirect and induced[5]) effect of the SSP on economic output was approximately $1.9 billion (which represents less than 1 percent of the nearly $350 billion[6] in overall economic activity in the 10-county region), $0.84 billion in earnings, and 14,700 jobs (NASA, 2007cc).

3.3.10 Solid Waste

3.3.10.1 Affected Environment

Nonhazardous refuse is taken to roll-off boxes at the Central Waste Collection Facility (Building 332) and then shipped to the City of Houston landfill (NASA, 2004a).

Approximately 93 metric tons of ACM were generated in 1996 from asbestos removal from buildings. Asbestos, an industrial solid waste, temporarily is stored in lined and covered roll-off boxes until being shipped to a landfill. Electrical equipment containing PCBs becomes industrial solid waste as electrical equipment is replaced. PCB-contaminated wastes currently are stored in Building 358.

3.3.11 Traffic and Transportation

3.3.11.1 Region of Influence

The transportation ROI for JSC is defined as the 10 counties of the Houston-Baytown-Sugarland, Texas, MSA–Austin, Harris, Brazoria, Liberty, Chambers, Montgomery, Fort Bend, San Jacinto, Galveston, and Waller counties (OMB, 2006). The majority (83 percent) of the JSC civil service and contractor employees live in Harris, Galveston, and Brazoria counties, while the rest live elsewhere in the Houston metropolitan area (NASA, 2007a).

3.3.11.2 Affected Environment

Transportation Routes. Autos and trucks reach the Clear Lake area on SR 3, State Highway 146, and I-45. NASA Parkway connects these roads with the main gate to JSC.

Access Roads to JSC. JSC is connected to the local roadway system by gates to NASA Parkway to the south, Space Center Boulevard to the north and east, and Saturn Boulevard to the west. The site adjoins homes and offices in the Clear Lake City development to the north and west. To the south are shops, offices, and homes in the City of Nassau Bay. Armand Bayou Nature Center is northeast of JSC. To the east are West Mansion and Clear Lake. The West Mansion once housed the Lunar and Planetary Institute of Rice University. Transportation to JSC for most employees is by private auto. The Center has gates on NASA Parkway to the south,

---

[5] Based on the "multiplier effect" using economic multipliers for the 10-county region from the U.S. Bureau of Economic Analysis.

[6] U.S. Bureau of the Census, 2002 Economic Census—Total sales, shipments, receipts, or revenue for all establishments (2-digit NAICS codes) in the 10-county JSC ROI.

Space Center Boulevard to the east and north, and Saturn Boulevard to the west. The transportation routes at JSC are shown in Exhibit 3-20.

Railroads. Railroads run parallel to SR 3 and State Highway 146. The Southern Pacific provides freight rail service to Seabrook and the Missouri-Kansas-Texas Railroad serves Webster. JSC does not have any direct rail service.

Airports and Ports. Bush Intercontinental Airport, Houston's major airport, is 60 km (38 miles) north of JSC. The William P. Hobby Airport, 24 km (15 miles) northwest of JSC, provides regular commercial air service by eight airlines. EF, which is 13 km (8 miles) north of the Center, is primarily a general aviation airport. Air freight service is available at all three airports.

The Port of Houston and the Port of Galveston serve ocean-going ships and provide worldwide cargo service. The Gulf Intracoastal Waterway and other barge canals accommodate smaller vessels.

Transit. The Metropolitan Transit Authority of Harris County provides "Park and Ride" bus service between Clear Lake City and downtown Houston on a staggered schedule and operates a shuttle to the Center.

3.4 Ellington Field

EF, the center of aviation-related operations for NASA's manned space program, is located 13 km (8 miles) northwest of JSC and 27 km (17 miles) southeast of downtown Houston, in Harris County, Texas (Exhibit 3-13). EF conducts aircraft operations for training astronauts and simulating aspects of manned space missions, including microgravity, remote sensing, and spacecraft operation.

The City of Houston owns the majority of the 1,900-acre airport and leases tracts of land to the State of Texas and several fixed-base operators. The Air National Guard operates a small parcel of property at EF. NASA pays a 6-cent-per-gallon cost for fuel dispensed to aid with the airfield maintenance. NASA also pays for paramedic and firefighting services at EF.

The airport is also a transportation hub for JSC employees and equipment, and in the past, the Shuttle, transported on a modified Boeing 747 carrier, has stopped at EF for transport to KSC. EF operations are conducted under the management of JSC. Directives issued for JSC also apply to EF.

3.4.1 Air Quality

3.4.1.1 Affected Environment

Regional Climate. The regional climate for the Houston area is provided in Section 3.3.1.

3. AFFECTED ENVIRONMENT

This page intentionally left blank.

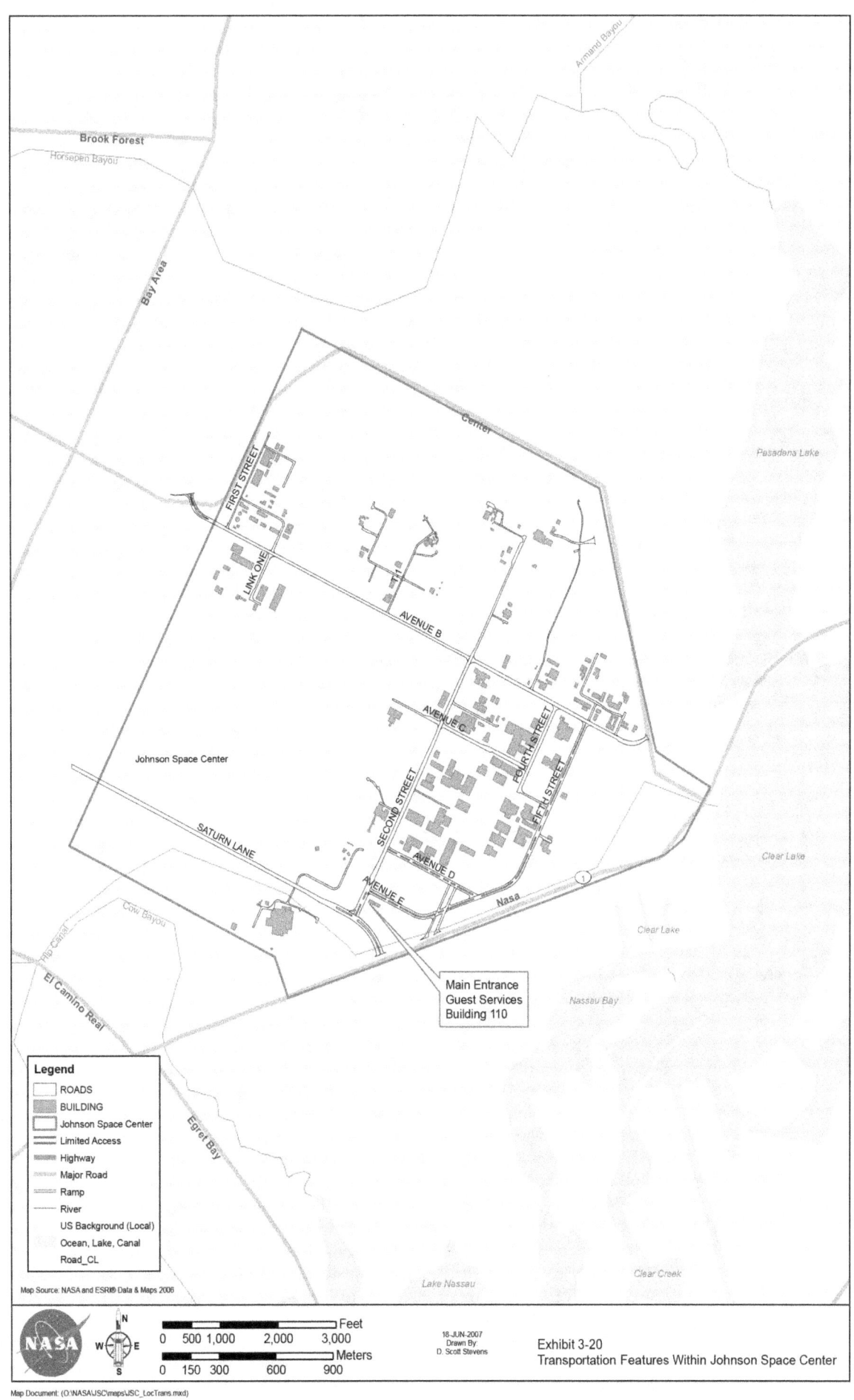

Exhibit 3-20
Transportation Features Within Johnson Space Center

This page intentionally left blank.

Emission Sources. The Houston-Galveston area is classified as a "moderate" ozone non-attainment area. Therefore, EF follows the NSR program and also must evaluate all new projects under the General Conformity rule. EF is categorized as a synthetic minor source of air emissions, with three registered PBR sources (NASA, 2007s). A synthetic minor permit was issued for these sources in April 2007.

Texas does not have any state-specific air quality standards; however, it does have a "Watch List" of HAPs. EF is on the watch list for benzene, styrene, and 1,3-butadiene (TCEQ, 2007).

Stationary sources of air pollutants at EF include aircraft engine testing, coatings of aircraft, fuel storage tank transfers (including fueling) and standing losses, paint stripping, degreasing, power generation, and fugitive emissions from chemical usage at various locations. The registered PBR sources are unenclosed abrasive blasting operations, unenclosed painting operations, and an aircraft corrosion control hangar (NASA, 2007s).

3.4.2 Hazardous and Toxic Materials and Waste

3.4.2.1 Affected Environment

Storage and Handling. NASA is regulated for the generation of hazardous wastes at EF, for which it holds an RCRA registration (#TX2800024067) as an LQG. EF stores small quantities of toxic substances inside buildings and in covered boxes that are capable of containing a spill of their contents. These areas have curbed concrete bases or internal steel structures to contain the toxic substances in case of spills (NASA, 2005b). There is one less-than-90-day waste accumulation area operated by NASA at EF, as well as chemical storage areas in each hangar and operating area. There are no major stockpiles of chemicals and the wastes routinely are disposed properly, according to the procedures developed by JSC (NASA, 2007s).

There are no underground storage tanks (USTs) associated with NASA's activities at EF (NASA, 2005b).

Waste Management. Wastes are stored at EF in the less-than-90-day accumulation yard and are then sent directly to an appropriate waste disposal site. The wastewater generated during the paint stripping activities at the Aircraft Tire and Wheel Maintenance Shop (Building 137) is the largest source of hazardous waste at EF. Operations at EF also generate large quantities of spent solvent and rags soaked with solvent and jet fuel (NASA, 2005b).

Occasionally, storm water and washwater become contaminated by hazardous materials; the water is collected from sumps and disposed as hazardous waste (NASA, 2005b).

Contaminated Areas. EF is not a CERCLA NPL site (NASA, 2007s). No contaminated areas at the NASA-operated areas at EF have been reported (NASA, 2005b).

Toxic Substances. Asbestos products are found in pipe lagging, boiler insulation, and fireproofing materials at EF. NASA has not surveyed all of its facilities to locate all ACMs. The Maintenance Hangar (Building 276) has two beams that have been sprayed with an asbestos-containing insulation. Additional facilities at EF may contain asbestos (NASA, 2005b). Although the paint currently used at EF is lead-free, some of the buildings may still contain LBPs (NASA, 2005b).

PCBs are present at EF only in old light ballasts. As these ballasts are replaced due to attrition, they are sent to the Hazardous Waste Storage Facility before disposal by a PCB disposal contractor (NASA, 2007s).

3.4.3 Health and Safety

3.4.3.1 Affected Environment

The following subsections outline EF's programs for protecting the health and safety of EF employees and the public. Noise hazards at EF are outlined in Section 3.3.8.

Hazardous Materials. Hazardous materials are used to maintain the aircraft used at EF. The hazardous materials used and the hazardous wastes generated at EF are discussed in Section 3.4.2. The implementation of work practices and control technologies minimizes employee exposures to hazardous materials. For spills that are too large to be handled by EF employees, a spill response team may be summoned from JSC or the Houston Fire Department.

Buildings at EF contain asbestos in the form of insulation, fireproofing materials, and other building materials. EF complies with the 29 CFR 1910.1001 standard for the protection of employees from asbestos exposure.

Hazardous Materials Transportation Safety. Hazardous materials such as fuels, chemicals, and hazardous wastes are transported in accordance with the DOT regulations for interstate shipment of hazardous substances (49 CFR 100 through 199).

Explosions and Fire Hazards. The use and storage of certain hazardous materials, including fuels, in aircraft operations presents a risk of explosions and fire hazards. The hangars at EF are equipped with automatic fire detection systems with sprinkler and foam systems. At EF, fire protection is contracted with the City of Houston Fire Department.

Aircraft Safety. Aircraft-related operations at EF are conducted in accordance with Federal Aviation Administration (FAA) regulations to protect the health and safety of the crew, the EF employees, and the public during taxis, takeoffs, flights, and landings.

3.4.4 Hydrology and Water Quality

3.4.4.1 Affected Environment

Surface Waters. EF is located 13 km (8 miles) northwest of JSC. It is part of the Armand Bayou watershed. Horsepen Bayou, a tributary of Armand Bayou, is located to the southeast (Exhibit 3-17). At EF, Horsepen Bayou is non-tidal. The Armand Bayou watershed drains about 140 km^2 (54 square miles) of southeastern Harris County. Armand Bayou itself flows into the northern end of Mud Lake, an estuary of Clear Lake. Clear Lake flows to the western side of Galveston Bay. Armand Bayou is a coastal preserve in the Galveston Bay NEP.

Storm water from NASA tracts at EF drains to the south into Horsepen Bayou via storm sewers, culverts, drainage ditches, and swales (NASA, 2005b).

Groundwater. Groundwater resources under EF are similar to those beneath JSC, as described in Section 3.3.6.

Water Quality. Water quality resources for EF are described in Section 3.3.6. Ground subsidence is an issue near EF. NASA uses the Houston municipal water supply, most of which comes from surface water (NASA, 2005b).

Regulated Water Discharges and Withdrawals. Treated water is supplied to the airport from the City of Houston's water main along State Highway 3.

Wastewater from NASA's EF operations includes sewage, rinse water, washwater, oil/water separator effluent, and washrack wastewater. Wastewater from EF is conveyed to a Sewage Treatment Plant (STP) owned by the Metro Central Advisory Committee southeast of the airport. The plant is operated by the Gulf Coast Waste Disposal Authority. Effluent from the plant flows into Horsepen Bayou south of the airport.

Impervious surfaces at EF generate runoff during rain events. Storm water discharges from industrial sources require discharge permits. A general permit from the TCEQ covers the discharges as long as the facility complies with the permit's conditions, including preparing a P2 Plan, monitoring effluent quality, and keeping records. NASA is covered under a general permit and complies with its conditions (TCEQ, 2007b).

3.4.5 Land Use

3.4.5.1 Affected Environment

NASA occupies 15 ha (37 acres) of EF on six separate tracts of land. Two tracts adjoin the apron of Runway 17R-35L. The tracts are fully developed; facilities include hangars, offices, warehouses, repair and maintenance facilities, fire suppression systems, and parking lots (NASA, 2005b). NASA land uses and other land uses adjacent to EF are discussed below. Land use planning for NASA's facilities at EF is performed by the Facilities Development Division of JSC's Center Operations Directorate.

NASA Land Uses. EF, the largest general aviation reliever airport in Houston by providing traffic relief to the two main airports in Houston, covers 750 ha (1,900 acres), of which 15 ha (37 acres) are NASA tracts (Turner, Collie and Braden, 1991). The largest tract (10 ha, or 23 acres) is at the southern end of the airport and contains most NASA activities and airplane parking. Adjoining this tract are small tracts used for fire protection and for auto parking.

Fifteen hundred m (1 mile) north of the southern tract is the second largest tract (4 ha, or 10 acres) containing the Maintenance Hangar (Building 990), Aircraft Operations Building (Building 993), Aircraft Maintenance Support Building (Building 994), and airplane parking. The two remaining tracts hold the Supply and Maintenance Warehouse (Building 380) and auto parking.

NASA Buildings and Other Structures. NASA has 22 buildings and 8 other structures at EF. Three hangars are used for aircraft maintenance (NASA, 2005b). One hangar is located in the north tract; this hangar has additions for shop machines and technical facilities. The two south hangars (Buildings 276 and 135) have additions for machine shops and office buildings. Nearby are warehouses and an office and warehouse building for purchasing, receiving, distribution, and shipping (NASA, 2005b).

Other NASA structures include five storage sheds, two gate houses, a deluge pump station, two deluge storage tanks, an airplane wash rack, an engine test complex, two special projects buildings, and a hazardous materials storage area.

Easements and Rights-of-Way. The NASA tracts have several sanitary sewer and other utility easements; these are shown in the Airport Master Plan (Hoyle, Tanner and Associates, 1987). NASA controls only parts of the airport roadway system and apron area (NASA, 2005b).

3.4.6 Noise

3.4.6.1 Affected Environment

The following subsections describe the two types of noise generated at EF, which include noise generated by engine testing and by aircraft operation. Most of the land surrounding EF is undeveloped. The closest sensitive receptor is a commercial development 200 m (670 ft) away.

Noise Generated by Engine Testing. The Engine Test Complex (Building 140) and the Sound Suppression Facility (Building 151) generate the most noise of the stationary sources at EF. These sources produce noise of variable duration and frequency (NASA, 2005b).

The Engine Test Complex tests engines outside of the airplane for up to 4 to 6 hours each day. Each engine is tested in idle, military thrust, and afterburner modes. During tests in the afterburner mode, noise levels as high as 142 dBA are generated in the building. Monitors have recorded noise levels of 90 dBA at a distance of 20 m (60 ft) from the facility. The nearest receptor to the Engine Test Complex is a commercial development 200 m (670 ft) to the southwest, beyond State Highway 3. Noise levels of 68 dBA at this receptor have been estimated when noise levels in the Engine Test Complex are at peak levels (NASA, 2005b).

The Sound Suppression Facility tests engines in the airplane after they are tested in the test complex. Tests are conducted twice per week for 30 minutes to 2 hours in the Sound Suppression Facility. Noise contours provided by the manufacturers of the engines indicate that noise levels of 90 dBA may extend 40 to 60 m (140 to 190 ft) from the test site. The nearest receptor from the Sound Suppression Facility is the same commercial development mentioned above, 400 m (1,300 ft) to the southwest. It is estimated that noise levels reaching 69 dBA would reach this receptor when an engine is being tested at its maximum output. Employees at these two facilities are required to wear hearing protection when tests are being conducted (NASA, 2005b).

Noise Generated by Aircraft. Noise sources at EF include the tactical jet operations conducted by the Texas Air National Guard (TxANG) and NASA. Noise levels at 75 dBA generated by NASA flight operations surround the runways; also, aircraft noise at this level extends over the airfield property boundary and encroaches on open areas beyond the runways. Noise levels of 65 dBA generated by NASA flight operations extend beyond the airport property into primarily undeveloped areas, but also reach surrounding residential and commercial communities. Noise levels of 65 dBA are higher than the noise generated by normal conversation, but are lower than the 85-dBA threshold that may cause hearing damage (NASA, 2005b).

3.4.7 Site Infrastructure

3.4.7.1 Region of Influence

The ROI for energy sources is defined as the SSP-related facilities and the energy sources for these facilities at EF.

3.4.7.2 Affected Environment

Potable Water Supply. See the Hydrology and Water Quality, Section 3.4.4.

Wastewater System. See the Hydrology and Water Quality, Section 3.4.4.

Storm Water. See the Hydrology and Water Quality, Section 3.4.4.

Energy Sources. NASA does not generate electricity at EF using natural gas. Electricity is generated on an emergency basis using diesel generators. EF does not provide electricity into the grid. NASA uses the Houston municipal electricity and natural gas supply at EF. The distribution systems are owned by the airport (NASA, 2007s).

3.4.8 Solid Waste

3.4.8.1 Affected Environment

Solid waste generated at EF is sent to JSC, where nonhazardous refuse is taken to roll-off boxes at the Central Waste Collection Facility and then shipped to the City of Houston landfill. Classified wastes (paper, microfilm, and microfiche) either are taken to the classified waste incinerator or to the Classified Waste Disintegrator Facility, then are landfilled as solid waste (NASA, 2005b).

3.4.9 Traffic and Transportation

3.4.9.1 Region of Influence

The transportation ROI for JSC is defined as the 10 counties of the Houston-Baytown-Sugarland MSA. Like JSC, EF is located in Harris County and is subject to the same commuting patterns.

3.4.9.2 Affected Environment

Transportation. EF is in the South Belt Ellington area, which includes areas along I-45 to the northwest and southeast of the airport and along the Sam Houston Parkway (South Belt) to the northeast and southwest. The airport is close to the Clear Lake area to the southeast, the City of Pasadena to the northeast, and South Houston to the north.

Access Roads to Ellington Field. Automobiles and trucks can reach the Clear Lake area on State Highway 3, State Highway 146, and I-45. NASA Parkway connects these roads with the main gate into EF.

Railroads. Railroads run parallel to State Highway 3 and State Highway 146. The Southern Pacific provides freight rail service to Seabrook and the Missouri-Kansas-Texas Railroad serves Webster. EF does not have any direct rail service.

Airports and Ports. Airports and ports in the Houston area are described in Section 3.3.12.

Transit. Transportation to EF for most employees is by private auto. The Metropolitan Transit Authority of Harris County provides "Park and Ride" bus service between Clear Lake City and downtown Houston on a staggered schedule and operates a shuttle to the EF Center.

3.5 El Paso Forward Operation Location

EPFOL is located at the EPIA in El Paso, Texas, on 2,833 ha (7,000 acres) of land (Exhibit 3-21). EPFOL operations are under JSC's management, and the management directives issued for JSC also apply to EPFOL.

EPIA is owned and operated by the City of El Paso and NASA leases land from the City. The lease between the City of El Paso and NASA, administered by the WSTF, ends in 2010. Biggs Army Airfield and the Fort Bliss Military Reservation are adjacent to the airport on the northern and eastern sides. The topographic features near EPIA are the Franklin Mountains to the west, desert terrain to the north and east, and the Rio Grande valley to the south. EPIA is classified as a medium-duty air traffic hub by the FAA and serves several airlines, air freight operators, NASA, and occasional military aircraft.

NASA operates its facilities according to state, city, FAA, and NASA rules and regulations. JSC supports EPFOL with regard to environmental issues such as permitting, inspections, and P2. WSTF supports EPFOL with regard to occupational health issues. EPFOL maintains two aircraft hangars at EPIA that have two distinct operations, supported by two different contractors. Hanger 1 is the Shuttle Training Aircraft (STA) Hangar, with two shifts operating at the hangar. Hangar 2 primarily is used for T-38 maintenance activities and avionics upgrades. There also are areas used to park aircraft and an aircraft washing area.

Astronauts fly T-38s from EF to EPFOL to prepare for flights in the STA. The astronauts are briefed at EPFOL for their training missions in the STA. A typical training mission in the STA is illustrated in Exhibit 3-22.

The aircraft maintenance crews are provided by the USAF from Holloman AFB through a contract administered at Tinker AFB. NASA also has the crews at Fort Bliss on hold for quick turnarounds on unexpected maintenance activities that are not performed routinely at EPFOL.

This page intentionally left blank.

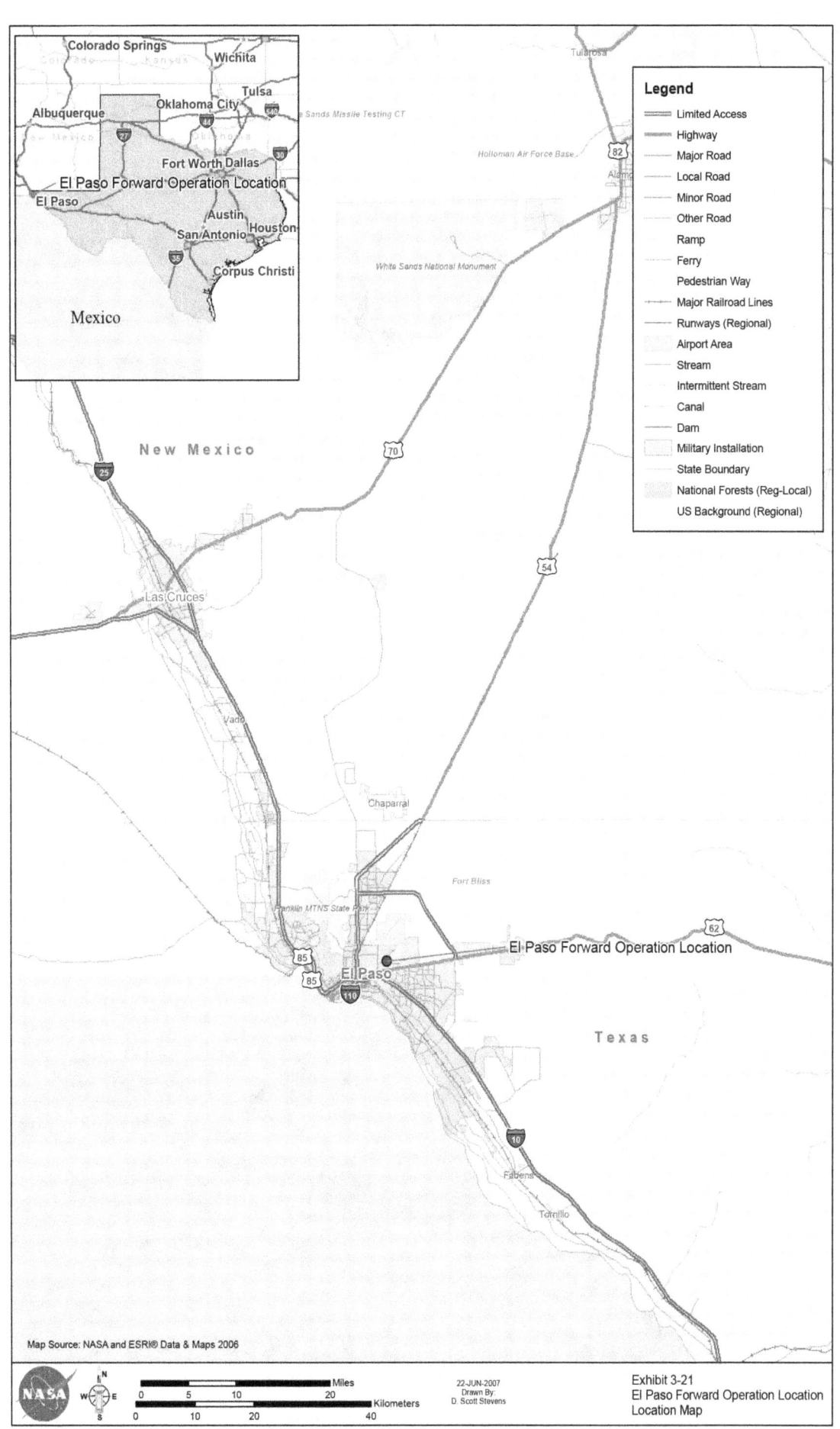

Exhibit 3-21
El Paso Forward Operation Location
Location Map

This page intentionally left blank.

EXHIBIT 3-22
Typical Astronaut Training Mission Activities Conducted at EPFOL

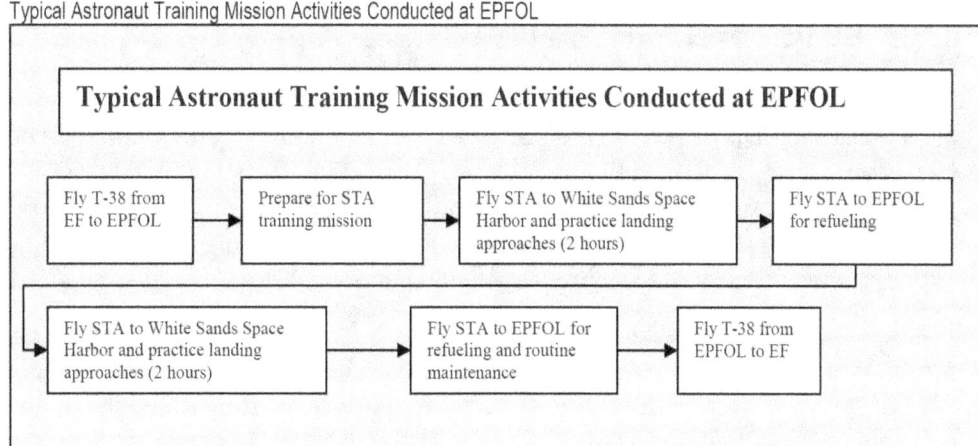

3.5.1 Air Quality

3.5.1.1 Affected Environment

Regional Climate. El Paso's climate is arid. The average daily maximum temperature over a year is 77.5 degrees Fahrenheit (°F) and the daily minimum is 49°F. The average monthly temperature is 63.3°F. Average rainfall per year is 8.8 inches. An average of 4.3 inches of snow falls per year, but snowfalls generally are gone within a few hours. Dust storms and sandstorms are common because natural vegetation is sparse (NASA, 2004c).

Emission Sources. EPA has designated El Paso County as a moderate non-attainment area for CO and PM-10. (On January 11, 2006, the State of Texas requested that EPA redesignate the area as attainment for the CO standard). Therefore, EPFOL follows the NSR program and also must evaluate new projects under the General Conformity rule. El Paso is in attainment for all other criteria pollutants (NASA, 2007s).

EPFOL is categorized as a minor source of air emissions, with one registered PBR source (the aircraft corrosion control hangar) and a few unregistered PBRs. The TCEQ only requires written notice when permitted sources are removed, and the permit then becomes void (NASA, 2007s).

Activities at EPFOL Hangar 1, the STA Hangar, include astronaut training, aircraft turn-around, unscheduled maintenance, and aircraft washing. Hangar 2 at EPFOL is the T-38 Hangar. Hangar 2 houses the corrosion prevention program, which provides corrosion prevention treatment to T-38 aircraft. This treatment involves physical grinding and applying primers, paints, and sealants in a paint booth.

Structural maintenance and avionics system upgrade operations also are performed in Hangar 2 (NASA, 2007s).

3.5.2 Hazardous and Toxic Materials and Waste

3.5.2.1 Affected Environment

Storage and Handling. EPFOL is categorized as a small-quantity generator (SQG) of hazardous waste, as defined by the State of Texas Waste Reduction Policy Act (WRPA) of 1991, generating between 100 and 1,000 kg of hazardous waste a month (NASA, 2007s). EPFOL is a sub-installation of JSC and, as such, environmental support is provided by the JSC Environmental Services Office (ESO) (NASA, 2004c).

Hangers 1 and 2 contain chemical storage areas operated by the USAF. Unused products and materials are returned to the USAF. There are no major stockpiles of chemicals and wastes routinely are disposed properly, according to the procedures developed by JSC (NASA, 2004c).

Waste Management. Each hangar contains less-than-90-day waste accumulation areas operated by the USAF.

Asbestos and Lead-based Paint. No asbestos is known to be present in SSP facilities at EPFOL.

LBP is a potential concern for the demolition of buildings and structures built before 1978. Demolition debris potentially containing LBP is subject to landfilling restrictions.

EPFOL does not release a quantity of pollutants high enough to trigger an Emergency Planning and Community Right-to-Know Act (EPCRA) Toxic Release Inventory (TRI) report (NASA, 2007s).

3.5.3 Health and Safety

3.5.3.1 Affected Environment

Astronauts fly T-38s from EF to EPFOL to prepare for flights in the STA. The astronauts are briefed at EPFOL for their training missions in the STA. At EPFOL, general maintenance on aircraft is conducted, including corrosion control, structural maintenance, and avionics work. The following subsections outline EPFOL's programs for protecting the health and safety of the employees at EPFOL, as well as the public. Noise hazards at EPFOL are outlined in Section 3.5.5.

Hazardous Materials. Hazardous materials are used to maintain the aircraft at EPFOL. The hazardous materials used and hazardous wastes generated are discussed in Section 3.5.2. The degree of exposure to hazardous materials is minimized by the implementation of work practices and control technologies. Spills occur infrequently at EPFOL. When spills do occur, they are immediately contained and recovered using rags or other absorbent material. The spill residue is transferred to an appropriately labeled drum for storage and transport offsite for disposal. Hazardous materials (HAZMAT) services are provided by the City of El Paso; employees may dial 911 for assistance with spills that are too large for EPFOL employees to manage.

The buildings at EPFOL were constructed after 1990; therefore, it is unlikely that any ACMs are present at the facility.

Hazardous Materials Transportation Safety. Hazardous materials such as fuels, chemicals, and hazardous wastes are transported in accordance with DOT regulations for the interstate shipment of hazardous substances (49 CFR 100 through 199).

Explosions and Fire Hazards. Using certain hazardous materials presents a risk of explosions and fires. At EPFOL, fire protection and rescue are provided by EPIA and the City of El Paso Fire Department. The Airport Fire Department responds within 7 minutes, as indicated by recent fire drills, but has limited capabilities for structural fires. The City of El Paso would provide a combined response for the structures at the NASA hangars.

Aircraft Safety. All aircraft-related operations at EPFOL are conducted in accordance with FAA regulations to protect the health and safety of the crew, the EPFOL employees, and the public during taxis, takeoffs, flights, and landings.

3.5.4 Hydrology and Water Quality

3.5.4.1 Affected Environment

Surface Waters. No waters of the U.S., such as rivers or arroyos, exist within the boundaries of EPIA. The airport is located in the Rio Grande River watershed.

In heavy rains, water drains to undefined drainages and pools in low areas. A retention basin is located in the eastern portion of EPIA. A water tank operated by the El Paso Water Utility Public Service Board is located in the southeastern portion of EPIA (NASA, 2004c).

Groundwater. The Hueco-Bolson aquifer underlies the EPFOL. Wells have been drilled into this aquifer in the vicinity of the eastern portion of EPIA. All of the wells are between 213 and 244 m (700 and 800 ft) deep.

The Hueco-Bolson aquifer is a thick sequence of Tertiary and Quaternary sediment formed in the faulting between area mountain ranges. The Hueco-Bolson consists of silt, sand, and gravel in the upper part and clay and silt in the lower part, with a combined thickness of approximately 274 m (9,000 ft). The aquifer contains fresh to slightly saline water (TCEQ, 2005).

Regulated Water Discharges and Withdrawals. Potable water is provided to the EPIA and NASA from El Paso Water Utilities.

Wastewater generated at EPFOL and discharged to the sanitary sewer system is directed to the El Paso Water Utilities' Public Service Board Haskell Street WWTP. The El Paso Water Utilities Public Service Board treats and discharges wastewater under NPDES Permit TX0026751, issued by EPA, and under Texas Pollutant Discharge Elimination System (TPDES) Permit WQ0010408-004, issued by the TCEQ (NASA, 2004a).

NASA has submitted a Notice of Intent (NOI) to be covered under a general permit for storm water discharges from industrial facilities and complies with its conditions (NASA, 2004a). The general permit from the TCEQ covers the discharges, as long as the facility complies with the permit's conditions, including preparing a P2 Plan, monitoring effluent quality, and keeping records.

3.5.5 Noise

3.5.5.1 Affected Environment

Aircraft operations are the primary source of noise at EPFOL. Normal operations at the EPFOL produce relatively low noise levels when compared to EPIA's flight operations, which produce the dominant noise levels in the vicinity. Most of the land immediately surrounding the site is undeveloped and does not contain sensitive site receptors (NASA, 2004c).

Activities that occur in Hangar 1 at EPFOL include flight checks, aircraft refueling, unscheduled maintenance, and aircraft washing. These activities generate low noise levels, below 85 dBA. Corrosion prevention and structural maintenance of aircraft are conducted in Hangar 2 as needed. These activities also generate low levels of noise. Hearing protection typically is not required (NASA, 2004c). No sensitive receptors are known to be located within the noise ROI.

3.5.6 Site Infrastructure

3.5.6.1 Region of Influence

The ROI for energy sources is defined as the SSP-related facilities and the energy sources for these facilities at EPFOL.

3.5.6.2 Affected Environment

Potable Water Supply. See "Hydrology and Water Quality," Section 3.5.4.

Wastewater System. See "Hydrology and Water Quality," Section 3.5.4.

Storm Water System. See "Hydrology and Water Quality," Section 3.5.4.

Energy Sources. NASA uses the City of El Paso's municipal electricity and natural gas supply at EPFOL. The distribution systems are owned by the airport (NASA, 2007s).

3.5.7 Solid Waste

3.5.7.1 Affected Environment

EPFOL is a small quantity generator of wastes and those wastes are regulated through the JSC industrial solid waste program. The location of waste disposal may change with waste type and facility audits.

3.5.8 Transportation and Traffic

3.5.8.1 Region of Influence

The transportation ROI for EPFOL is defined as El Paso County, Texas. El Paso County is also the El Paso MSA, made up of the City of El Paso in the northwestern portion of the county and the surrounding area (OMB, 2006).

3.5.8.2 Affected Environment

Transportation. I-10 is south of EPFOL. U.S. Highway 180 and one major arterial, Airway Boulevard, connect the project site to I-10. U.S. 180, to the south of the project site, runs east-west and has three lanes in each direction.

Access Roads to EPFOL. The primary connection from the project site to U.S. 180 is through Airway Boulevard and Terminal Drive (Exhibit 3-23). The secondary access to the project site from U.S. 180 is through American Drive, Boeing Drive, and Air Way Boulevard. Airway Boulevard links the project site to I-10.

Railroads. Southern Pacific Railroad runs west of the project site. Freight service to the international airport is provided by this railroad. EPFOL does not have any direct rail service.

Airports. EPIA is adjacent to and north of the project site. Cidad International Airport is approximately 24 km (15 miles) south of the project site. Air freight services are available at these airports.

This page intentionally left blank.

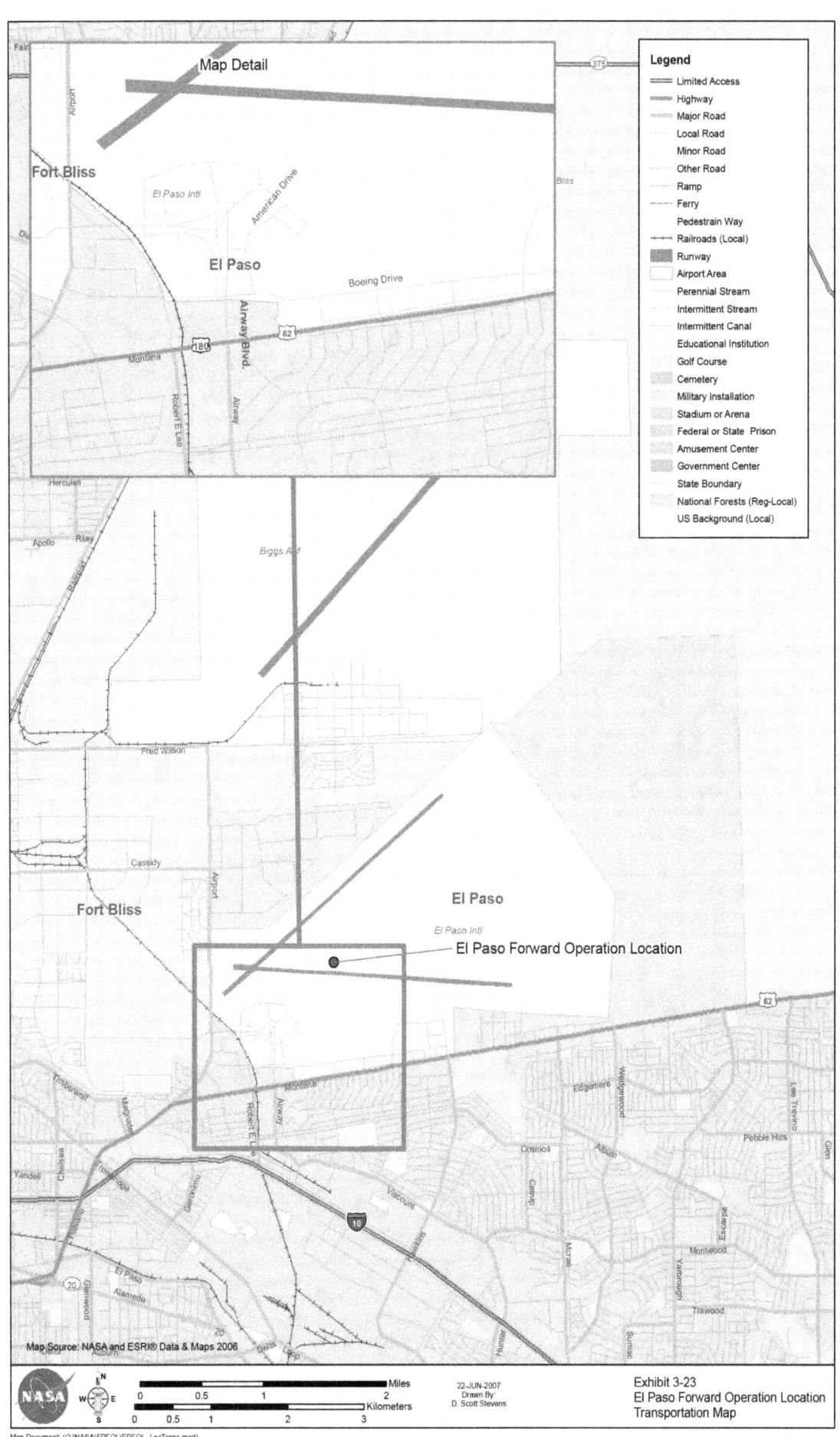

Exhibit 3-23
El Paso Forward Operation Location
Transportation Map

This page intentionally left blank.

3.6 Stennis Space Center

NASA's SSC is located near the Gulf of Mexico in western Hancock County, Mississippi, approximately 89 km (55 miles) northeast of New Orleans, Louisiana, and approximately 48 km (30 miles) west of Biloxi and Gulfport, Mississippi (Exhibit 3-24). The facility is situated at 30.38 north latitude and 89.60 west longitude at its center point. In May 1962, the federal government acquired approximately 56 km^2 (13,800 acres) that constitute the "Fee Area," or the confines within the gates of SSC. In this area, NASA, along with numerous federal and state agencies, has constructed administrative, research, remote sensing, and propulsion testing facilities. The latter activity is restricted to NASA and is the major function of the Center. SSC has been named as NASA's program manager for propulsion testing, and many new programs are envisioned. Because of SSC's proximity to MAF, the socioeconomic analysis for SSC in this document is included in the analysis for MAF.

Rocket testing operations necessitated the development of a Buffer Zone for safety and acoustic considerations. A perpetual restrictive easement of 506 km^2 (125,001 acres) was acquired, which extends 9.6 km (6 miles) in all directions of the Fee Area. The majority of the Buffer Zone is located in Hancock County, Mississippi, although portions extend into Pearl River County, Mississippi; and St. Tammany Parish, Louisiana. The region is bounded on the east and west by the Pearl River and Jourdan River watersheds, respectively. Currently, the government owns 30.6 km^2 (6,808 acres) of the Buffer Zone, with the remainder being held by individuals or corporations. Provisions of the restrictive easement prohibit maintenance or construction of dwellings and other buildings suitable for human habitation. The predominant land use in the Buffer Zone includes sand and gravel mining, timber production, livestock production, and recreational pursuits such as hunting and fishing.

Several communities are situated just outside the Buffer Zone including Pearlington, Waveland, Bay St. Louis, Kiln, and Picayune, Mississippi; and Slidell and Pearl River, Louisiana. There are 12.1 km (7.5 miles) of canals inside the Fee Area available to transport material within SSC. The SSC canal system links to the East Pearl River through a canal lock system. The East Pearl River links SSC to the national waterway transportation system. It is 33.8 km (21 miles) from the main canal to the Gulf Intracoastal Waterway. The canal system provides a means of transporting large rocket engines, propellants, and other heavy equipment and materials to the facility.

3. AFFECTED ENVIRONMENT

This page intentionally left blank.

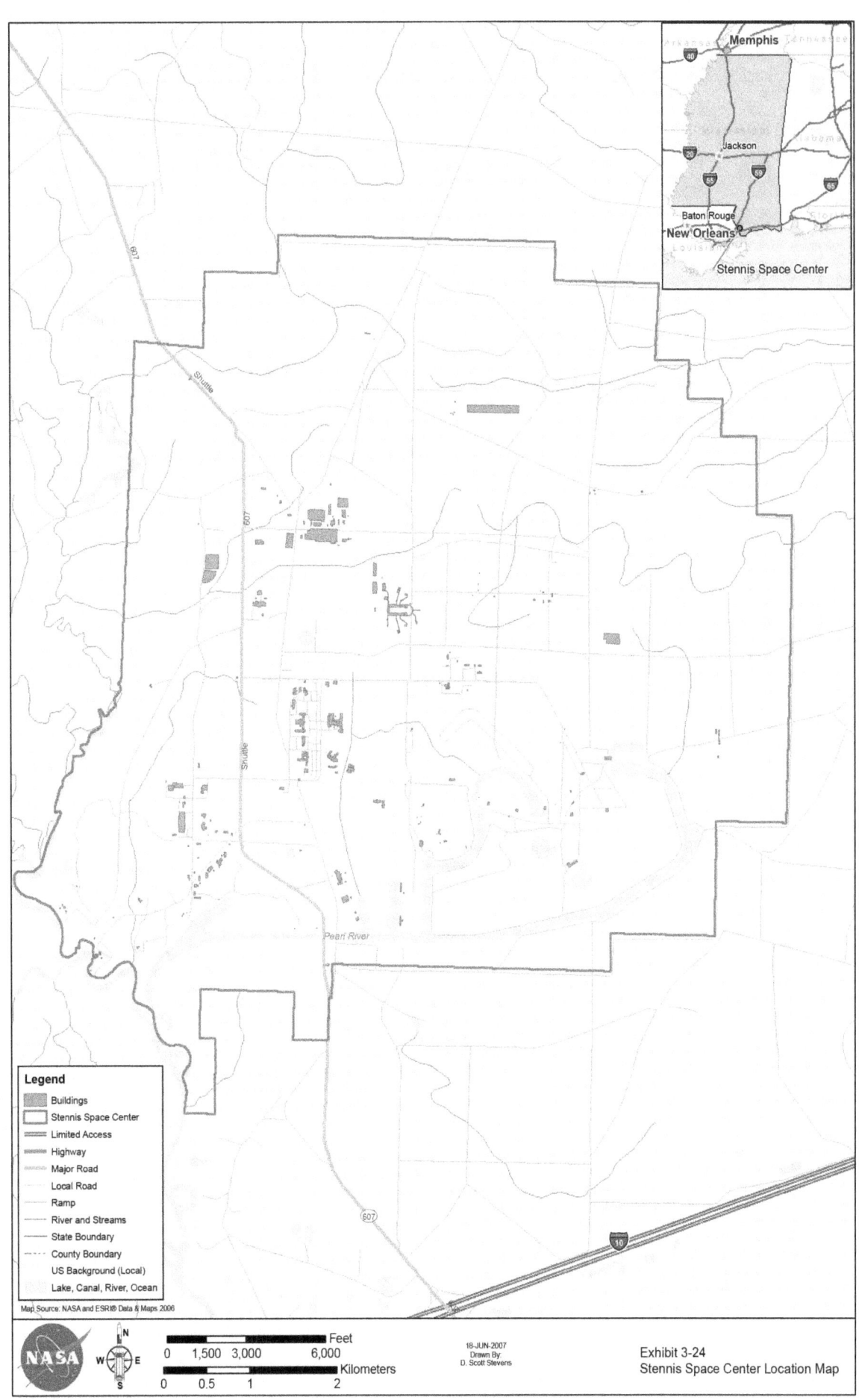

This page intentionally left blank.

The SSC's Center Operations Directorate, NASA Environmental Management, is responsible for permitting, compliance, and monitoring NASA activities and many of the resident activities that may affect the environment. Sitewide environmental and industrial hygiene programs are provided by NASA to tenant agencies as part of the shared-pool operations; however, NASA does not accept responsibility for tenant or contractor compliance. Each resident organization has its own environmental personnel responsible for the organization's environmental compliance for permitting, NEPA, etc. When a resident organization wants to change a discharge for which SSC holds a permit or to perform an activity onsite, the organization is required to fill out a Preliminary Environmental Survey (PES) to evaluate whether its discharge requires pretreatment or if the action requires NEPA documentation. Resident organizations typically are required to obtain their own air permits. If a resident agency's wastewater discharge exceeds the SSC permit limits, the resident agency must obtain its own water permit. SSC Environmental Management reviews the PES to determine if the resident organization activities will affect SSC and requires the resident agency to follow its own NEPA regulations. The Facility Operating Services (FOS) contractor can dispose of tenant waste or respond to a spill on a fee-per-service basis (NASA, 2007s).

3.6.1 Air Quality

3.6.1.1 Affected Environment

Regional Climate. SSC lies in a humid subtropical region, based on the Köppen-Geiger system of climate classification. The climate typically lacks a dry season. The climate is temperate and rainy with hot summers (NASA, 2005a).

The average annual temperature at SSC is about 66°F. The average seasonal temperatures are 53°F in the winter, 65°F in the spring, 79°F in the summer, and 64°F in the fall (NASA, 2005a).

On average, there are only 84 clear days per year. For the rest of the year, it is typically partly cloudy for 114 days and cloudy for 167 days. It is frequently foggy from mid-October to May. Heavy fogs that limit surface visibility to 1/4 mile or less occur an average of 42 days per year, usually during late night and early morning hours (NASA, 2005a).

Rainfall averages about 60 inches per year, but varies by plus or minus 20 inches per year (NASA, 2005a).

Prevailing surface winds are from the south and southeast through two thirds of the year and from the north for the rest of the year, while upper level winds generally prevail from the west and southwest. The hurricane (tropical cyclone) season runs from June to November. Cyclone intensity ranges from weak to large and intense, with maximum wind speeds approaching 200 mph. The Gulf Coast averages one tropical cyclone per year; approximately two thirds of these are hurricane force with winds greater than 74 mph (NASA, 2005a).

Emission Sources. The ambient air quality of the three southern Mississippi counties (Hancock, Harrison, and Jackson) is considered to be in attainment for PM_{10}, $PM_{2.5}$, ozone, CO, sulfur dioxide, nitrogen oxides (NOx), and lead (NASA, 2007b). Mississippi does not have any additional state-specific air quality standards.

SSC is a minor source of HAPs. As a result, maximum achievable control technology (MACT) standards do not apply. Currently, SSC operates under Title V Operating Permit #1000-00005, originally issued in February 1998 and renewed in 2003. On March 27, 2001, SSC obtained a PSD permit (NASA, 2005a).

NASA operates more than 40 diesel fuel-burning generators and engines, including four 1,500-kilowatt (kW) generators and ten 3,475-kW engines that support the deluge water system for the A1, A2, and B1/B2 Test Stands. NASA also operates a Fuel Dispensing Facility, an HCFC Recovery Facility, an Abrasive Blast Facility, a Rocket Testing Facility, and Flare Stacks (NASA, 2007s).

NASA maintains an ODS phase-out plan at SSC. Since the implementation of this plan in 1993, NASA has reduced its use of chlorofluorocarbons (CFCs) and methyl chloroform and terminated its use of halons. NASA operates a computerized refrigerant database system (Refrigerant Compliance Manager [RCM]) at SSC to track and maintain information about all cooling systems onsite. This system enables NASA to monitor refrigerant usages, leak rates, and other data at SSC to minimize the release of ODSs to the environment (NASA, 2007s).

3.6.2 Biological Resources

3.6.2.1 Affected Environment

Vegetation. Four major plant community types have been identified in the SSC area. These community types, generally identified by the predominant type of vegetation, are as follows: Pine Flatwoods, Bottomland hardwood, Pitcher plant bogs and swamps, and Grasslands and marshes. Pine Flatwoods account for the majority of the vegetation in the undeveloped portions of SSC and in the surrounding Buffer Zone. The dominant species in these communities are slash pine interspersed with some cypress, loblolly pine, swamp tupelo, red maple, and sweet gum. Oak species occur in locations that are more elevated with better drainage. The understory in these communities includes holly species, sweet bay gallberry, yaupon, wax myrtle, grasses, and cane. Bottomland hardwood communities occur in low, poorly drained soils, which may have standing or slowly moving water. The dominant species in these communities are black gum, swamp tupelo, and pond cypress. The understory includes ash species, black willow, red maple, poison ivy, and honeysuckle. Few grass or forb (herbs other than grass) species occur in these communities. Pitcher plant bogs are unique to the coastal plain of the southeastern

U.S. and occur in low-lying, poorly drained areas with acidic soil. The few mature trees, if any are present, are generally cypress or longleaf pine species. These communities occur where the area is burned regularly, which prevents transition to forest or bottomland hardwood-type communities. Prominent herbaceous species in Pitcher plant bogs include orchids, sundews, pitcher plants, pipeworts, and yellow-eyed grass (NASA, 2005a).

Wetlands. Large portions of the Fee Area and Buffer Zone are considered jurisdictional wetlands by the USACE (Exhibit 3-25). SSC maintains four areas to provide for wetland mitigation to compensate for the filling of jurisdictional wetlands during construction activities in the Fee Area. NASA and SSC Environmental Management coordinate with the USACE for activities that affect wetlands and for mitigation activities (NASA, 2007s). On SSC, portions of Bayou LaCroix, Mulatto Bayou, and the Pearl River in the Buffer Zone are designated as below the watermark of ordinary high tides (NASA, 2007s).

Floodplains. Documented floodplains at SSC are a 100-year floodplain along the East Pearl River at the western edge of the Fee Area and 100-year floodplains along the Wolf Branch and the Lion Branch of Catahoula Creek in the northeastern portion of the Fee Area (Exhibit 3-26). The majority of SSC is in an area of minimal flooding, and there is little development in the documented floodplains at SSC (NASA, 2005a). NASA and SSC Environmental Management coordinate with the USACE for activities that affect floodplains (NASA, 2007s).

The Pearl River, extending through the Buffer Zone, and the Jourdan River from the confluence of Catahoula Creek to the Bay of St. Louis are Inventory Rivers listed on the Nationwide Rivers Inventory (NRI). These rivers are protected under the Wild and Scenic Rivers Act (NASA, 2007s).

Wildlife. SSC provides diverse terrestrial habitats including grasslands, forests, and wetlands for a variety of wildlife species. A complete list of the wildlife documented on SSC is provided in the ERD (NASA, 2005a).

Protected Species and Habitats. The Pearl River, which is used for SSC barge traffic, has been identified as an excellent example of a large Gulf Coastal Plain river with extensive swamplands. The river supports numerous endangered, threatened, and rare species. The only plant species at SSC listed as endangered by the USFWS (and also as critically imperiled by the Louisiana Department of Wildlife and Fisheries [LDWF]) is the Louisiana quillwort (*Isoetes louisianensis*). None of the surveys conducted in the early 1990s or in 1998 found any evidence of the existence of Louisiana quillwort on SSC (NASA, 2005a). Exhibit 3-27 lists the federal and Louisiana or Mississippi state-listed wildlife species that have ranges within SSC (NASA, 2005a; LDWF, 2004; Mississippi National Heritage Program [MNHP], 2002).

3. AFFECTED ENVIRONMENT

This page intentionally left blank.

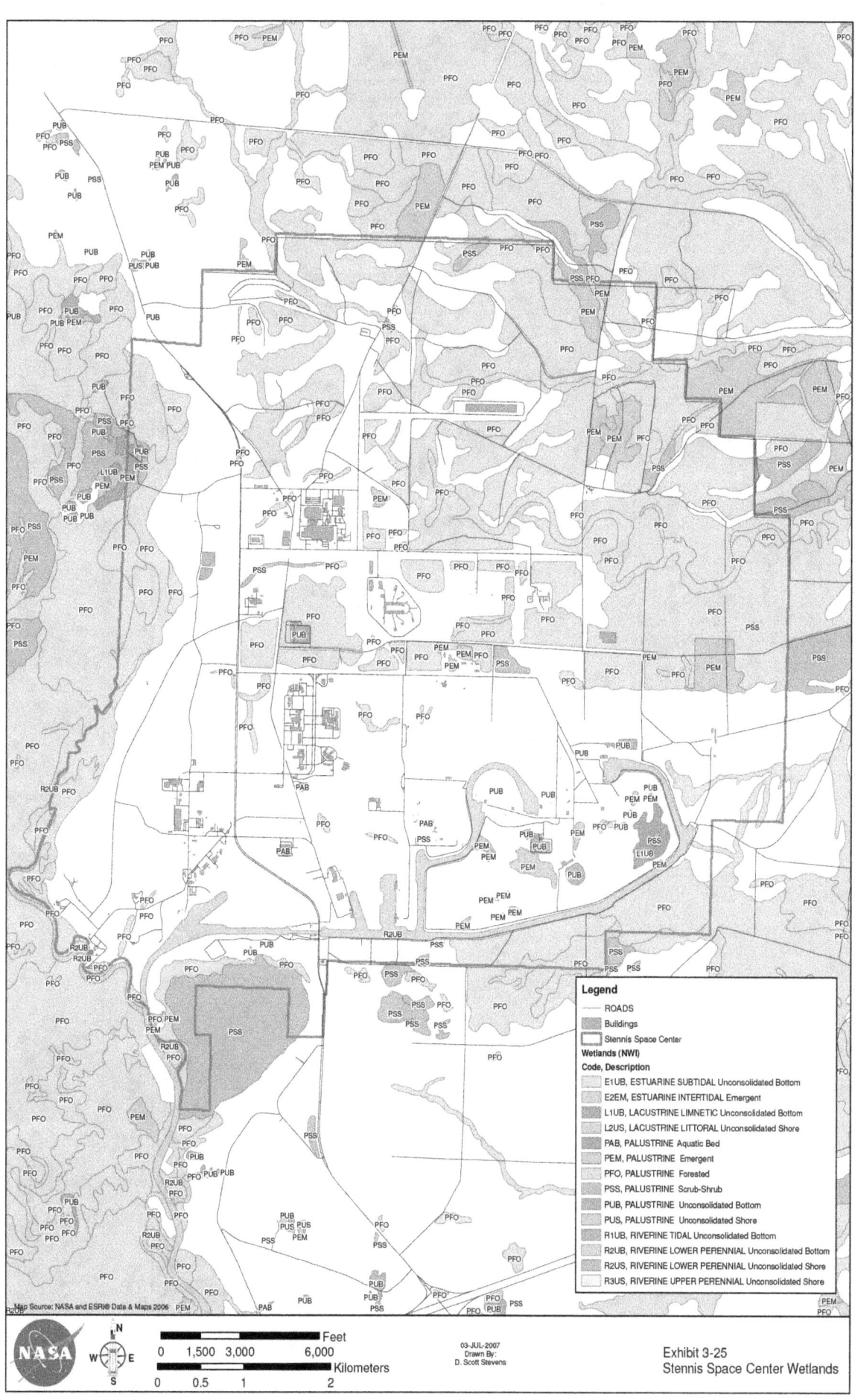

Exhibit 3-25
Stennis Space Center Wetlands

This page intentionally left blank.

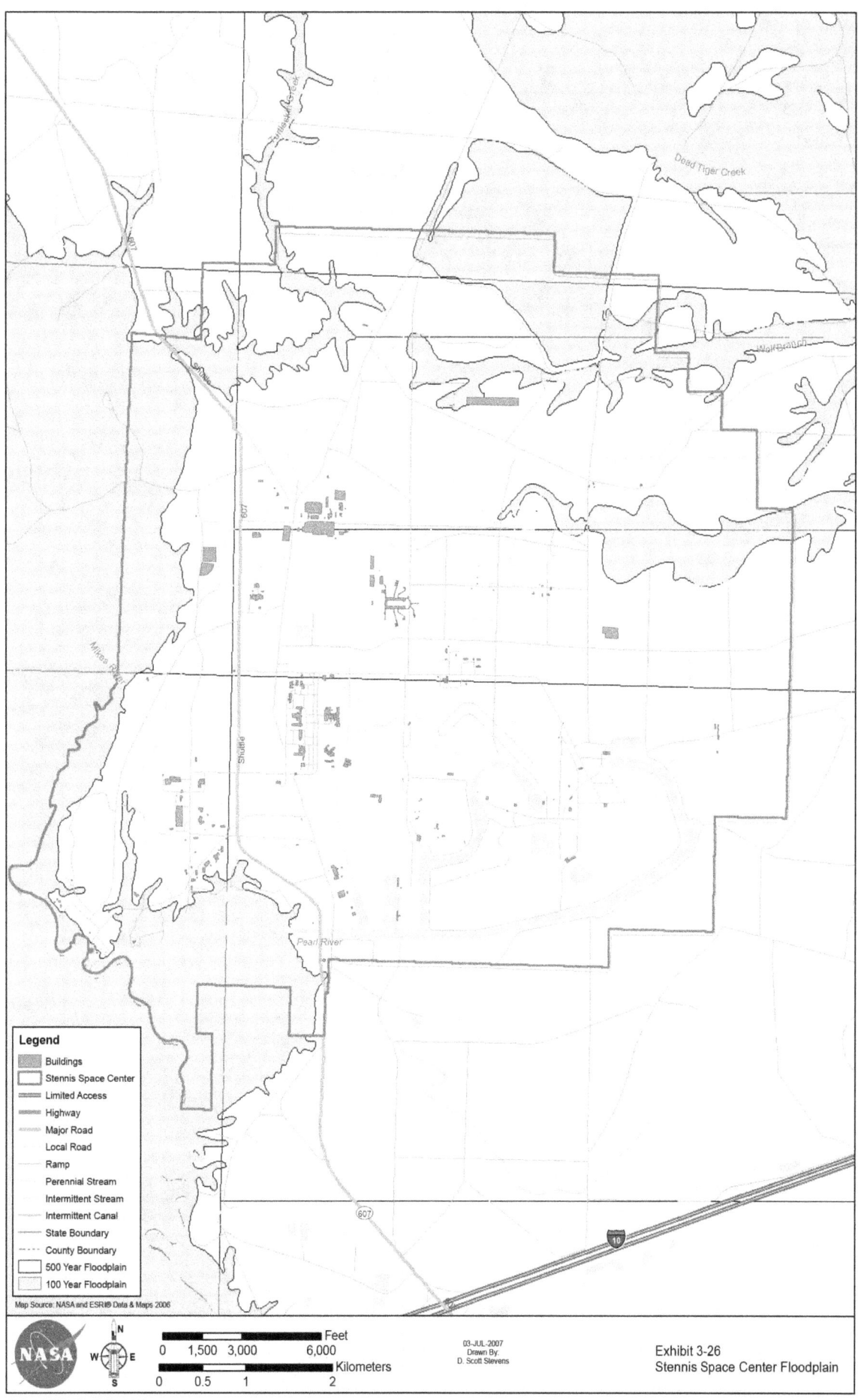

Exhibit 3-26
Stennis Space Center Floodplain

This page intentionally left blank.

EXHIBIT 3-27
Federal- and State-listed Wildlife Species with Ranges that Include SSC

Scientific Name	Common Name	Level of Protection State Louisiana/ Mississippi	Federal
Fish			
Acipenser oxyrhynchus desotoi	Gulf sturgeon	T/E	T
Reptiles and Amphibians			
Drymarchon couperi	Eastern indigo snake	-/E	T
Gopherus polyhemus	Gopher tortoise	T/E	T
Birds			
Picoides borealis	Red-cockaded woodpecker	E/E	E
Falco peregrinus	American peregrine falcon	T/T	-
Mammals			
Felis concolor coryi	Florida panther	E/E	E

Notes:
E = Endangered
T = Threatened

Plants. The Mississippi Department of Wildlife, Fisheries and Parks (MDWFP) and the LDWF list 142 plant species on SSC as special concerns because they are known or suspected to occur in low numbers (NASA, 2005a). (Note that the Mississippi and Louisiana state rankings are assigned by each state's Natural Heritage Program, which may result in inconsistencies in species' rankings from state to state.) A total of 52 of these plant species are listed as critically imperiled because of their extreme rarity (five or fewer occurrences or few remaining individuals or acres) or because of some other factor(s) making them vulnerable to extirpation (NASA, 2005a). The ERD lists the special concern plant species identified at SSC (NASA, 2005a).

Wildlife. The SSC ERD (NASA, 2005a) lists the MDWFP and LDWF endangered or threatened wildlife species and wildlife species of special concern for Hancock County and/or St. Tammany Parish. Ecological surveys found no evidence of the existence of the gopher tortoise, eastern indigo snake, red-cockaded woodpecker, peregrine falcon, or Florida panther (NASA, 2005a). The following wildlife species were identified in the western portion of the Fee Area: American alligator (*Alligator mississipiensis*), ringed map turtle (*Graptemys oculifera*), alligator snapping turtle (*Macroclemys temminckii*), and Louisiana black bear (*Ursus americanus luteolus*) (NASA, 2005a).

3.6.3 Cultural Resources

3.6.3.1 Affected Environment

Archaeological Resources. There are four NRHP-eligible archaeological sites within the boundaries of SSC, one of which has been nominated but not yet listed. The remains of Gainesville, a small logging town, have been excavated partially and nominated to the NRHP. The remains of Logtown have been determined potentially eligible for listing in the NRHP and an evaluation of the site is ongoing. Napoleon, the first European settlement in Hancock County, and the area around Bayou LaCroix are known to have been inhabited by the Southern Band of the Choctaw Nation since prehistory. These two sites also have been determined to be potentially eligible for NRHP listing, and further evaluation is planned. To protect the resources from vandalism, it is customary not to publish the exact locations of the sites. All of these sites are discussed in the *SSC Historic Preservation Plan*, which must be followed if any type of development or other ground disturbance were to occur in these areas (NASA, 2005a:123-28; NASA, 2007s:159-60).

Historic Resources. The Rocket Propulsion Test Complex, also formerly called the National Space Technology Laboratories, has been designated as an NHL. The complex is made up of Building 4120 (A-1 Test Stand), Building 4122 (A-2 Test Stand) and Building 4220 (B1/B2 Test Stand). NASA, the ACHP, and other consulting parties negotiated a Programmatic Agreement in 1989 to address potential alterations to these NHL properties. The Programmatic Agreement and the SSC Historic Preservation Plan will be followed if there should be any modifications to these resources (NASA, 2005a:127-28; NASA, 2007s).

All NHL properties are listed automatically in the NRHP. The NHL designation recognizes properties that exemplify important trends in U.S. history (NPS, 2007e). These properties are significant both for their contributions to the Apollo Program and to the SSP. A survey of 48 facilities was conducted to determine their potential eligibility to the SSP. On the basis of this survey, no additional properties were considered eligible for listing on the NRHP either as individually significant or as contributing to a historic district (NASA, 2007s).

Exhibit 3-28 shows the properties that are listed on the NRHP at SSC.

3.6.4 Hazardous and Toxic Materials and Waste

3.6.4.1 Affected Environment

Storage and Handling. NASA maintains LQG status under RCRA Subtitle C at SSC for generating hazardous waste and having it transported offsite for treatment, storage,

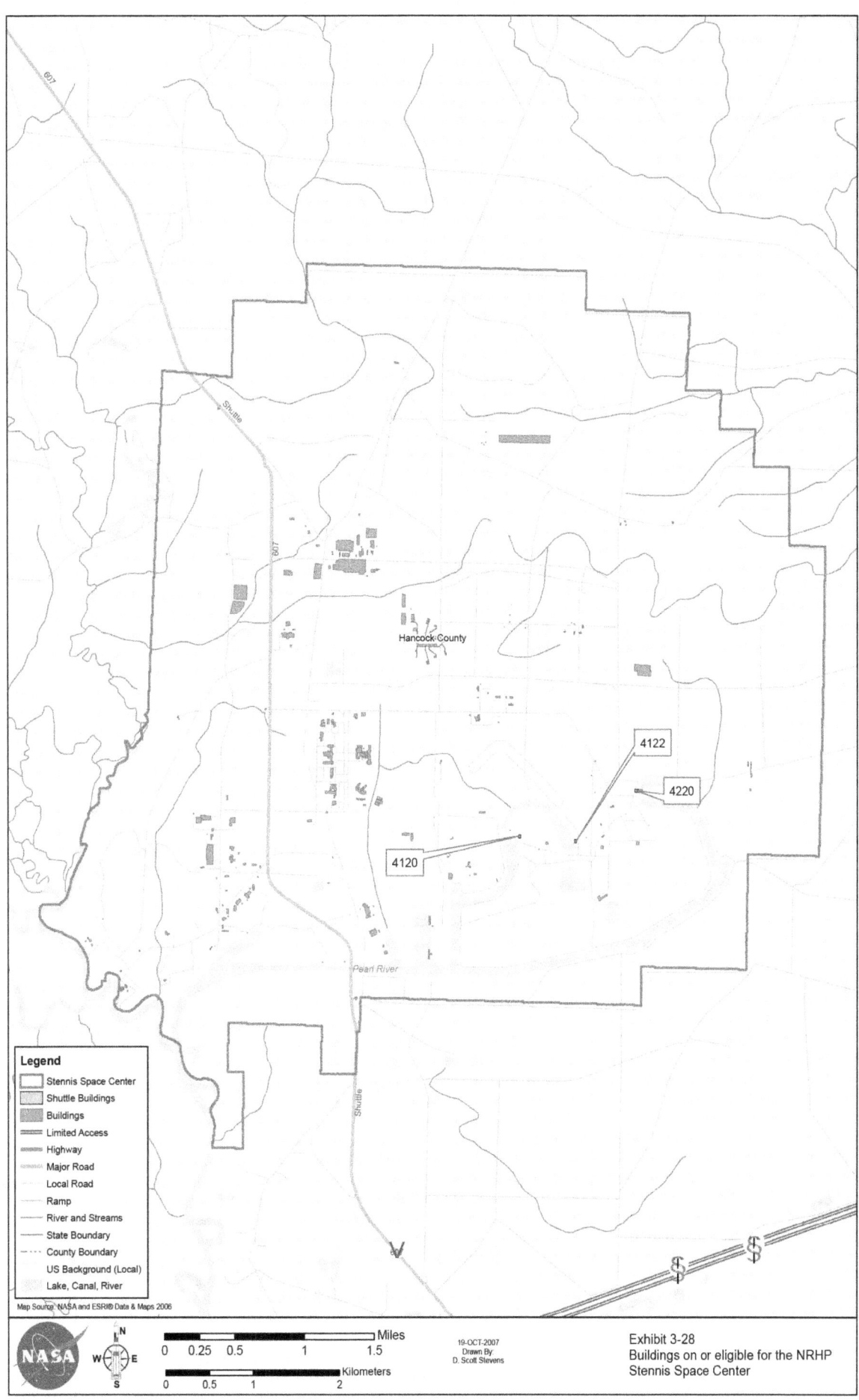

This page intentionally left blank.

or disposal (NASA, 2007s). There are no SSP-related material stockpiles or waste accumulation areas at SSC (NASA, 2007s).

Federal regulations require UST owners to reduce the risk of spills by providing quick release detection and spill cleanup. The NASA SSC Environmental Management must be notified before the installation, reactivation, or removal of any tank. In the event of a spill, the procedures for reporting, investigation, and cleanup are provided in the Integrated Contingency Plan (NASA, 2007s). Currently, USTs are used by SSP for storing gasoline, diesel, and waste oil. All of the tanks have been upgraded to meet or exceed the regulatory standards.

Waste Management. The following operational processes or activities generate hazardous wastes at SSC, in addition to facilities-related wastes from construction and routine maintenance:

- R&D and analytical testing–spent solvents, reaction products, unused or expired reagents, acids, bases, and test sample wastes
- Aerospace testing, cleaning, and maintenance–spent cleaning solutions, dyes, and photographic wastes
- Equipment cleaning and degreasing–alkaline cleaners and nitric acid

Hazardous wastes generated at SSC must be shipped offsite for treatment, storage, or disposal within 90 days from the start date of accumulation at the Accumulation Area Building. Hazardous waste disposal is handled through NASA's hazardous waste contractor at SSC. For NASA and its onsite contractors, when the specified waste limit is reached at a satellite accumulation area, a completed Waste Removal Form is submitted to the FOS contractor for environmental services for timely removal of the wastes. For resident agencies, the FOS contractor either coordinates for the disposal of the waste, or tells the generator how to dispose of the waste.

The *Hazardous Materials, Hazardous Waste, and Solid Waste Plan* provides guidance about the proper handling, compliance, and disposal of hazardous materials, hazardous wastes, universal wastes, and nonhazardous solid wastes (NASA, 2007s).

Contaminated Areas. SSC has not been listed as an NPL facility (NASA, 2007s). Following CERCLA processes, through an MOA with the Mississippi Department of Environmental Quality (MDEQ), NASA is investigating areas that may have been affected by historical releases as a proactive measure. EPA does not monitor the investigation or cleanup. NASA conducted Preliminary Assessments of 40 potential sites at SSC. Twenty-six potential sites were classified as clean or as having localized contamination. NASA conducted cleanup activities at the potential sites that had localized contamination. Of the 40 potential sites originally identified, 30 are potential sites for which actions are not recommended, 1 is a potential site that probably will not require any action, 1 is a long-term monitoring (LTM) site, the landfill at SSC is an LTM site, 7 are cleanup sites, and 1 is a potential cleanup site. Active remediation is being conducted at the 7 cleanup sites, which are referred to as

Cleanup Areas A through G. Five of the Areas–B, C, D, E, and G–are associated with SSP activities. Pump-and-treat systems located at Areas B, C, D, and E are used to remediate groundwater contaminated with trichloroethene (TCE) and its degradation products. Groundwater from Site G is extracted and transported to the treatment unit at Area B. Areas A and F are not associated with SSP activities. Hundreds of monitoring wells have been installed to monitor cleanup progress. Monitoring is expected to continue for the next 30 years. The monitoring results are submitted to the MDEQ semiannually (NASA, 2007s).

Toxic Substances. SSC has completed a survey of buildings for asbestos. Asbestos is present in buildings at SSC; it is removed as buildings are renovated and is disposed in the onsite hazardous solid waste landfill. An *Asbestos Hazard Control Plan* provides guidance for the proper handling and disposal of asbestos (NASA, 2007s).

LBP was used on the SSC Test Stands and in some other locations onsite. As this paint is removed, it is disposed as hazardous waste. A *Lead Hazard Control Plan* provides guidance for the proper handling and disposal of lead and lead-containing property (NASA, 2007s).

NASA completes TRI reporting and Tier II reporting for SSC annually. In 2005, NASA reported 10 chemicals on the EPCRA reports for SSC–Diesel Fuel #2, gasoline, propane, LH2, LOX, sodium hydroxide, hydrogen peroxide (35-percent), sulfuric acid, nitric acid, and chlorine (NASA, 2007s; NASA, 2002d).

To ensure the proper disposal of wastes, the NASA SSC Environmental Management must be contacted if any action causes a disturbance of PCBs, asbestos, or other substances regulated under TSCA. If it is unknown whether a waste is regulated by TSCA, the NASA SSC Environmental Management also must be contacted (NASA, 2007s).

As of May 2006, there were 16 pad-mounted transformers that had PCB contents of 50 ppm or greater in use at SSC. All pole-mounted transformers that had a PCB content of 50 ppm or greater have been removed. Some pole-mounted transformers at SSC may contain low levels (about 2 ppm or less) of PCBs in their fluid. Fluorescent lighting fixtures equipped with PCB-containing ballasts are replaced, upon failure, with non-PCB ballasts. The fixtures are disposed in accordance with state and federal regulations (NASA, 2007s).

3.6.5 Health and Safety

3.6.5.1 Affected Environment

Rocket testing for the SSME is conducted at SSC. The Buffer Zone surrounding SSC minimizes the potential effects that accidents or emergencies occurring in the SSC would have on the surrounding areas. The following subsections outline NASA's programs for protecting the health and safety of employees at SSC and the public. Noise hazards at SSC are outlined in Section 3.6.8.

Hazardous Materials. Hazardous materials, including fuels, are used to test the SSMEs at SSC. The hazardous materials used and hazardous wastes generated are discussed in Section 3.6.4. The degree of exposure to hazardous materials is minimized by the implementation of work practices and control technologies. SSC Procedural Requirement (SPR) 8715.1 requires that SSC and its contractors develop procedures for working with the following (NASA, 2007s):

- Triethylaluminum (TEAL) and triethylborane (TEB)
- Cryogenics, including LH2, LOX, liquid nitrogen (LN2), and liquid helium (LHe)
- Pressure systems
- Explosives
- Rocket Propellant (RP)-1 or any hydrocarbon fuels
- Hydrogen peroxide propellants

Additionally, SPR 8715.1 requires SSC and its contractors to comply with the "Oxygen Standard" (American Society for Testing and Materials [ASTM] Manual 36, *Manual for Safe Use of Oxygen and Oxygen Systems: Guidelines for Oxygen System Design, Materials Selection, Operation, Storage and Transportation*) when liquid and gaseous oxygen systems are used to protect the health and safety of its workers. SSC and its contractors also are required to comply with the *Guide to Safety of Hydrogen and Hydrogen Systems,* American National Standards Institute (ANSI)/ American Institute of Aeronautics and Astronautics (AIAA) G-095-2004, when operating liquid and gaseous hydrogen systems (NASA, 2007s).

Hazardous Materials Transportation Safety. Hazardous materials such as fuels, chemicals, and hazardous wastes are transported in accordance with the DOT regulations for the interstate shipment of hazardous substances (49 CFR 100 through 199).

Explosions and Fire Hazards. Using certain hazardous materials, including fuels, in SSME testing operations presents a risk of explosions and fire hazards. Fire protection at SSC is provided on a 24-hour-per-day, year-round basis for all areas and activities at SSC by the SSC Fire Department. Other services are fire prevention inspections, stand-by duty for LOX and LH2 transfers, explosive and engine tests, basic and refresher fire-fighting training for full-time firemen and officers, and assistance to the contractor in establishing fire-fighting training programs to qualify the contractor's personnel in the use of fire-fighting equipment (NASA, 2005a).

3.6.6 Hydrology and Water Quality

3.6.6.1 Affected Environment

Surface Waters. The East Pearl River flows along the southwestern boundary of SSC and the Jourdan River flows in a southeasterly direction through the eastern portion of the Buffer Zone surrounding SSC. Two tributaries to the East Pearl River and two tributaries to the Jourdan River are located within SSC. Mikes River and Turtleskin Creek are in the East Pearl River Basin. The East Pearl River drains to Lake Borgne and eventually to the Mississippi Sound. Two intermittent tributaries, the Lion and Wolf branches, drain offsite to Catahoula Creek in the Jourdan River Basin. The Jourdan River drains to the Bay of St. Louis and eventually to the Mississippi Sound.

Approximately 13.7 km (8.5 miles) of constructed canals are located in the southeastern portion of SSC. The canals are connected to the East Pearl River through a lock system. A spillway and overflow from the main access canal drains into Devils Swamp, which discharges into Bayou LaCroix and the Bay of St. Louis to the Mississippi Sound (NASA, 2005a). The water resources are shown in Exhibit 3-29.

Groundwater. Several aquifers occur in Hancock County. The area is underlain by southward-tipping Miocene and Pliocene age sands. Within these water-bearing sands, one freshwater unconfined aquifer is near the surface. Ten or more freshwater aquifers, confined by discontinuous clay layers, occur at depth. The sequence of alternating sand and clay layers is part of the Coastal Lowlands Aquifer System. The fresh water-bearing zone is 600 to 900 m (2,000 to 3,000 ft) thick beneath SSC. Individual aquifers range from 30 to 140 m (100 to 450 ft) in thickness. The aquifers have plentiful supplies of fresh water (NASA, 2005a).

Water Quality. The State of Mississippi assigns one or more of five use designations to streams and has developed criteria to protect those uses:

- Public Water Supply
- Shellfish Harvesting
- Recreation
- Fish and Wildlife
- Ephemeral Stream

The Pearl and the Jourdan Rivers are classified as recreation waters. The fish and wildlife use is assigned to any stream not specifically designated in Mississippi's water quality standards (MDEQ, 2003). Thus, all other streams on SSC are assigned the fish and wildlife use.

Both the Pearl and Jourdan Rivers are listed on the NRI. NRI rivers possess one or more "outstandingly remarkable" natural or cultural values considered to be of more than local or regional significance (NPS, 2007f).

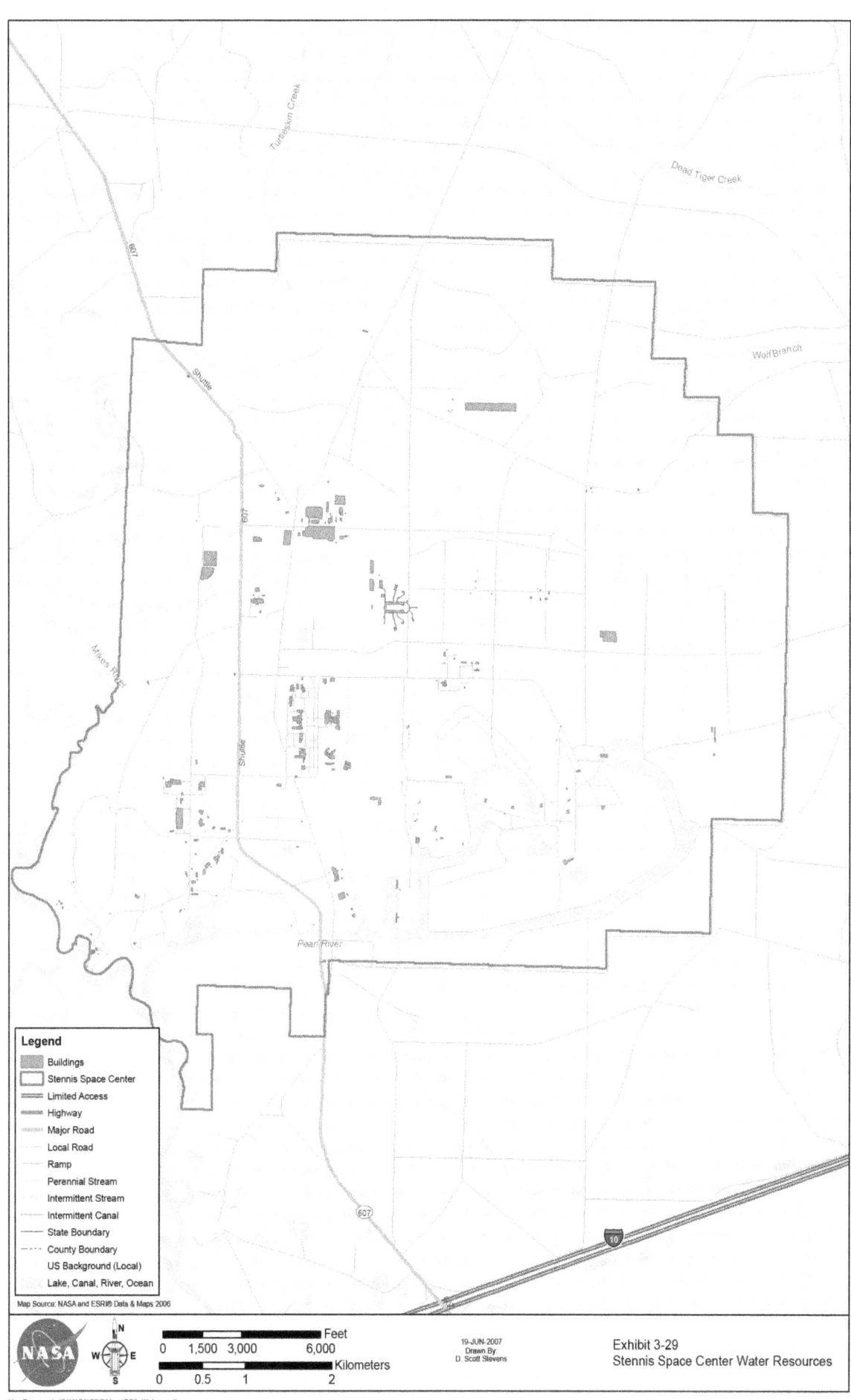

This page intentionally left blank.

Background surface water quality information is limited; however, discharge stations are maintained by the U.S. Geological Survey (USGS) on the Pearl River approximately 40 km (25 miles) northwest of SSC. A USGS monitoring station on the West Pearl River also measures flow and is located approximately 11.3 km (7 miles) west of the SSC. The surface waters in the streams of the area generally are suitable for most uses. USGS analyses indicate that the water in freshwater streams is generally soft and slightly acidic (5 to 7 pH units), with low concentrations of dissolved solids. Dissolved solids concentrations increase in the Pearl and Jourdan Rivers with the movement of saltwater during high tides (NASA, 2005a).

Regulated Water Discharges and Withdrawals. Wastewater is discharged from SSC under NPDES Permit #MS0021610. Five outfalls discharge water from various sources at the facility. These outfalls contain deluge water (that is, water used to cool the test facility flame deflectors), and sanitary wastewater that has been treated by biological lagoons or rock reed filters. Water samples are collected from each outfall and analyzed for specific contaminants at a frequency specified in SSC's permit. Monitoring includes biochemical oxygen demand (BOD), fecal coliform bacteria, total suspended solids (TSS), total dissolved solids (TDS), conductivity, metals, and nutrients. All outfalls are released to the Access Canal or Mikes River, then on to the East Pearl River (NASA, 2007s). Sewage treatment systems at SSC consist of 4 permitted treatment facilities and 57 lift stations (NASA, 2005a).

SSC water supplies include both ground and surface water sources. Industrial water is used for deluge water for the test stands, cooling water, and fire control. The Access Canal is the primary source of industrial water at SSC. SSC is permitted by the State of Mississippi to withdraw water from the East Pearl River into the elevated portion of the Access Canal. Industrial water also is supplied by three groundwater wells that range in depth from 205 to 571 m (672 to 1,873 ft). These wells are maintained as a back-up system for the surface water withdrawal system (NASA, 2007c).

Groundwater is used as the drinking water source at SSC under a licensed public water supply. Drinking water comes from water-bearing zones about 427 m (1,400 ft) deep and is drawn from three onsite potable water wells that range in depth from 437 to 464 m (1,434 to 1,524 ft). Well information is listed in Exhibit 3-30.

EXHIBIT 3-30
SSC Groundwater Well Use Permits

Permit No.	Well Use	Depth		Normal Discharge		Maximum Discharge	
		Meters	Feet	Million Liters/Day	mgd	Liters/ Min.	Gal./ Min.
MSGW01907	Industrial Water	570.9	1,873	3.2	0.84	13,248	3,500
MSGW01908	Industrial Water	516.6	1,695	4.5	1.2	18,925	5,000
MSGW01909	Industrial Water	204.8	672	4.5	1.2	18,925	5,000

Permit No.	Well Use	Depth		Normal Discharge		Maximum Discharge	
		Meters	Feet	Million Liters/Day	mgd	Liters/ Min.	Gal./ Min.
MSGW01910	Drinking Water	466.3	1,530	1.5	0.4	2,271	600
MSGW01911	Drinking Water	451.4	1,481	0.4	0.1	2,271	600
MSGW01912	Drinking Water	437.1	1,434	0.75	0.2	2,839	750

Notes:
gal. = Gallons
mgd = Million gallons per day
min. = Minute
Source: NASA, 2005a

Storm water discharges are covered under MDEQ's Land Disposal Storm Water General NPDES Permit MSR500068. The land disposal storm water permit is applicable to the operation of the SSC nonhazardous waste landfill, which allows storm water associated with industrial activity to be discharged into state waters. An SWP3 also was developed to identify potential sources of pollution that could affect the quality of storm water discharges associated with industrial activities (NASA, 2005a).

3.6.7 Land Use

3.6.7.1 Region of Influence

The ROI includes all lands within the SSC Fee Zone and Buffer Zone boundaries.

3.6.7.2 Affected Environment

Fee Area. The master plan for the SSC facilities, which is updated on an ongoing basis, establishes controls and criteria to guide future growth and development in the Fee Area (NASA, 2005a). The plan is used as a general tool to guide orderly site growth and expansion, and not as a detailed outline for design purposes.

The following 14 land use categories describe the general land uses in the SSC Fee Area, where the various types of operational or support activities are conducted (NASA, 2005a):

- Component and small propulsion system testing
- Medium propulsion system testing
- Large propulsion system testing
- Engineering and administration
- Test support
- Maintenance, supply, and security
- Utility
- Waterways and canals

- Recreation
- Mississippi Army Ammunition Plant (MSAAP)
- Mitigation area
- Landfill
- Restricted
- Open areas

Buffer Zone. NASA maintains control of the SSC Buffer Zone through a perpetual easement that prohibits the maintenance or construction of buildings suitable for human habitation. As stated in Section 3.6, the purpose of the Buffer Zone is to provide an acoustical and safety protection zone for NASA's testing operations.

The majority of land in the Buffer Zone is used for commercial pine forests. Besides commercial forestry, other uses in the Buffer Zone include wildlife management areas, nature preserves, cattle grazing, limited cropland, and small mineral operations. McLeod Park and Stennis International Airport are areas classified for special or unique land use in and along the perimeter of the Buffer Zone. McLeod Park is a 426-acre recreational facility along the banks of the Jourdan River. The park is operated by Hancock County and is open throughout the year for public camping and day use. Stennis International Airport is a county-run airfield located partially within the Buffer Zone. In addition, there is a small industrial park located adjacent to the airfield (NASA, 2005a).

SSC received approval in May 1996 from the USACE to mine sand and clay from a 10-acre area. This mining operation complies with the Mississippi Surface Mining and Reclamation Act and the Mississippi Air and Water Pollution Law, which regulate the disposal of wastewater generated from mining operations. In January 1992, SSC received permission to move mineable materials within the facility without a mining permit, as long as the material remained on NASA property (NASA, 2005a).

The Fee Area and Buffer Zone at SSC occupy approximately 36 percent of the Hancock County land base.

Other Areas within SSC. In addition to NASA and its support contractors, the Center has facilitated the establishment of outside operations involving federal and state agencies at SSC. Following is a list of the major facilities at SSC:

- Naval Meteorological and Oceanography Command
- Naval Research Laboratory
- MSAAP Industrial Complex
- National Data Buoy Center
- Mississippi Laboratories of the Southeast Fisheries Center
- USGS
- EPA

- Technology Transfer Offices
- Mississippi Space Commerce Initiative
- The Naval Small Craft Instruction and Technical Training School (NAVSCIATTS)
- Navy's Special Boat Unit 22

3.6.8 Noise

3.6.8.1 Affected Environment

The major sources of noise at SSC are generated by vehicle traffic, cooling towers, and indoor manufacturing operations (NASA, 2005a).

Because of the nature of static rocket engine testing, noise, and to a smaller extent, vibrations, have always been taken into consideration at SSC. The land area required for SSC and its Buffer Zone was based on acoustic environment calculations made for the NOVA first stage rocket engine. NASA determined that it was necessary to purchase all land within a 125-dB acoustical boundary and to prohibit human habitation within a 110-dB acoustical boundary. A perpetual restrictive easement on 506 km^2 (125,001 acres) was acquired for the Buffer Zone, which extends 9.6 km (6 miles) in all directions of the SSC fence line (NASA, 2005a). The closest sensitive receptors to SSC include a day care facility 0.5 km (0.3 mile) southwest of SSC and a private school 1.9 km (1.2 miles) north of SSC. The noise levels at SSC decrease to acceptable levels outside the buffer zone and are not detectable by the sensitive receptors outside the buffer zone. Sensitive receptors within 3.2 km (2 miles) of the outside borders of the buffer zone include 12 schools and several residential subdivisions.

Background Noise Levels. Generally, noise levels at SSC are low. The following continuous sources of noise at the facility have been identified:

- Diesel generators
- Pumps
- Boilers
- Automotive traffic

The effects of the generators, pumps, and boilers are minimal because these sources are contained within structures that minimize the noise levels. SSC maintains a hearing protection program to ensure that workers exposed to 8-hour TWA SPLs of 85 dBA and 90 dBA are monitored and provided with hearing protection, as required by OSHA regulations. Traffic noise is highest during the morning and evening while employees are commuting to and from work (NASA, 2005a).

One-hour noise measurements were recorded at SSC at four locations in the Fee Area in 1974 when no rocket tests were being conducted. The results of these measurements were all below 45 dBA. In addition to NASA's measurements, background noise levels measured along I-10 at the Highway 607 interchange range from 60 dBA to a peak noise level of 75 dBA, depending on traffic levels (NASA, 2005a).

3.6.9 Site Infrastructure

3.6.9.1 Region of Influence

The ROI is defined as the SSP-related facilities and the energy sources for these facilities at SSC.

3.6.9.2 Affected Environment

Potable Water Supply. See "Hydrology and Water Quality," Section 3.6.6.

Wastewater System. See "Hydrology and Water Quality," Section 3.6.6.

Storm Water System. See "Hydrology and Water Quality," Section 3.6.6.

Energy Sources. Dual overhead 110-kilovolt (kV) transmission lines normally supply electricity to SSC. The lines are owned and operated by the Mississippi Power Company; an alternate power service is available from the Louisiana Power and Light Company. The High Pressure Industrial Water (HPIW) facility also houses an emergency back-up electrical power generation facility for the test complexes (NASA, 2005a).

Natural gas is purchased from United Gas Pipeline Company and supplied to the SSC facilities through 8 miles of pipeline and 2 miles of branch line. A pressure-reducing and metering system supplies the gas to SSC at 100 pounds per square inch gauge (psig) (NASA, 2005a).

Natural gas is used as fuel for emergency back-up generators, flare stacks, and laboratory use in the Engineering and Administration Building and the Environmental Laboratory (NASA, 2005a). SSC operates under a Title V operating permit (#1000-00005) originally issued in February 1998 and renewed in 2003. This permit covers back-up electricity generation on the facility (NASA, 2007s).

3.6.10 Solid Waste

3.6.10.1 Affected Environment

Nonhazardous solid waste generated within the Fee Area at SSC is disposed onsite in a Class A solid waste landfill under the authority of Permit #SW02401B0376. The 2005 average quantity of solid wastes accepted for disposal in the landfill was approximately 208,000 pounds per month. The groundwater at the landfill is monitored per the requirements in the Solid Waste Permit, issued in February 2005 (NASA, 2007c).

3.6.11 Traffic and Transportation

3.6.11.1 Region of Influence

The transportation ROI for SSC is defined as the counties in which at least 50 SSC employees live–Hancock, Harrison, and Pearl River counties in Mississippi and St. Tammany Parish in Louisiana. Together, these counties accounted for 86 percent of the civil servants and contractors working at SSC in 2007 (NASA, 2007a).

3.6.11.2 Affected Environment

Transportation Routes. I-10 and I-59, U.S. Highway 90, and Mississippi Highway 607 serve the SSC area. I-10 is the primary corridor linking Biloxi, Gulfport, Bay St. Louis, and other coastal cities with New Orleans. It is located approximately 5 km (3 miles) south of SSC. I-59 joins I-10 near Slidell, Louisiana, and extends northeastward to Hattiesburg, Mississippi, and on into Alabama, passing about 8 km (5 miles) from the northwestern corner of SSC.

Access Roads to SSC. Direct access to and through SSC from I-10 and I-59 is provided by Mississippi Highway 607 (Exhibit 3-31). The highway is closed to the general public within the Fee Area and checkpoints exist at both entrances to SSC. Highway 607 connects with U.S. 90 approximately 14.5 km (9 miles) southeast of SSC.

Airports. Two airports, Gulfport-Biloxi International Airport located in Gulfport, Mississippi, and the Louis B. Armstrong International Airport in New Orleans, Louisiana, provide nationwide access to the site. Commuter air services are proposed to accommodate personnel directly between SSC and the NASA Shared Services Center (NSSC) site to Washington D.C. Local access to SSC and the proposed NSSC site averages less than 48 km (30 miles), with a commuting time of less than 35 minutes.

3.7 Michoud Assembly Facility

MAF is located 16 miles east of downtown New Orleans in southeastern Louisiana (Exhibit 3-32). The site is about 161 km (100 miles) north of the mouth of the

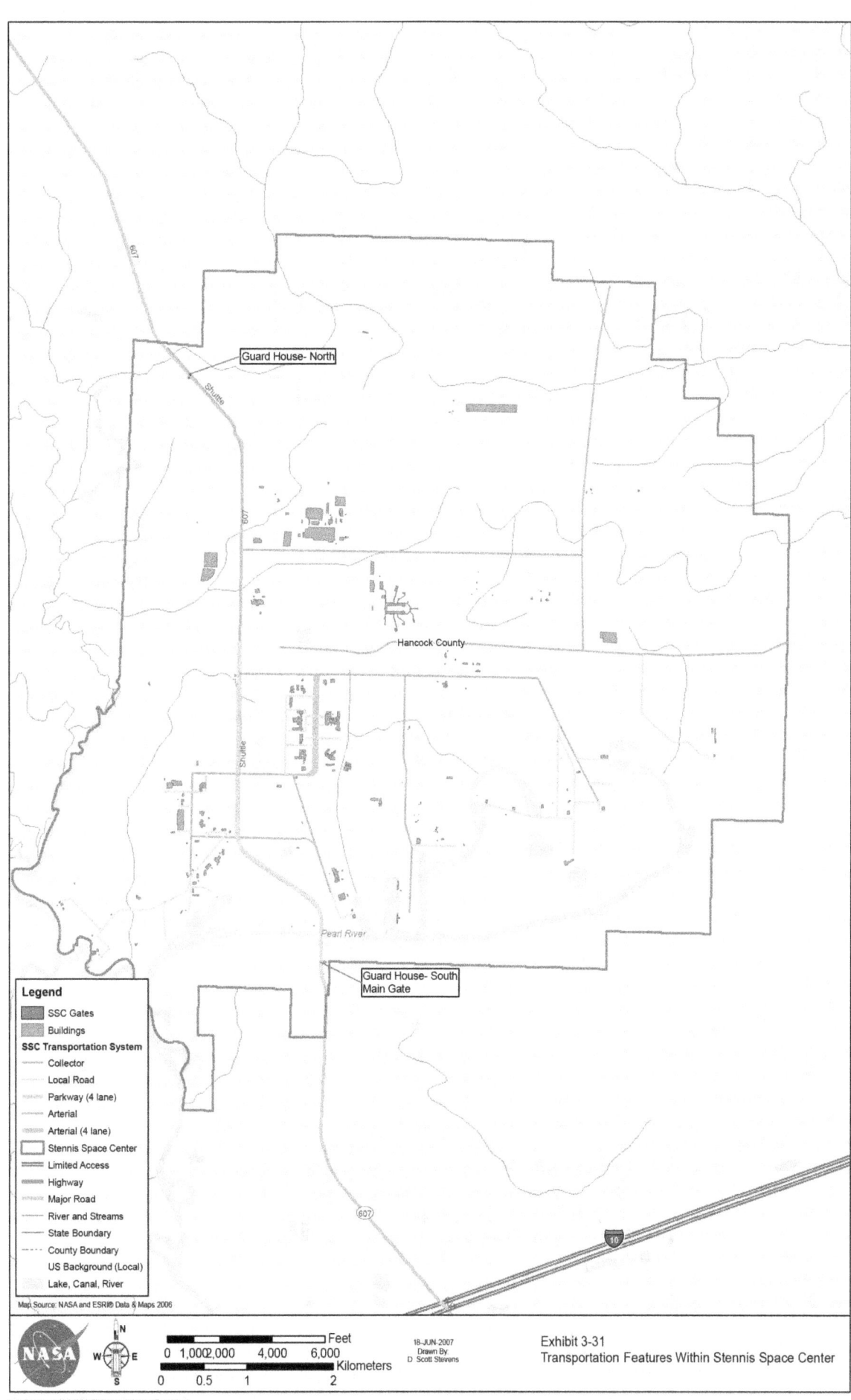

Exhibit 3-31
Transportation Features Within Stennis Space Center

This page intentionally left blank.

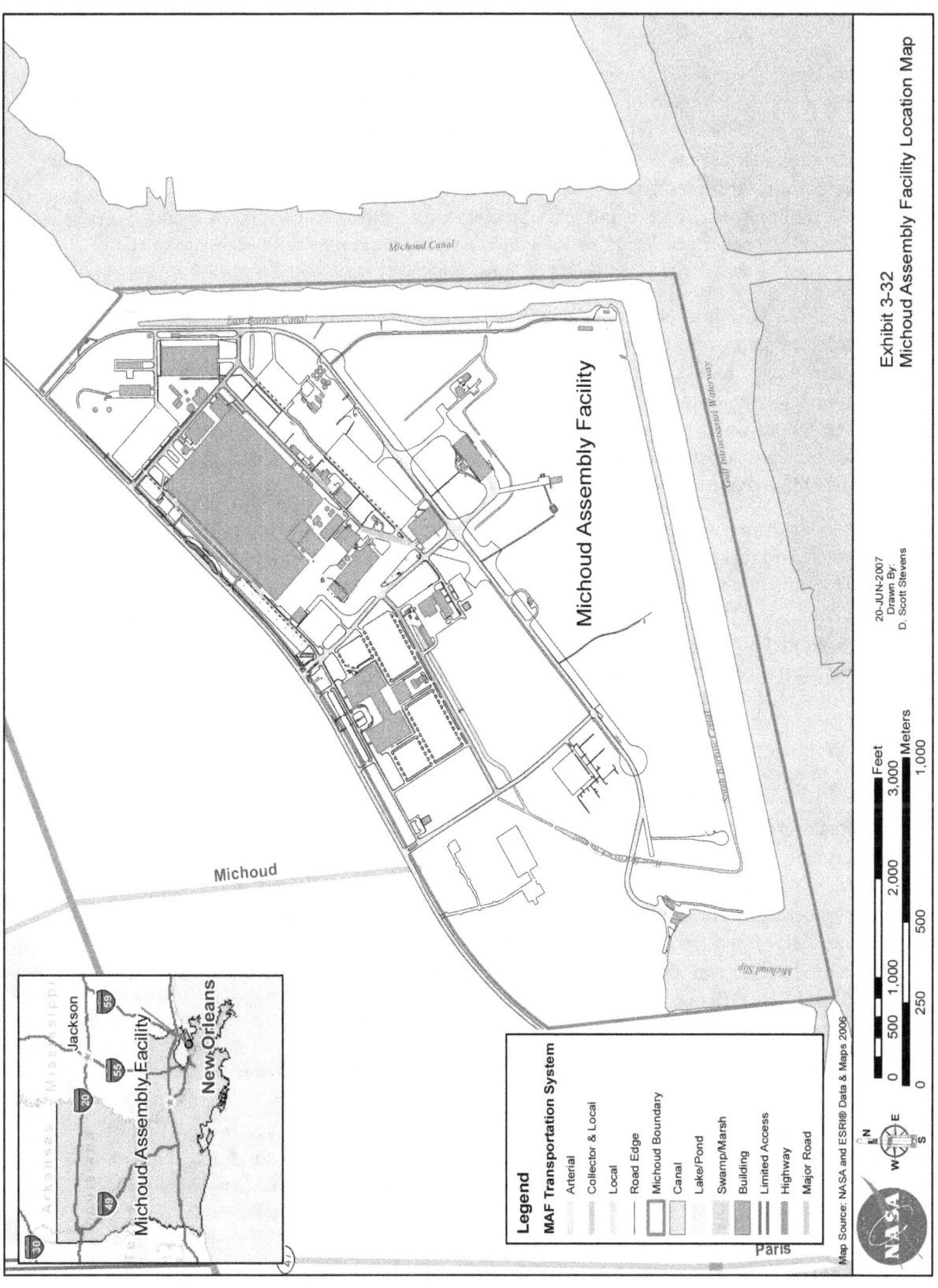

Exhibit 3-32
Michoud Assembly Facility Location Map

Mississippi River in an industrialized area. MAF occupies 338 ha (832.5 acres) owned by NASA, 20 percent of which is devoted to buildings, roads, and parking; the remaining 80 percent is vacant land consisting mostly of mowed grass lands and canals. MAF has 41 buildings. Approximately 60 percent of the buildings onsite are devoted to manufacturing activities, 20 percent are used for offices, and the remaining 20 percent are used as storage and support facilities. MAF has 2.3 million square feet (ft^2) of manufacturing space, 895,000 ft^2 of office facilities, and 703,000 ft^2 of support facilities, as well as a deep-water port with access to the Gulf of Mexico. MAF also is occupied by numerous tenants.

NASA's MAF is a satellite organization of MSFC in Huntsville, Alabama. NASA has a Maintenance and Base Operations Management (MOM) contract with Lockheed Martin to provide services such as spill response; training; audits; environmental management system (EMS) implementation; inspections; chemical management; hazardous waste and solid waste management; and medical services such as ambulances, doctors, and nurses; fire and crash rescue; security; and public affairs. The MOM contract expires in December 2008.

MAF's primary mission is to support the design and assembly of the ET, the liquid-fuel-carrying component of the Space Shuttle. Activities conducted at MAF include system engineering, engineering design, manufacturing, fabrication, assembly and testing, and total automatic checkout and computer data reduction. The nominal six ETs-per-year production cycle is 24 months and has two primary processes–Structural Fabrication and TPS Operations. Exhibit 3-33 shows the manufacturing flows for both processes.

3.7.1 Air Quality

3.7.1.1 Affected Environment

Regional Climate. MAF is located in the New Orleans metropolitan area. The region's climate is subtropical and humid, primarily because the area is virtually surrounded by water. Afternoon thunderstorms from mid-June through September keep temperatures from rising much above 90°F. An occasional winter storm can bring a northerly flow of cold continental air into the area, resulting in a sudden drop in temperature. Monthly mean temperatures range from 43.4 to 85.8°F. Precipitation in the New Orleans area averages nearly 62 inches annually. Weather hazards in the area include fog, thunderstorms, tornadoes, and hurricanes (NASA, 2001b).

Emission Sources. Utility emission sources at MAF include fuel storage tanks (gasoline and No. 2 fuel oil) and the operation of fuel-burning equipment including boilers, generators, pumps, and compressors. Primary production process emission sources at MAF include solvent cleaning of metal, preparation and application of high-performance primers and coatings, application of cryogenic foam insulation, and miscellaneous small-quantity usage of adhesives and cleaners. Emission sources related to the preparation and application of the Superlight Ablator (SLA) at

EXHIBIT 3-33
Nominal Six ETs-per-Year Production Cycle

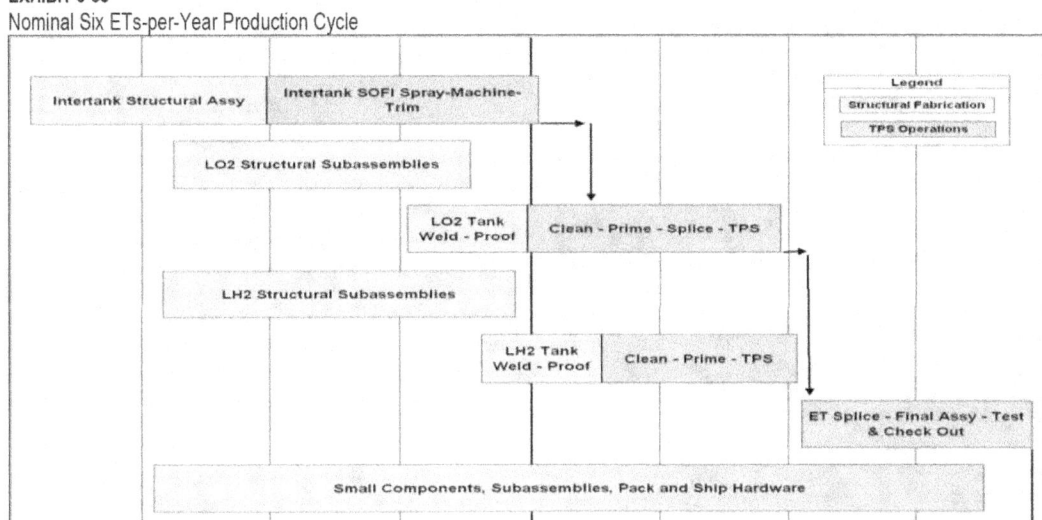

MAF include storage tanks, fume hoods, and thermal oxidizers. The groundwater recovery system consists of two caternary countercurrent air stripper towers in series. The emissions from the air stripper towers are routed to one of two parallel horizontal carbon absorption units (NASA, 2001b).

Currently, the U.S. Department of Agriculture (USDA) has two emergency generators that are covered under the NASA MAF site utilities air permit (NASA, 2007s). Under the Title V program, MAF is categorized as a synthetic minor source because of boiler emissions. NASA currently has four air permits issued by the state for emission sources at MAF (NASA, 2007s). MAF is located in Orleans Parish, which is classified as attainment for all NAAQS (EPA, 2007a). Therefore, MAF follows the PSD program, instead of the more stringent NSR program.

3.7.2 Biological Resources

3.7.2.1 Affected Environment

Vegetation. Nearly all naturally occurring vegetation on MAF has been altered (NASA, 2001b). A large potion of MAF has been cleared and developed for buildings, parking lots, and industrial operations. Undeveloped areas on MAF are primarily maintained lawn, with common weeds, some shrubs, and a few scattered trees (ARCADIS, 2004a). Habitat types present at MAF include urban, agriculture-cropland-grassland, wetland barren, non-vegetated urban, and water (Exhibit 3-34).

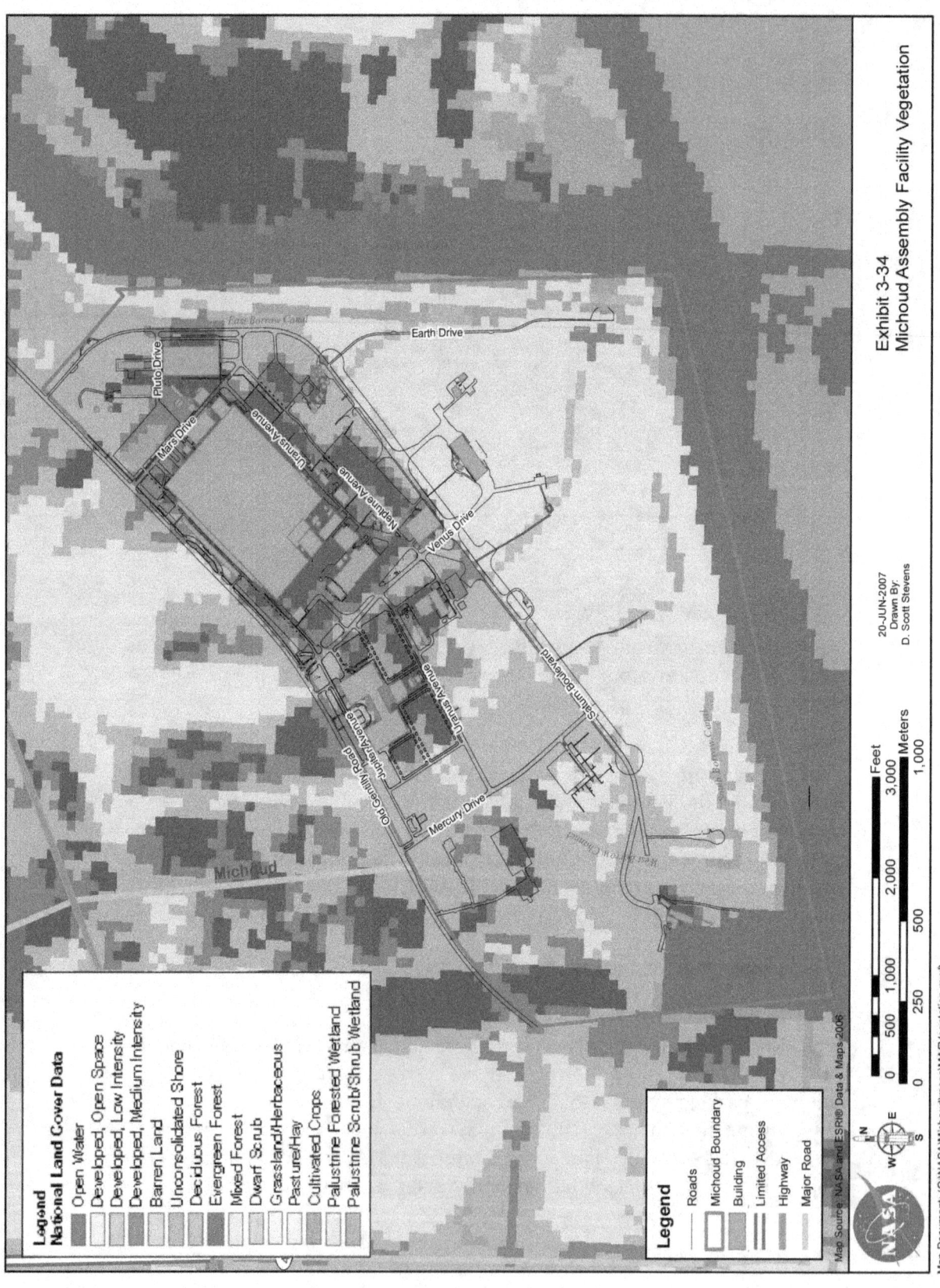

Exhibit 3-34
Michoud Assembly Facility Vegetation

Wetlands. A 2004 wetland survey delineated approximately 110 ha (272 acres) of wetlands on MAF (Exhibit 3-35), including the undeveloped areas in the southern portion of the property (ARCADIS, 2004b). Approximately 183 ha (452 acres) of MAF are not considered wetlands, including the spoil bank and levee area and the open grass-covered areas located adjacent to buildings, parking lots, and roads (ARCADIS, 2004b).

Floodplains. FEMA has delineated floodplain areas on FIRMs. According to the FIRMs prepared by FEMA in March 1984, MAF is designated in Ponding Area 32, Zone A1, and Zone B (FEMA, 2007). The floodplain map for MAF is provided in Exhibit 3-36.

Wildlife. As noted previously, MAF is in an area that has been altered extensively by human development, and all habitats on MAF have been modified previously (NASA, 2001b). MAF contains little natural habitat and is dominated by highly developed and industrialized areas consisting of buildings and parking lots. MAF is limited in species diversity and wildlife numbers because of the relatively small size of the facility, limited habitat types, and current land use conditions (NASA, 2001b). With the exception of a few species of common reptiles, amphibians, birds, and mammals, no fauna are known to regularly inhabit MAF (NASA, 2001b).

In April 2004, a 3-day baseline biological inventory was conducted at MAF (ARCADIS, 2004a). During this inventory, 1 amphibian species, 5 reptile species, 100 bird species, and 5 mammal species were identified (ARCADIS, 2004a). A complete list of the species identified during the survey is located in the *Biological Survey for Sensitive Flora and Fauna, MAF* (ARCADIS, 2004a).

Protected Species and Habitats. Although many listed species could occur in the bayous near MAF and at the Bayou Sauvage National Wildlife Refuge, which is within 3.2 km (2 miles) of the MAF property boundary, the lack of suitable habitat on MAF makes their presence onsite unlikely (NASA, 2001b).

The Louisiana Natural Heritage Program (LNHP) lists 21 rare elements found in Orleans Parish (LNHP, 2006). Of these, 5 rare elements may be found in the vicinity of MAF; in 2000, observations were made of a waterbird nesting colony, manatees in the surrounding waterbodies, pallid sturgeon in the surrounding waterbodies, and Gulf sturgeon critical habitat near the site. In 1987, a live oak forest natural community was observed near MAF (ARCADIS, 2004a).

Plants. No federally protected plant species are listed for Orleans Parish (USFWS, 2003). LNHP maintains a database of rare plants, and because of the highly disturbed nature of the site, rare plants whose distribution lies outside of Orleans Parish are not likely to be found on MAF. The 2004 biological survey found no rare plants on MAF (ARCADIS, 2004a).

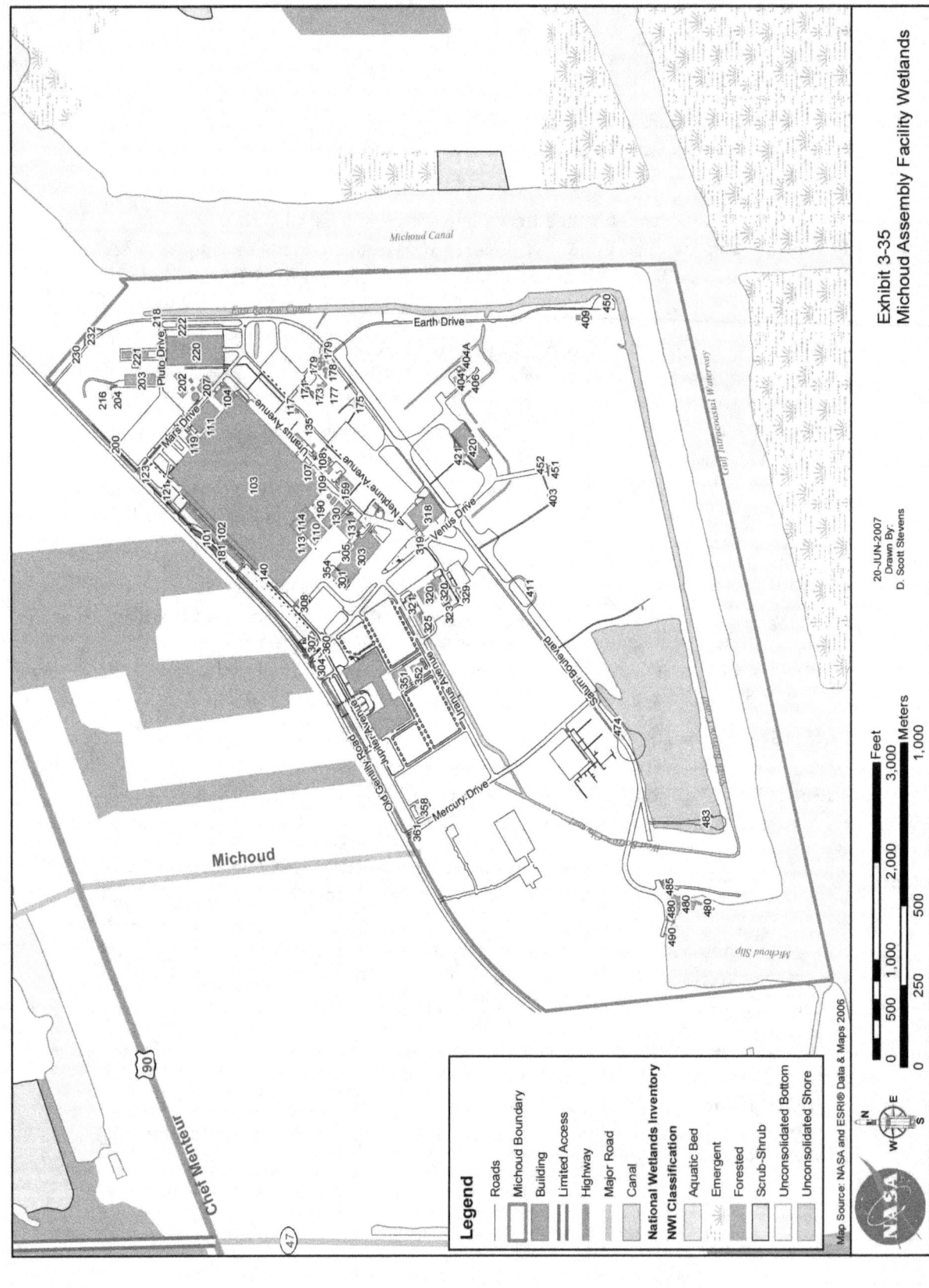

Exhibit 3-35
Michoud Assembly Facility Wetlands

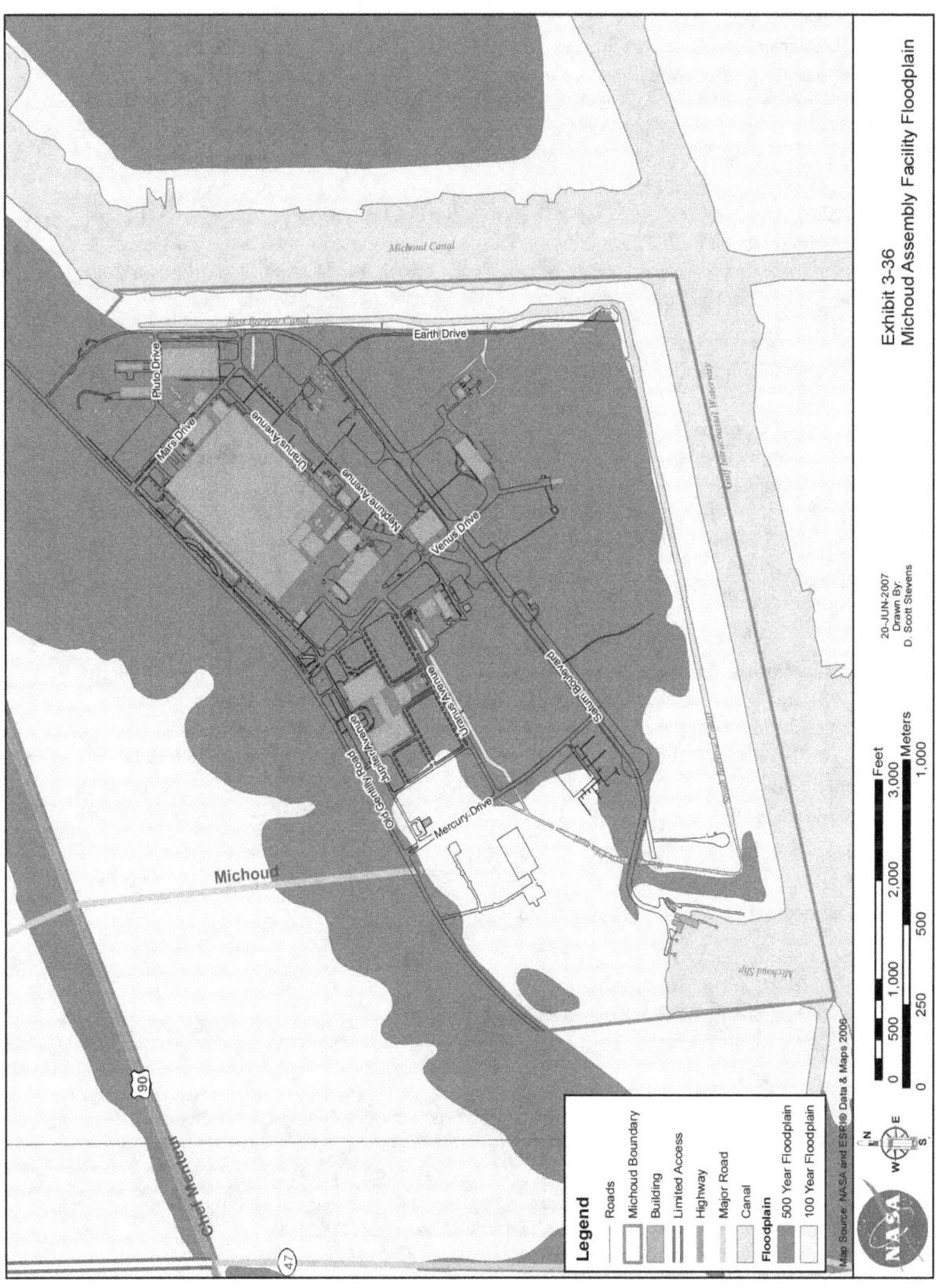

Exhibit 3-36
Michoud Assembly Facility Floodplain

Threatened and Endangered Species. The USFWS has identified five threatened or endangered species in Orleans Parish. Threatened species include the Gulf sturgeon (*Acipenser oxyrhynchus desotoi*). Endangered species in Orleans Parish include the brown pelican (*Pelecanus occidentalis*), West Indian manatee (*Trichechus manatus*), and pallid sturgeon (*Scaphirhychus albus*) (USFWS, 2003). Brown pelicans may occur on MAF as transients (NASA, 2001b). The brown pelican is known to occur near MAF (ARCADIS, 2004a). The West Indian manatee has been reported a few times in the Michoud Canal (ARCADIS, 2004a). The Gulf sturgeon and pallid sturgeon have been reported from areas adjacent to Orleans Parish, but it is not likely that suitable aquatic habitat exists at MAF for these species (ARCADIS, 2004a).

The brown pelican (*Pelecanus occidentalis*) was the only state and/or federally listed threatened and endangered species found during the 2004 survey. Five bird species classified as Louisiana state species of concern including Cooper's hawk (*Accipiter cooperii*), Caspian tern (*Sterna caspia*), lark sparrow (*Chondestes grammacus*), grasshopper sparrow (*Ammodramus savannarum*), and Henslow's sparrow (*Ammodramus henslowii*) were identified during the survey (ARCADIS, 2004a). No state and/or federally listed threatened and endangered amphibian, reptile, or mammal species were detected.

3.7.3 Cultural Resources

3.7.3.1 Affected Environment

Archaeological Resources. Archaeological surveys were conducted at or near MAF in 1981 and 1999. Two archaeological sites were identified on MAF property–one a sugar house from the 19th-century Michoud Plantation and the other a World War II-era brick building–but neither of these sites was determined to be NRHP eligible. There are no known sites within the boundaries of MAF that are listed or eligible for listing in the NRHP (NASA, 2001b; NASA, 2007s).

Historic Resources. Eight resources recently were surveyed to assess their eligibility to the NRHP for their association with the SSP. The following buildings at MAF have been determined as being eligible for listing in the NRHP as SSP-significant:

- VAB (Building 110)
- HB Addition (Building 114)
- Acceptance and Preparation Building (Building 420)
- Pneumatic Test Facility Structure (Building 451)
- Pneumatic Test Facility Control Room (Building 452)

Building 103, with 68 major tools, was considered ineligible, but it has been proposed that before the next NASA program, a record of the tooling be made so that the knowledge is not lost (TRC, 2007). Building 110, previously considered eligible as contributing to the Apollo Program, also has been determined to be significant to the SSP context (TRC Garrow Associates, 2001).

Exhibit 3-37 shows the properties at MAF that potentially are eligible to be listed on the NRHP.

3.7.4 Hazardous and Toxic Materials and Waste

3.7.4.1 Affected Environment

Storage and Handling. MAF has an RCRA Part B permit for a Hazardous Waste Storage Facility (Building 159, for wastes in containers) and three hazardous waste solvent aboveground storage tanks (ASTs). The RCRA permit is effective from 2006 to 2016 (NASA, 2007s).

All USTs at MAF were closed in 1995. The facility currently does not have any active USTs onsite (NASA, 2007s).

MAF has approximately 200 ASTs. The tanks are used for solvent storage, chemical supply, petroleum storage, and ET processing. The tanks' uses and emergency spill responses are included in MAF's *Spill Response Plan* (NASA, 2007s).

Waste Management. MAF is classified as an LQG of hazardous waste. Approximately 40 percent of the solid and hazardous waste streams come from the SSP ET processing (NASA, 2007s). The primary waste streams generated include solvents, various sludges, photographic wastes, batteries, paint wastes, and corrosive liquids.

Contaminated Areas. NASA MAF is not listed as a CERCLA NPL site, but is involved with several RCRA corrective action projects (NASA, 2007s). In 1982, NASA initiated a groundwater detection monitoring program at MAF. TCE contamination was identified and led to the initiation of a groundwater compliance monitoring program, established in 1984. A groundwater treatment system is in place for treating the contaminated groundwater (NASA, 2007s). In December 1987, NASA received a Federal Part B Hazardous Waste Permit, LA4800014587, for activities at MAF, which included provisions for conducting an RCRA Corrective Action Program (CAP). The CAP portion of the permit identified 24 SWMUs (NASA, 2007s).

The RCRA CAP identified 24 SWMUs. In 1993, one additional SWMU was added, the SWMUs were grouped into 14 areas of concern (AOCs), and the Phase I RCRA Facility Investigation (RFI) began. The RCRA permit includes these SWMUs and AOCs and includes provisions for completing the closure of two RCRA surface impoundment operating units at the site, as a result of these requirements and the completion of an RFA by an EPA contractor in 1986 (NASA, 2007s).

In May 2000, the Louisiana Department of Environmental Quality (LDEQ) approved no further action (NFA) for 8 of the 14 AOCs at MAF: A, C, I, J, K, L, M, and N. The remaining six AOCs (B, D, E/F/G, and H) remain in the CAP at MAF. AOCs B, D,

This page intentionally left blank.

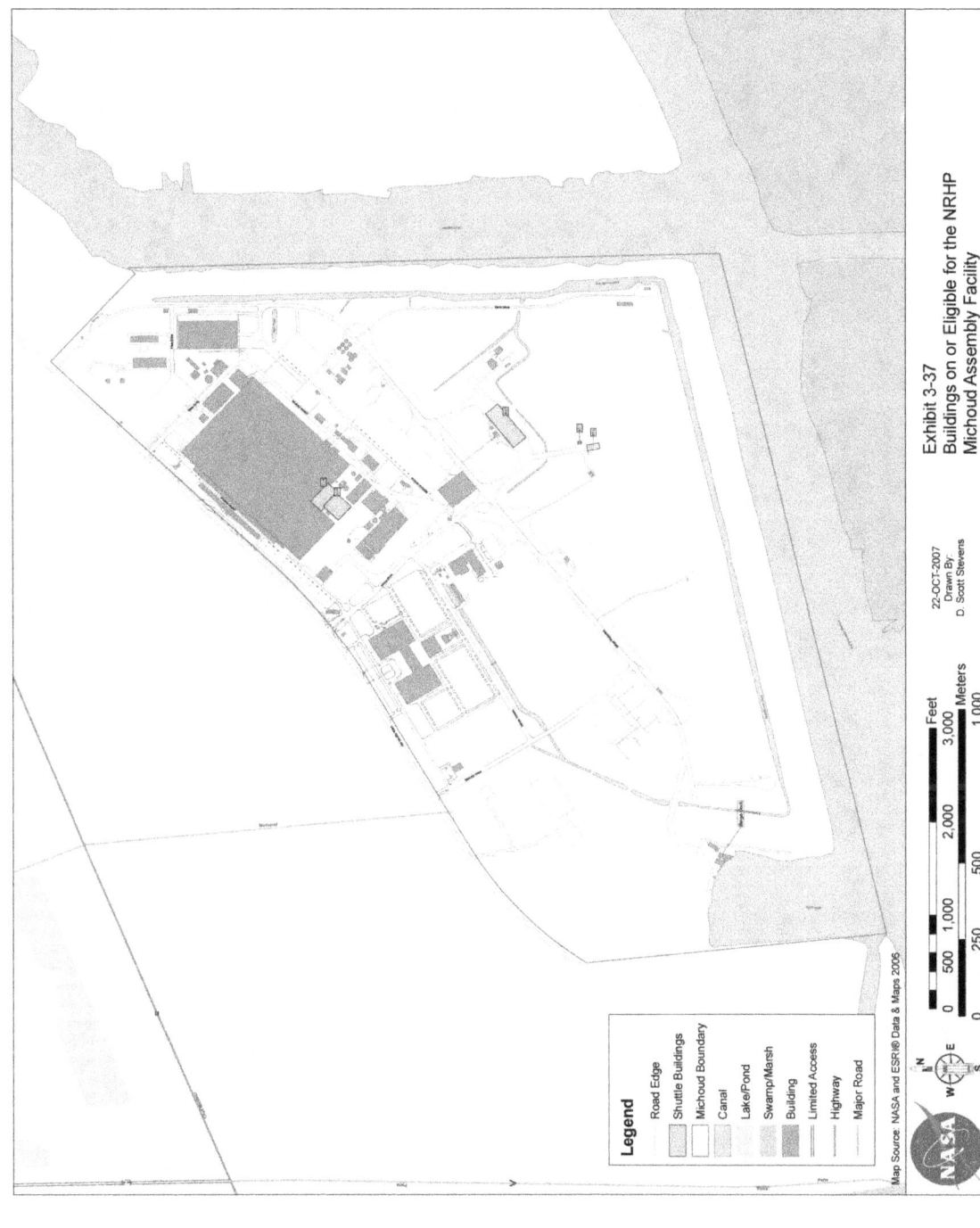

Exhibit 3-37
Buildings on or Eligible for the NRHP
Michoud Assembly Facility

This page intentionally left blank.

E, F, and G are being investigated and remediated as sources of TCE in the groundwater. AOC H is being investigated and remediated for PCBs, chromium, and polynuclear aromatic hydrocarbons (PAHs) in the sediment.

Three treatment systems are in operation or being pilot tested at AOC B: a horizontal recovery well, a dense non-aqueous phase layer (DNAPL) zero valent iron trench, and an air stripper.

Asbestos and Lead-based Paint. Several MAF buildings contain asbestos; however, construction and maintenance projects that involve asbestos removal are evaluated as they occur, and removal and disposal are performed per the applicable state and federal requirements (NASA, 2007s).

LBP is a potential concern for the demolition of buildings and structures built before 1978. Demolition debris potentially containing LBP is subject to landfilling restrictions. Maintenance activities could have created the potential for localized lead contamination in soils in areas around those older buildings or structures.

3.7.5 Health and Safety

3.7.5.1 Affected Environment

As mentioned previously, MAF manufactures ETs for the Space Shuttle. The following subsections outline MAF's programs for protecting the health and safety of MAF employees, as well as the public. Noise hazards at MAF are outlined in Section 3.7.8.

Hazardous Materials. Hazardous materials are used to manufacture ETs for the SSP at MAF. The hazardous materials used and hazardous wastes generated at MAF are discussed in detail in Section 3.7.4. The degree of exposure to hazardous materials is minimized by the implementation of work practices and control technologies, which include, but are not limited to, the following:

- Ventilated, controlled access work areas
- Controlled use and restricted access work areas and associated standard operating procedures (SOPs)
- Regular monitoring to ensure that exposure levels do not exceed the OSHA standard thresholds

The implementation of these work practices and control technologies minimizes employee exposure to hazardous materials. Risks associated with hazardous materials are managed under NPD 1820.1B. Hazardous material spills or releases are to be cleaned up by employees in the shops where the spills are generated. If spills are too large or cannot be managed, employees must call the MAF Leak Line to provide notification that a cleanup is needed. The Leak Line is monitored 24 hours every day by onsite personnel.

Several buildings at MAF contain asbestos. MAF uses a licensed subcontractor for projects that require asbestos removal.. Construction projects that involve asbestos removal are evaluated as they occur, and removal and disposal are performed per 29 CFR 1910.1001, OSHA's standard for the protection of employees from asbestos exposure.

Hazardous Materials Transportation Safety. Hazardous materials such as fuels, chemicals, and hazardous wastes are transported in accordance with the DOT regulations for the interstate shipment of hazardous substances (49 CFR 100 through 199).

Explosions and Fire Hazards. Using certain hazardous materials in manufacturing operations presents a risk of explosions and fire hazards. The main buildings at MAF have individual fire suppression and alarm systems. The MAF fire station is located in Building 320. Firefighting services and equipment are provided by the plant protection contractor. Fire water is drawn from the Borrow Canal and stored onsite (Lockheed Martin, 2006b). Because of damage from Hurricane Katrina, most of the fire stations near MAF are out of service and are located just outside MAF's gate. The nearest city fire station that is in service is located 18.4 km (11.5 miles) from the facility (New Orleans Fire Department [NOFD], 2006).

3.7.6 Hydrology and Water Quality

3.7.6.1 Affected Environment

Surface Waters. MAF is within the coastal area of southeast Louisiana. It is surrounded by major water bodies including Lake Ponchartrain, Lake Borgne, and the Mississippi River, as shown in Exhibit 3-38. The natural hydrologic regime in the vicinity of MAF has been modified substantially by human activity. Canals collect and store runoff from precipitation and from developed areas. Natural and artificial levees protect the area from flooding, tidal flushing, and storm surges. Surface drainage at MAF comes from within the levee system. Precipitation and groundwater seepage are pumped over the levees to keep the ground surface above the water table.

There is no natural surface drainage system within 305 m (1,000 ft) of MAF and no streams or rivers pass through the property. The facility is bounded on the west by the Michoud Slip, on the south by the Gulf Intracoastal Waterway (GIWW) and Mississippi River Gulf Outlet (MRGO), and on the east by the Michoud Canal. To the north is a dredged extension of Maxtent Lagoon, a drainage canal that is pumped to the GIWW. A marsh is located south of MAF and the GIWW. These waters are estuarine and influenced by tidal action from the Gulf of Mexico.

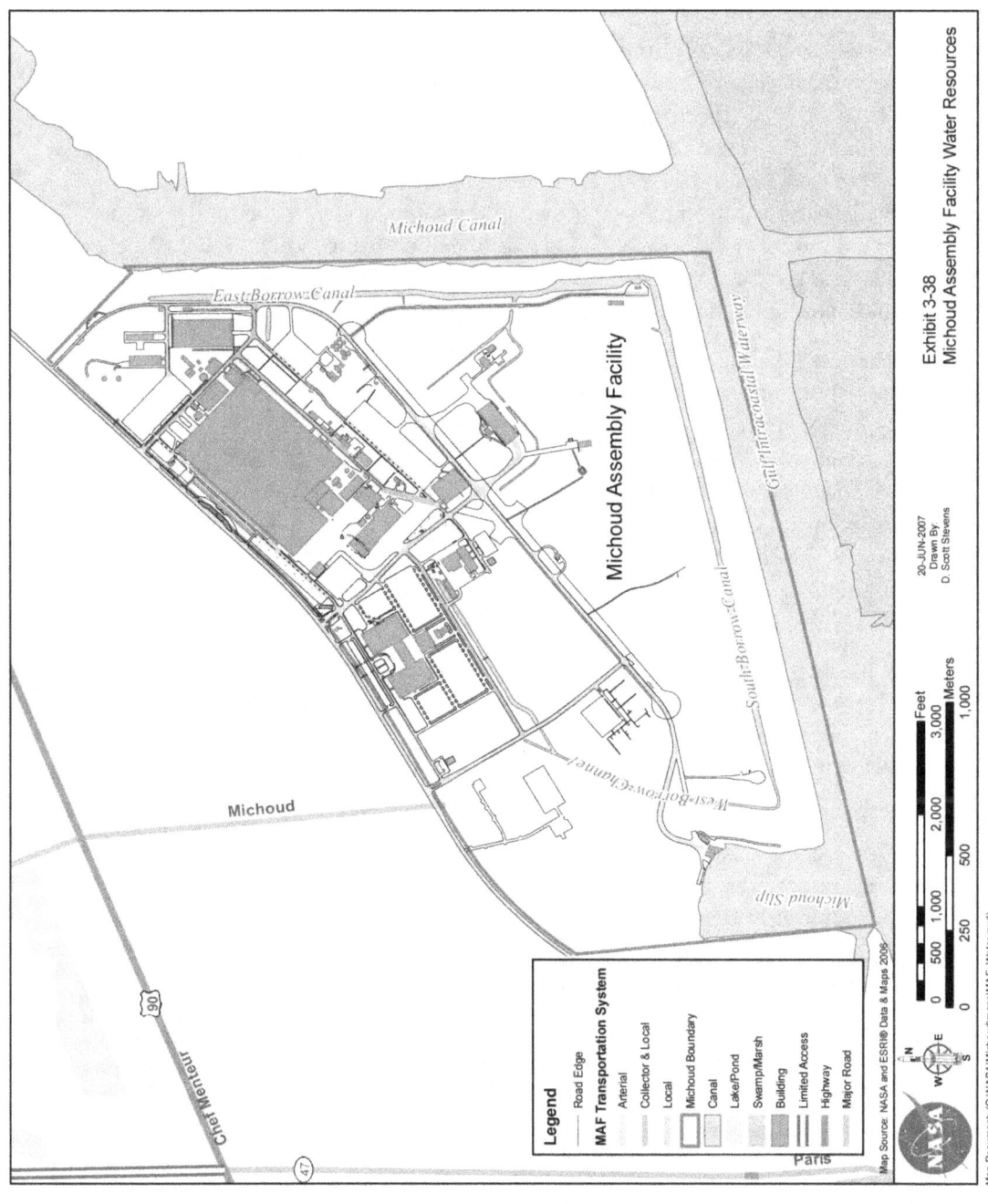

Exhibit 3-38
Michoud Assembly Facility Water Resources

The storm water drainage system at MAF is composed of open water ditches, catch basins, and underground pipes that direct storm water to a Borrow Canal located inside the levees. Both surface water drainage and industrial wastewater are diverted to the Borrow Canal, which is located along the southern, eastern, and western boundaries. A pump station is used to remove surface water from the Borrow Canal. The Borrow Canal is approximately 1.5 to 2.4 m (5 to 8 ft) deep and 9 to 15 m (30 to 50 ft) across, with a volume of approximately 8 million gallons (30 million L) of water. Its water level is generally maintained at -1.2 m (-4 ft) below mean sea level (msl). The bottom of the canal is about -3.2 m (-10.5 ft) msl. The MAF discharge pump station is located at the southeastern corner of the site and pumps water from the Borrow Canal over the flood protection levee into Michoud Canal. This is a Louisiana Pollutant Discharge Elimination System (LPDES)-permitted outfall (001) (NASA, 2001b).

Groundwater. MAF is underlain by four aquifer systems. The Surficial Aquifer is present at the land surface to a depth of 6.1 m (20 ft). It is connected hydrologically to the Borrow Canal and onsite subsurface drainage and sewer piping. Across MAF, the groundwater level is approximately 0.3 to 1.2 m (1 to 4 ft) below ground surface. The Surficial Aquifer is composed of clay and peat layers with thin discontinuous sand lenses and fibrous peat zones that create horizontal and vertical flow pathways. The aquifer has a natural upward vertical gradient that is suppressed through pumping. Relatively fresh to brackish water is present in this aquifer.

The Shallow Aquifer is a semi-confined layer located between 6 and 15 m (20 and 50 ft) below the ground surface. It is separated from the Surface Aquifer by a thick clay layer. The aquifer system is comprised of sandy silt and silty sand interbedded with layers of fine sand. This aquifer contains relatively stagnant fresh to brackish water. A thick clay layer (15 to 18 m [50 to 60 ft]) is located immediately beneath this aquifer.

The 700-foot Sand Aquifer extends from 137 to 183 m (450 to 600 ft) below the ground surface. It is the only known formation south of Lake Pontchartrain to produce significant quantities of fresh water. The aquifer is composed of fine- to medium-grained silty sand and forms a continuous layer under the New Orleans area. The 700-foot aquifer is pumped heavily for industrial use.

The 1,200-foot Sand Aquifer is the lowermost aquifer under MAF. Water in the aquifer is highly mineralized, with a high chloride content. It is composed of layers of sand, silty clay, and clay (NASA, 2001b).

Water Quality. Water quality in surface waters at MAF is regulated by the LDEQ. Louisiana surface waters are designated according to eight classifications based on their potential uses. Numeric criteria are assigned to protect those uses, as noted below (LDEQ, 2004):

- Primary contact recreation (designated swimming months of May through October, only)
- Secondary contact recreation (all months)
- Fish and wildlife propagation
- Drinking water source
- Outstanding natural resource
- Agriculture
- Oyster production
- Limited aquatic and wildlife

Minimum water quality standards have been established by LDEQ for these classifications.

The segment of the GIWW immediately adjacent to the MAF has been assigned four designated uses–primary contact recreation, secondary contact recreation, fish and wildlife propagation, and oyster propagation. Michoud Canal and Michoud Slip are not specifically listed in the Louisiana water quality standards, but they fall under the same standards as the adjoining portion of the GIWW.

Regulated Water Discharges and Withdrawals. MAF currently discharges wastewater and storm water under LPDES Permit #LA0052256. An industrial wastewater treatment facility treats manufacturing process-related wastewater before discharge. Several shuttle-related processes at MAF generate wastewater. Internal and ET cleaning, hydrostatic testing, tank priming, and small component cleaning and heat treating, among others, generate approximately 26 million gallons (94 million L) of wastewater per year. Wastewater is pumped to holding tanks and treated before discharge into Mars Canal, an onsite canal that connects to the Borrow Canal. Utility wastewater (boiler blowdown, steam condensate, and chilled water) is discharged directly to the storm water system and is not treated. Storm water is conveyed by a discharge system to the Borrow Canal. Discharges occur in batches as water is pumped from the Borrow Canal. Each discharge is approximately 210,000 gallons (795,000 L), with discharges occurring 4 to 5 times per week (NASA, 2001b). Water from this discharge location is tested for chemical oxygen demand (COD), pH, total organic carbon (TOC), TCE, and residual chlorine (EPA, 2007h).

Sanitary waste is collected in a separate system of sewer lines and is serviced by the City of New Orleans' Publicly Owned Treatment Works (POTW).

Potable water is obtained from the Sewerage and Water Board (S&WB) of New Orleans for a supply pipe located offsite. Groundwater is not used at MAF.

Orleans Parish, which includes MAF, is entirely within a Coastal Zone Management (CZM) boundary and must comply with CZM policies. All activities occurring in coastal water bodies and wetlands outside the flood protection levees require a coastal use permit. Activities in upland areas also must have a permit if they are expected to have direct or significant effects on coastal waters or wetlands. Existing pumping of water for drainage purposes is excluded. The CZM program is administered by the Louisiana Department of Natural Resources (LDNR). MAF has a CZM permit for general site activities (#C20010555) (NASA, 2001b).

3.7.7 Land Use

3.7.7.1 Affected Environment

MAF is an area zoned for industrial use (NASA, 2001b). Buildings at MAF are dedicated to a variety of activities that are compatible with the industrial land use designation including manufacturing, hazardous waste storage, laboratory services, storage, and miscellaneous support. Buildings devoted to miscellaneous support activities are the Facility Operations Building (Building 320), Cafeteria and Equipment Building (Building 351), Maintenance Building (Building 301), Barge Docking Area (Building 480), and a wide range of facilities including drainage pump stations, small shops, and generator rooms. No land on MAF has been designated as unique farmland by the USDA's Soil Conservation Service (SCS).

Easements, Rights-of-Way, and Agreements. Easements have been reserved along the drainage ditch west of Mercury Drive to permit dredging. Another easement is reserved along the original property line to the west to allow for any necessary repairs to the signal line that serves the barge dock.

3.7.8 Noise

3.7.8.1 Affected Environment

The following subsections describe the three types of noise generated at MAF, which include noise generated by vehicle traffic, cooling towers, and indoor manufacturing operations.

The closest sensitive receptors to MAF include a day care facility located 0.5 km (0.3 mile) southwest of MAF and a private school 1.9 km (1.2 miles) north of MAF. The noise levels at MAF decrease to acceptable industrial levels at the property line and are not detectable by the sensitive receptors.

Noise Generated by Vehicle Traffic. Vehicle traffic at MAF is generated by employees commuting to and from work daily and from shipping and receiving. Approximately 30 trucks visit MAF daily for shipping and receiving activities. Traffic noise peaks at 74 dBA at a distance of 30 m (100 ft) from the building and 70 dBA at a distance of 60 m (200 ft) from the building (NASA, 2001b).

Noise Generated by Cooling Towers. MAF operates cooling towers that generate noise levels ranging from 85 dBA to a peak noise level of 107 dBA at a distance of 1 m (3.3 ft) from the building. Noise levels from the cooling towers range from approximately 61 dBA to a peak noise level of 83 dBA at a distance of 15 m (50 ft) from the building (NASA, 2001b).

Noise Generated by Manufacturing Operations. MAF conducts a variety of manufacturing operations in support of the production of the SSP ET. These activities include, but are not limited to, solvent cleaning, surface coating, surface preparation, metal working and ablating, and applying cryogenic foam insulation. MAF maintains a hearing protection program to ensure that workers exposed to 8-hour TWA SPLs of 85 dBA and 90 dBA are monitored and provided with hearing protection, per the OSHA regulations under 29 CFR 1910.95 (NASA, 2001b).

3.7.9 Site Infrastructure

3.7.9.1 Region of Influence

The ROI is defined as the SSP-related facilities and the energy sources for these facilities at MAF.

3.7.9.2 Affected Environment

Potable Water Supply. See "Hydrology and Water Quality," Section 3.7.6.

Wastewater System. See "Hydrology and Water Quality," Section 3.7.6.

Storm Water System. See "Hydrology and Water Quality," Section 3.7.6.

Energy Sources. Electrical power, purchased from Entergy, Inc., is transmitted from two offsite substations to the MAF at 115-kV to two onsite master substations (one rated at 40 mega volt amps [MVA] and one rated at 20 MVA) located near the northeastern and northwestern corners of Manufacturing Building 103. From the master substations, the power voltage is reduced to 13.8 kV and distributed to 60 secondary low-voltage substations via a radial loop feeder system throughout the site. The two master substations also are internally looped, which allows both to be independently or jointly operated and permits routine preventive maintenance and continuous maintenance (NASA, 2001b).

Steam, used for manufacturing processes and for heating, ventilating, and air conditioning (HVAC) requirements is generated at the steam plant in the Boiler House (207) east of Building 103 and in the Incinerator Building (105) south of Building 103. The Building 105 boiler is capable of backing up the Building 207 boilers as required. The steam generated in the Boiler House is used to support more than one hundred 40-ton to 80-ton HVAC units and numerous production-related heating requirements such as LH2 tank washes and rinses (NASA, 2001b).

Natural gas used for process and HVAC heating requirements is supplied by Entergy, Inc., at 80 psig through a 10-inch gas main that feeds an onsite natural gas plant. The natural gas piping supplies gas to five steam boilers in Building 207, one steam boiler in Building 105, and steam and hot water boilers in Buildings 105, 175,

303, 318, 320, 351, and 421. Natural gas also is supplied to thermal oxidizers serving Buildings 110, 114, 141, and 318. The remaining natural gas lines are run to miscellaneous laboratory taps, unit heaters, emergency generators, a heat treat furnace, and food service equipment throughout the facility (NASA, 2001b).

3.7.10 Socioeconomics

3.7.10.1 Region of Influence

The economic ROI for MAF is defined as the 11 counties and parishes shown in Exhibit 3-39, consisting of the 7 parishes of the New Orleans-Metairie-Kenner, Louisiana, MSA[7] (Jefferson, Orleans, Plaquemines, St. Bernard, St. Charles, St. John the Baptist, and St. Tammany parishes); 1 adjacent parish (Tangipahoa); and 3 additional counties in Mississippi (Hancock, Harrison, and Pearl River). Approximately 88 percent of the MAF employees live in the New Orleans MSA. Nearly half (43 percent) of the MAF employees live in Saint Tammany Parish alone (Personal Communication, 2007c; OMB, 2006).

Because a substantial number of MAF and SSC employees live in the same Louisiana parishes and Mississippi counties, the MAF ROI was expanded to include the counties of Hancock, Harrison, and Pearl River, Mississippi (Personal Communication, 2007c ; Personal Communication, 2007e). Collectively, 54 percent of the MAF and SSC employees live in the 7 parishes of the New Orleans MSA and 35 percent live in the 3 additional Mississippi counties.

Estimated employment and expenditures at SSC that are directly involved in the SSP are included in the description of the SSP's impact on the combined MAF and SSC ROI, provided under the Socioeconomics section for MAF in this document.

3.7.10.2 Affected Environment

Hurricane Katrina. After Hurricane Katrina in August 2005 and Hurricane Rita in September 2005, the City of New Orleans was shut down for a month and many of its residents relocated to other cities. Severe damage and flooding of residences and businesses occurred. Recovery is underway in parts of the city, but large areas remain nearly or completely empty.

During the hurricanes, 94 percent of MAF workers' homes were destroyed or damaged; several buildings at MAF suffered wind damage, but the facility was able to avoid major flooding. Personnel from MAF, MSFC, and other NASA locations worked to quickly restore operations, assisted by the USACE. Nine weeks after Hurricane Katrina, MAF was restored to full operations and most of its employees had returned (Lockheed Martin, 2006b).

---

[7] An MSA is an area, defined by the OMB for federal statistical purposes, consisting of a core area with 50,000 or more population and adjacent counties that have a high degree of social and economic integration, measured by commuting patterns, with that urban core (OMB, 2006). The core city of the MSA is New Orleans (Orleans Parish), along with Metairie and Kenner in Jefferson Parish and Slidell in St. Tammany Parish.

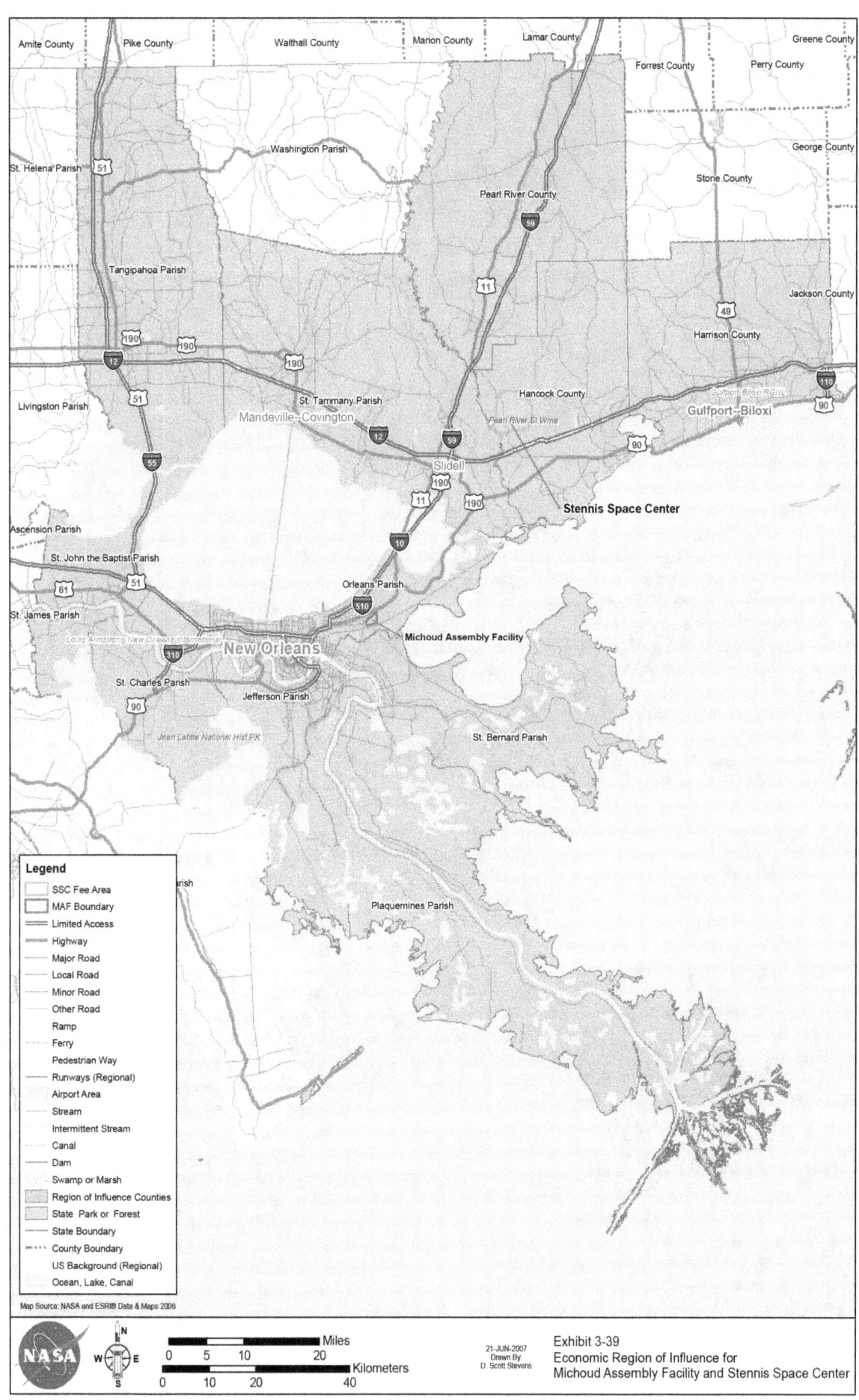

Exhibit 3-39
Economic Region of Influence for
Michoud Assembly Facility and Stennis Space Center

This page intentionally left blank.

According to the April 2007 Katrina Index, "… indicators suggest that the rebuilding of essential infrastructure is basically stalled, housing indicators remain mixed at best, but economic indices suggest a notable strengthening of the economy in New Orleans and in the metro area as a whole" (Greater New Orleans Community Data Center [GNOCDC], 2007).

Population. Approximately 1.4 million people lived in the ROI as of July 2006. At that time, almost 1 year after the hurricanes, the population was approximately 50 percent of the July 2005 levels in the city of New Orleans (Orleans Parish) and 81 percent in the 11- parish ROI as a whole. The parishes and counties that absorbed the relocated population (Tangipahoa Parish, St. John the Baptist Parish, and Pearl River County) continued to house much of it.

By 2008, the population of the city of New Orleans is expected to reach nearly 60 percent of its pre-Katrina level. The density of population in the hardest-hit neighborhoods (such as Lakeview, Gentilly, and New Orleans East, where much of the housing is unusable) could remain much lower than under pre-Katrina conditions indefinitely (U.S. Census Bureau, 2007b; Lockheed Martin, 2006b).

Regional Employment and Economic Activity. The New Orleans MSA historically has relied on five major economic activities: 1) maritime and port-related industries; 2) oil, gas, and related industries; 3) tourism, centered on the history, music, food and culture of the City of New Orleans; 4) ship-building and boat-building; and 5) aerospace manufacturing, centered on MAF as one of the largest industrial employers in the metropolitan area. Together, the shipbuilding and aerospace industries accounted for 25 percent of all manufacturing employment (NASA, 2001b).

Since the hurricane, many of these industries are struggling to recover, although employment and income in the construction industry have grown, due to recovery activities. In January 2006, 6 months after the hurricanes, the port was operating at 50-percent capacity, while the oil and gas industry had recovered 80 percent of its capacity and the military and aerospace industries were operating at nearly full capacity. About 60 percent of the city's small businesses had closed or relocated. As of April 2007, 92 percent of the major hotels and 46 percent of the retail food establishments in the New Orleans metro area were open. The number of air passengers arriving monthly was about 70 percent of the pre-Katrina levels (Lockheed Martin, 2006b; GNOCDC, 2007).

In 2006, the total labor force of the ROI was nearly 804,000 people. The overall unemployment rate of 5.2 percent was slightly lower than that for the State of Louisiana and considerably lower than that for the State of Mississippi, but higher than the national unemployment rate of 4.6 percent. As a result of the hurricanes, the labor force in the New Orleans-Metairie-Kenner MSA declined by 20 percent from 2005 to 2006. By April 2007, the labor force in the city and the MSA had returned to approximately 75 percent of the pre-Katrina levels (GNOCDC, 2007).

Economic Contribution of the Space Shuttle Program. Initiatives at MAF will continue to play an important role in the recovery of the manufacturing and technology industries in New Orleans. In February 2007, the Governor of Louisiana signed an MOU with the director of NASA MSFC. The MOU documents the shared goal of building on the existing public-private commercial partnerships involved in technical research and development activities at MAF, by using undeveloped areas of the facility (Advocate, 2007). The Michoud Master Plan envisions the development of vacant land at MAF to provide additional office and light industrial, manufacturing, industrial testing, and port and harbor operations by 2025 (Lockheed Martin, 2006b).

In 2006, MAF employed approximately 2,000 persons, primarily Lockheed Martin Space Systems Company personnel. More than 1,500 FTEs, consisting of 1,495 contractor FTEs and 15 federal civil service FTEs, worked directly on the SSP in 2006. The remainder worked in hurricane recovery, plant operations, security, project support, and other indirectly related functions. In addition, non-NASA tenants at the MAF complex together employ about 1,250 workers (Lockheed Martin, 2006b; NASA, 2007s; NASA, 2007a).

In 2006, nearly 4,700 workers were employed at SSC, of whom approximately 1,900 (40 percent) were NASA civil federal civil service and contract personnel. Only about 5 percent of that total (10 civil service and 230 contractor FTEs) worked on the SSP (NASA, 2007aPersonal Communication, 2007e).

In FY 2006, the SSP put nearly $74.3 million (including civil service and prime contractor salaries and non-payroll procurements) into the regional economy. Those expenditures translated into additional economic output, jobs, and income in supporting industries throughout the combined MAF and SSC ROI. The total (direct plus indirect and induced[8]) effect of the SSP on economic output was approximately $205 million (less than 1 percent of the nearly $85 billion[9] in overall economic activity in the 11-county region in 2002), $107 million in personal income, and 1,800 jobs (NASA, 2007cc). Equivalent post-Katrina economic data are not yet available; MAF and SSC probably provide a somewhat greater percent contribution to the economy of this still-recovering region at present.

3.7.11 Solid Waste

3.7.11.1 Affected Environment

The nonhazardous solid waste streams generated at MAF includes asbestos, blasting media, soil, rock, sand, spray-on foam insulation (SOFI), construction and factory

---

[8] Based on the "multiplier effect" using economic multipliers for the 11-county region from the U.S. Bureau of Economic Analysis.

[9] U.S. Bureau of the Census, 2002 Economic Census–Total sales, shipments, receipts, or revenue for all establishments (2-digit NAICS codes) in the 11-county ROI for MAF.

debris, and creosote pilings. Non-RCRA solid wastes are collected and sent to an offsite landfill (NASA, 1978).

3.7.12 Traffic and Transportation

3.7.12.1 Region of Influence

The ROI for MAF is defined as the 6 parishes and counties in which at least 50 MAF employees live–Jefferson Parish, Orleans Parish, St. Bernard Parish, and St. Tammany Parish in Louisiana; and Hancock and Pearl River counties in Mississippi. Together, these counties account for 87 percent of MAF employees. Nearly half (42 percent) of the MAF employees live in St. Tammany Parish alone (NASA, 2007a).

3.7.12.2 Affected Environment

Transportation. I-10 and I-510, U.S. Highway 90, and Mississippi 11 serve the MAF area. Direct access to and through MAF from I-10 and I-510 is provided by Mississippi Highway 90.

Access Roads to MAF. The MAF facility is accessed either by taking I-10 east to Paris Road and then south to Old Gentilly Road, or by driving east on Chef Menteur Highway (U.S. 90) to Old Gentilly Road or Paris Road. MAF is responsible for a major portion of the vehicular traffic on Old Gentilly Road, because it is the single public thoroughfare outside the facility. The main arterial roads at MAF include Mercury Drive, Venus Drive, and Jupiter Boulevard. Local roads include Uranus Avenue (south of Building 103) and Pluto Drive (Exhibit 3-40). Collector roads at MAF include Uranus Avenue (south of Building 350) and Mars Drive. The major transportation routes, I-510, and I-10, also are accessible.

Railroads. Six major truck line railroads serve the New Orleans area, providing single carrier access to virtually all of the America's major markets. In addition to the rail carriage of freight, 34 Amtrak trains carry passengers to or from New Orleans each week. In the past, rail spurs served the MAF complex from the main Seaboard System Rail Road; however, these lines are no longer serviceable.

Airports. Three airports serve the New Orleans area. Domestic and international commercial air transportation services are provided by the New Orleans International (Moisant) Airport, 32 km (20 miles) to the west of MAF. The New Orleans Airport, located 8 km (5 miles) to the northwest on Lake Pontchartrain, is devoted exclusively to private and corporate plane usage. Alvin Calendar Naval Air Station, located 24 km (15 miles) to the southwest of MAF, serves as a training area for the air reserve units of the Navy, Air Force, Marines, Coast Guard, and National Guard.

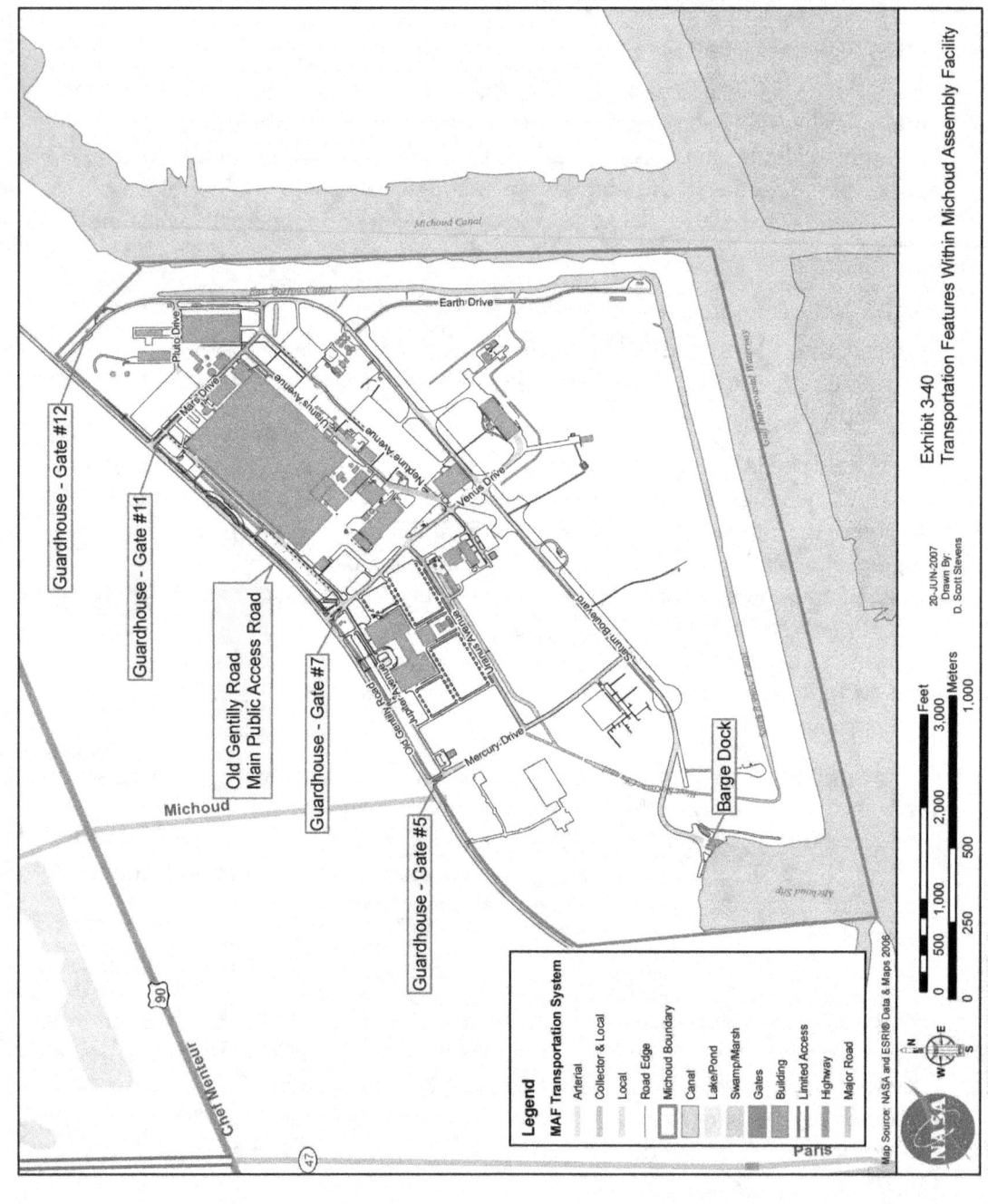

Exhibit 3-40
Transportation Features Within Michoud Assembly Facility

3.8 Marshall Space Flight Center

MSFC is located in north-central Alabama on approximately 7.4 km^2 (1,841 acres) within Redstone Arsenal (RSA) (Exhibit 3-41). MSFC is approximately 1,600 km (100 miles) north of Birmingham, Alabama; 100 miles south of Nashville, Tennessee; and 180 miles west of Atlanta, Georgia. The irregularly shaped property is roughly 4.8 km (3 miles) long on its north-south axis and 3.2 km (2 miles) wide on its east-west axis.

MSFC is centrally located within RSA, which is owned by the DA. The DA granted irrevocable use and occupancy of the lands and facilities known as MSFC to NASA for a term of 99 years beginning on July 1, 1960, and ending on June 30, 2059. The DA granted NASA full control and responsibility for MSFC land and facilities; however, the DA retained access rights to all major utility lines, railroad tracks, and main roads for the purposes of operating, maintaining, modifying, and extending the utilities, railroad tracks, and roads. A substantial portion of RSA, including most of the lands to the south and west of MSFC, is a part of the Wheeler National Wildlife Refuge (WNWR). Approximately 0.7 km^2 (180 acres) of the WNWR extend onto property controlled by MSFC.

RSA occupies 153 km^2 (38,309 acres) in the southwestern portion of Madison County, Alabama. RSA is roughly 160 km (10 miles) long on its north-south axis and 9.6 km (6 miles) wide on its east-west axis. The southern boundary of RSA is formed by the Tennessee River. The City of Huntsville surrounds RSA on the eastern, northern, and most of the western sides.

MSFC is NASA's principal propulsion research center. Its scientists, engineers, and support personnel play a major role in the National Space Transportation System, managing the SSMEs, SRBs, RSRMs, and ETs. In addition, MSFC will be a significant contributor to several of NASA's future programs, including the development of advanced propulsion systems and planetary observatories, and research on a variety of space science applications.

Six project offices for major development projects and six directorates that embody the institutional capabilities of MSFC carry out NASA's missions. The directorates are Safety and Mission Assurance; Science and Mission Systems; Shuttle Propulsion, Engineering, and Center Operations; Ares Projects Office; Office of Human Capital; Contractors; Diversity and Equal Opportunity; Procurement; and Strategic Analysis and Communication.

This page intentionally left blank.

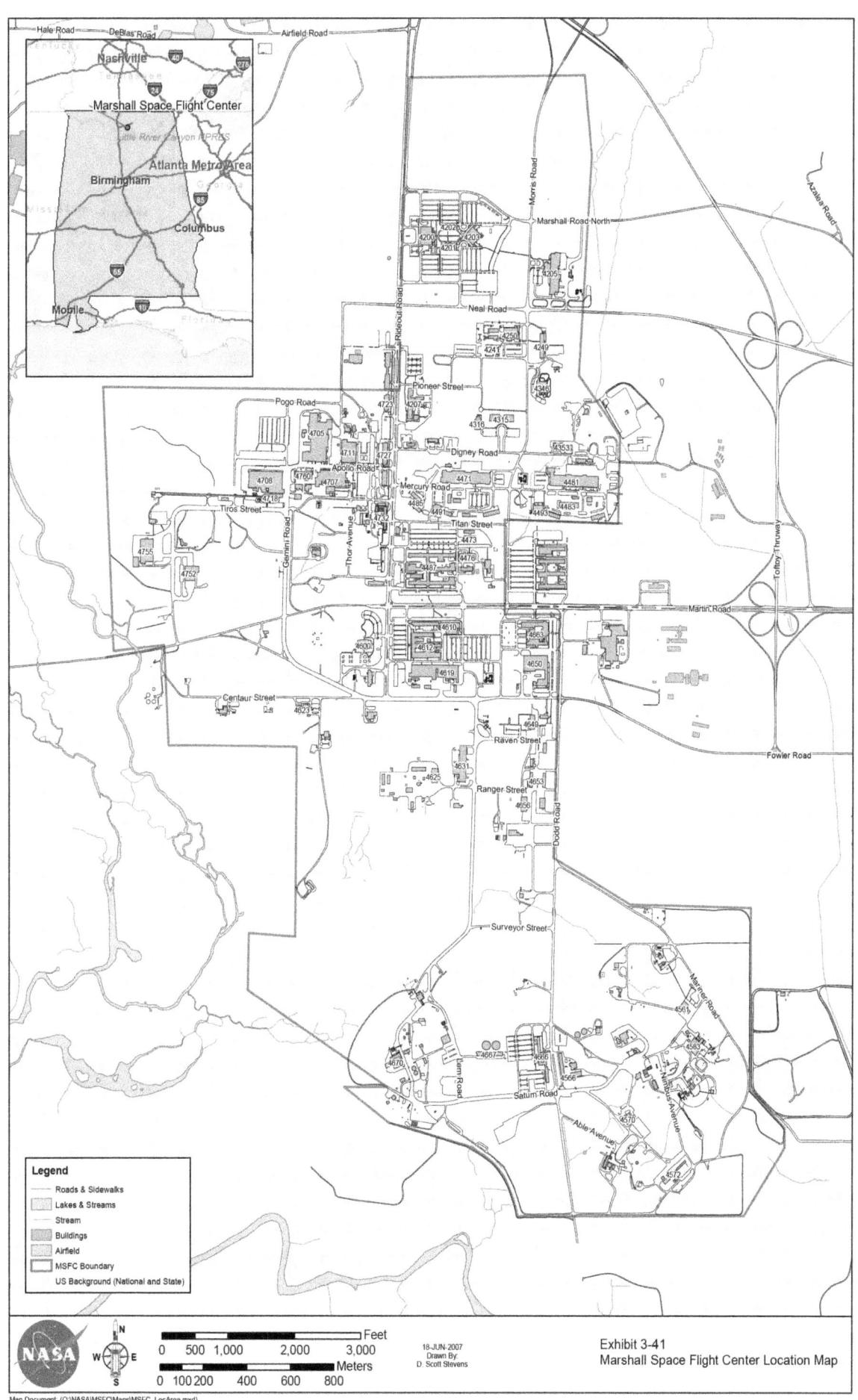

Exhibit 3-41
Marshall Space Flight Center Location Map

This page intentionally left blank.

3.8.1 Air Quality

3.8.1.1 Affected Environment

Regional Climate. The Huntsville, Alabama, area has a temperate climate. Summers are characterized by warm and humid weather with frequent thunderstorms. The City of Huntsville is almost surrounded by the foothills of the Appalachian Mountains (NASA, 2002a).

Cold air masses from the continent are predominant over the area during the winter season, but at times, mild air from the Gulf of Mexico spreads northward to Huntsville or beyond, and may persist for several days in succession. The contrast between air masses frequenting the region in winter provides a potential source of energy for producing extensive periods of low cloudiness and rain, the result being that 4 months, December through March, account for about 43 percent of the normal annual precipitation (NASA, 2002a).

Precipitation is mostly in the form of rain, but snow can be expected to some extent each winter, and seasonal totals have ranged from less than 1 inch to more than 20 inches (NASA, 2002a).

Temperatures frequently rise to 90°F or higher in the summers, but reach 100°F only on rare occasions. During the fall, the weather is usually dry. The air masses are cooler in the lower levels and the thunderstorm activity of summer decreases sharply. A major departure from the relatively dry weather of fall is an occasional rainy spell of one or more days associated with a decaying hurricane drifting northward from the Gulf of Mexico (NASA, 2002a).

Emission Sources. MSFC is classified as being in attainment for all NAAQS. This ranking causes the facility to follow the PSD program, instead of the more stringent NSR program (NASA, 2007s). Alabama does not have any additional state-specific air quality standards.

MSFC is a major source of HAPs, and therefore falls under the applicable NESHAPs. The amount of regulated ACMs at MSCF meets the threshold of the Asbestos NESHAP, which requires that asbestos surveys and remediation must take place before all remediation and demolition projects (NASA, 2002a).

The facility currently is operating under Title V Permit #0108900014 for the emission of HAPs at MSFC (NASA, 2007s). MSFC uses steam and electricity to heat the facility buildings. Steam used for heating and processes at MSFC is generated by three plants that have multiple boilers and a number of dispersed individual module boilers.

There are nine paint booths and two surface coating operations at MSFC. Routine maintenance on the buildings and equipment and the application of protective coatings account for almost all of the spray painting operations at MSFC.

There are three surface coating operations onsite at MSFC in which protective finishes are applied to various parts for testing under simulated flight or lift-off conditions.

MSFC performs experiments in the development, testing, and management of propulsion engines and launch vehicle systems.

MSFC also has permitted degreasers and storage tanks.

3.8.2 Biological Resources

3.8.2.1 Affected Environment

Vegetation. Land cover classes on MSFC consist of deciduous forest, pine forest, pine-deciduous forest, mowed fields, wetlands, and developed land (Exhibit 3-42). Approximately 50 percent of MSFC is developed, consisting of industrial complexes, office buildings, and parking lots and roads. Another 20 percent of MSFC is mowed fields. About 27 percent of MSFC is forested land cover, including mixed hardwoods, hardwood-pine vegetation communities, pineland, and hardwood swamps. Twenty-two percent of the forested land cover is pine-deciduous forest vegetation. Six percent of MSFC is wetlands (NASA, 2002a).

Wetlands. Approximately 45.8 ha (113.2 acres) of wetlands are found on MSFC (Exhibit 3-43). Wetlands in MSFC are palustrine systems and are scrub-shrub, forested, emergent, or open water systems. A 1993 survey found a total of 24 individual jurisdictional wetlands within MSFC property.

None of the facilities used by the SSP are located in wetlands.

Floodplains. FEMA has delineated floodplain areas on FIRMs (Exhibit 3-44). The FIRMs prepared by FEMA and updated by NASA for MSFC in February 2006 indicate that a significant portion of MSFC is located within the 100-year floodplain.

Building 4202 is not within the 100-year floodplain. Buildings 4436, 4625, and Trailer 236 are within the 100-year floodplain.

Wildlife. Approximately 0.7 km2 (180 acres) of the WNWR extends onto property controlled by MSFC, constituting 0.09% of the total MSFC land. There are examples of birds, mammals, reptiles and amphibians, but as a whole, MSFC has relatively low wildlife diversity (NASA, 2002a). A complete list of species can be found in the MSFC ERD (NASA 2002a).

Protected Species and Habitats. The ERD for MSFC (NASA, 2002a) lists all of the species potentially occurring at MSFC that receive protection from the federal government and the State of Alabama. The USFWS, the Alabama Department of Conservation and Natural Resources (ADCNR), and the Alabama Natural Heritage Program (ANHP) have rankings for wildlife and plant species, except that the State of Alabama does not have a list of officially threatened or endangered plants

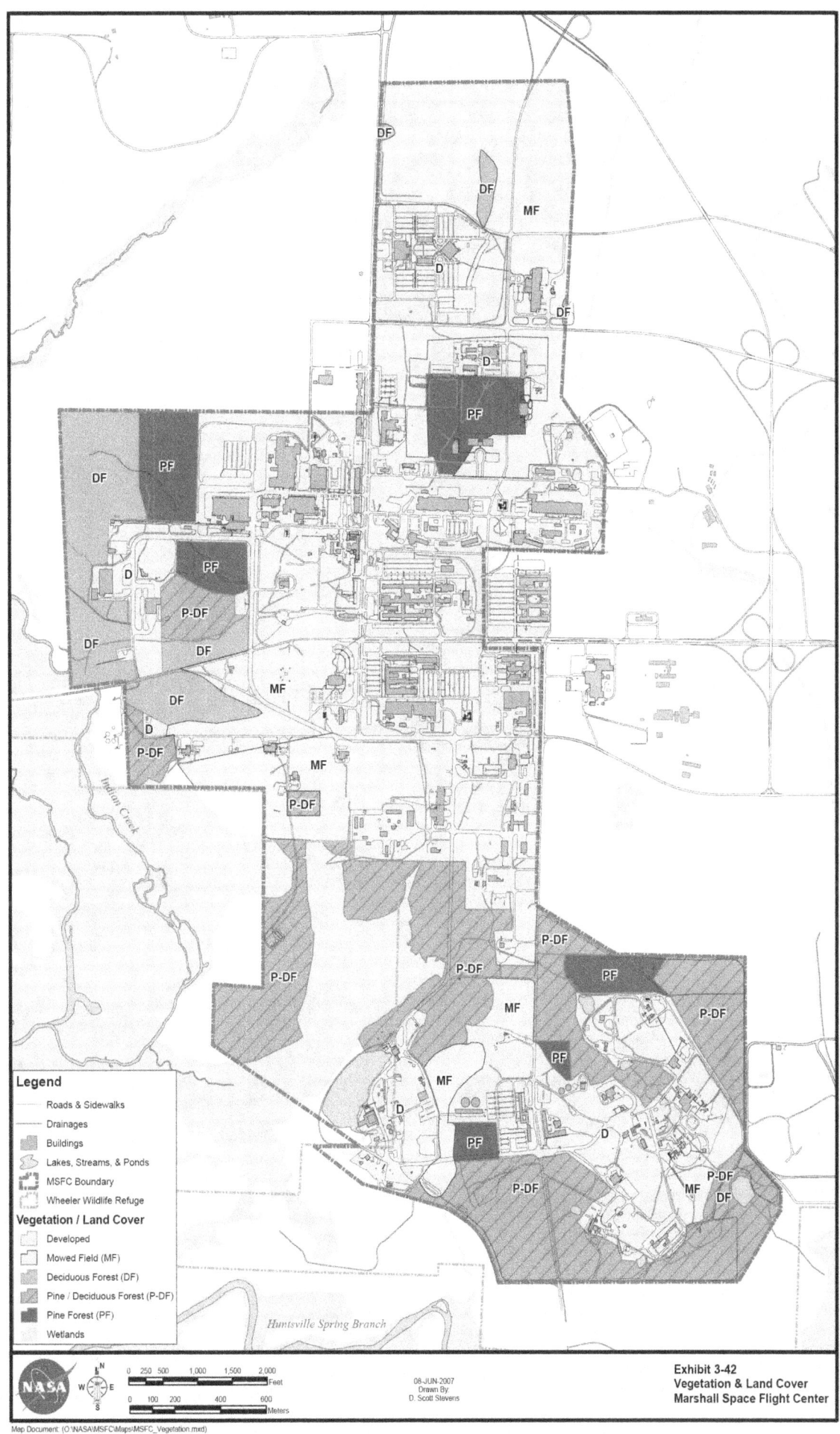

Exhibit 3-42
Vegetation & Land Cover
Marshall Space Flight Center

This page intentionally left blank.

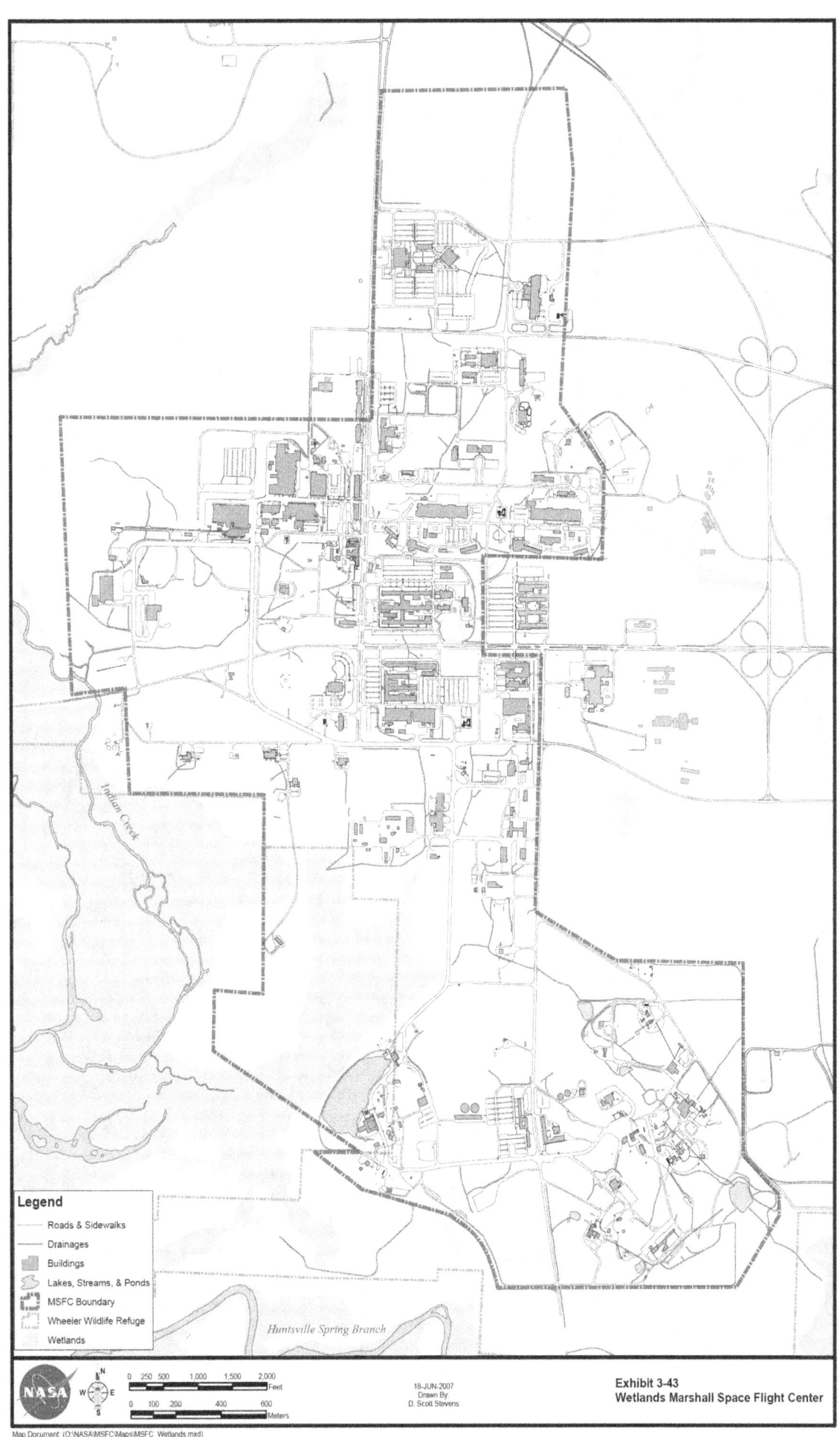

3. AFFECTED ENVIRONMENT

This page intentionally left blank.

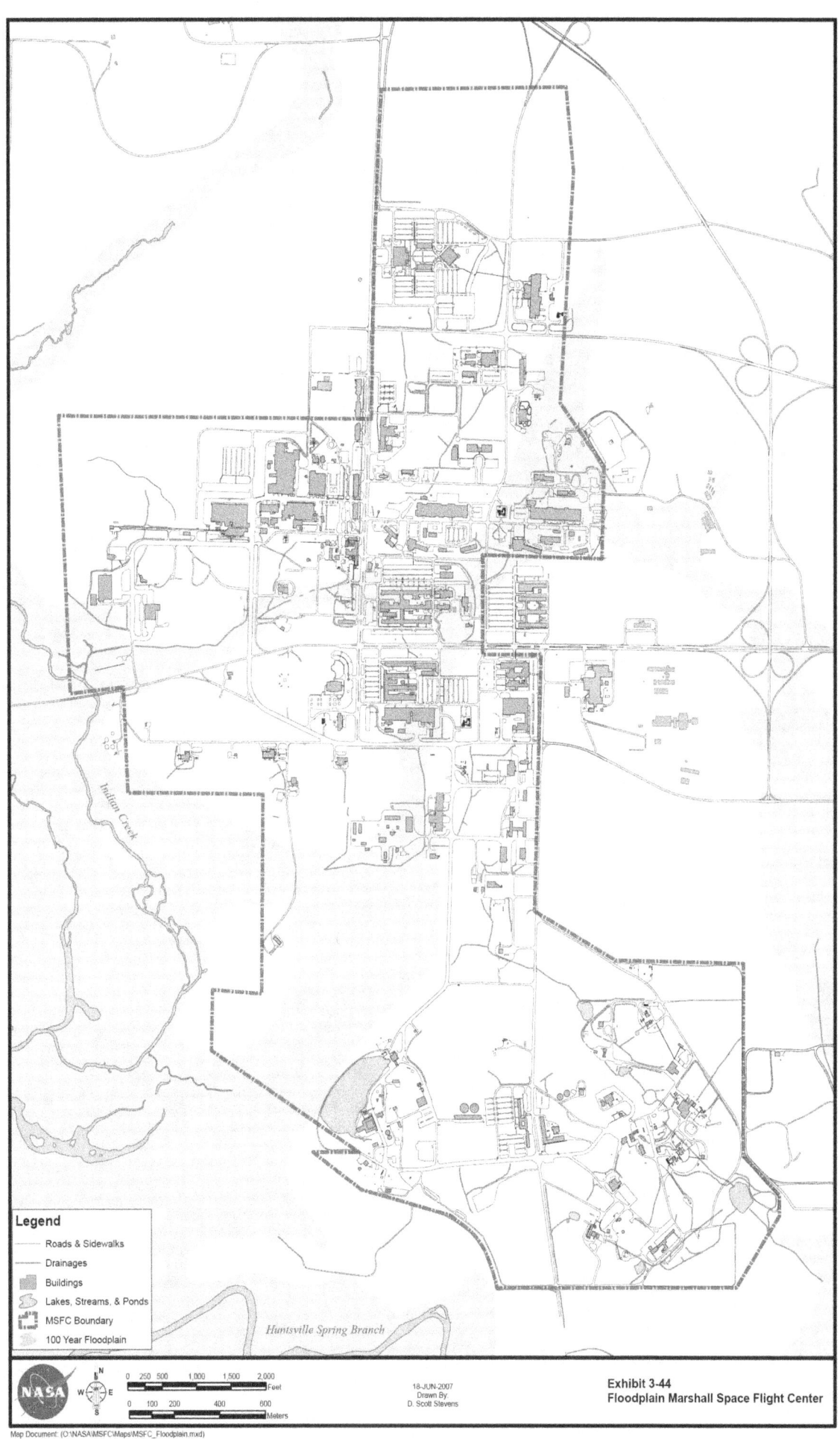

This page intentionally left blank.

(ADCNR, 2006; ANHP, 2006; NASA, 2002a). Exhibit 3-45 lists the federal- and state-protected species that have potential suitable habitat available at MSFC; these are the protected species most likely to be found on MSFC. In addition, NASA requires the proper best management practices and storm water pollution prevention measures be in place during construction activities and for operations conducted around the Center to aid in protecting habitats. The requirements are detailed in specifications for each construction project, MWI 8550, and MPR 8500.1.

Seven sensitive areas have been identified on RSA, and one of these, the Williams Spring Ecological Sensitive Area, is located within MSFC's boundaries. None of the facilities used by the SSP are located in the Williams Spring Ecological Sensitive Area.

3.8.3 Cultural Resources

3.8.3.1 Affected Environment

Historic Resources. A total of 40 properties were surveyed and evaluated for significance related to the SSP context. The SSME–Hardware Simulation Lab (HSL) Block II Facility (Building 4436), the Huntsville Operations Support Center (HOSC) (Building 4663), and the Office and Wind Tunnel Facility (Building 4732) are individually eligible under the SSP context. The newly eligible Test Area HD has five individually significant buildings:

- Test Facility (TF) 116 (TF 116, Building 4540)
- Structural Dynamic TF (Building 4550)
- Test & Data Recording Facility (Test Stand [TS] 115, Building 4583)
- Advanced Engine TF (AETF) (Building 4583)
- Control Facility (Building 4674)

Eight buildings are considered as contributing to the district, but are not individually significant:

- Transient Pressure TF (Building 4515)
- Solid Propulsion TF (Building 4520)
- TF 500 (Building 4522)
- TF 300 (TF 300, Building 4530)
- Test Stand Control Building (Building 4541)
- Hot Gas TF (Building 4554)
- Test Control and Service Building (Building 4561)
- Advanced Propulsion Research Facility (Building 4570) (Archaeological Consultants, Inc., 2007d)

EXHIBIT 3-45
Federal- and State-Protected Species with Potential Habitat at MSFC

Scientific Name	Common Name	Level of Protection State	Federal
Fish			
Etheostoma tuscumbia	Tuscumbia darter	S2/SP	-
Birds			
Pandion haliaetus	Osprey	SP	-
Haliaeetus leucocephalus	Bald Eagle	SP	-
Butorides striatus	Green-backed Heron	S2	-
Vireo solitaries	Solitary Vireo	S2	-
Falco sparverius paulus	Southeastern American kestrel	S3B	-
Mammals			
Neotoma floridana magister	Eastern wood rat	S3	-
Myotis grisescens	Gray bat	S2/SP	E
Myotis sodalist	Indiana bat	S2/SP	E(CH)
Myotis septentrionalis	Northern Long-eared Myotis	S2	-
Microtus ochrogaster	Prairie Vole	S2	-
Corynorhinus rafinesquii	Rafinesque's big-eared bat	S2/SP	-
Crustaceans			
Palaemonaias alabamae	Alabama cave shrimp	S1/SP	E
Reptiles			
Terrapene Carolina	Eastern Box Turtle	SP/S5	-
Amphibians			
Aneides aeneus	Green Salamander	SP/S3	-
Plants			
Trillium pusillum var. pusillum	Alabama dwarf trillium	S2	-
Hottonia inflate	Featherfoil	S2	-
Ophioglossum engelmannii	Limestone Adders Tongue	S3	-
Silphium asteriscus	Southern rosinweed	-	C
Prenanthes barbata	Barbed rattlesnake-root	S1/S2	-

EXHIBIT 3-45
Federal- and State-Protected Species with Potential Habitat at MSFC

Scientific Name	Common Name	Level of Protection State	Federal
Juglans cinerea	Butternut	S1	-
Silphium brachiatum	Cumberland rosinweed	S2	-
Leavenworthia crassa var. crassa	Fleshy-fruit glade cress	S1	C
Armoracia lacustris	Lake cress	S1	-
Marshallia mohrii	Mohr's Barbara buttons	S3	T
Silene ovata	Ovate catchfly	S1	-
Leavenworthia exigua var. lutea	Pasture glade cress	S1	-
Rudbeckia triloba var. pinnatiloba	Pinnate lobed black-eyed susan	S2/S3	-
Apios priceana	Price's potato bean	S1	T
Carex purpurifera	Purple sedge	S2	-
Silene regia	Royal catchfly	S2	-
Quercus boyntonii	Running post oak	S1	-
Leavenworthia alabamica var. brachystyla	Short-styled glade cress	S2	-
Cypripedium kentuckiense	Southern lady's-slipper	S1	-
Aureolaria patula	Spreading false-foxglove	S1	-
Rudbeckia heliopsidis	Sun-facing coneflower	S2	-
Delphinium exaltatum	Tall larkspur	SH	-
Astragalus tenneseensis	Tennessee milk-vetch	S1/S2	-
Xyris tennesseensis	Yellow-eyed grass	S1	E

Notes:
This list will change if species are extirpated or extinct.

Key:
FEDERAL:

- C Candidate species. Species is ready for proposal.
- CH Critical Habitat has been designated.
- E A species that is in danger of extinction throughout all or a significant part of its range, other than a species of the Class Insecta determined by the Secretary (of the Department of Interior) to constitute a pest whose protection under the provisions of the Endangered Species Act would present an overwhelming and overriding risk to man.
- PS An infraspecific taxon or population has federal status but the entire species does not – status is in only a portion of the species range.
- T Any species that is likely to become an threatened species within the foreseeable future throughout all or a significant portion of its range.

EXHIBIT 3-45
Federal- and State-Protected Species with Potential Habitat at MSFC

		Level of Protection	
Scientific Name	**Common Name**	**State**	**Federal**

STATE-Alabama Natural Heritage Program

- S1 Critically imperiled in Alabama because of extreme rarity (5 or fewer occurrences or very few remaining individuals or acres) or because of some factor(s) making it especially vulnerable to extirpation from Alabama.
- S2 Imperiled in Alabama because of rarity (6 to 20 occurrences or few remaining individuals or acres) or because of some factor(s) making it very vulnerable to extirpation from Alabama.
- S3 Rare or uncommon in Alabama (on the order of 21 to 100 occurrences).
- SB Regularly occurring, migratory and present only during the breeding season. A rank of S3B indicates a species uncommon during the breeding season (spring/summer) in Alabama.
- SH Of historical occurrence, perhaps not verified in the past 20 years, and suspected to be still extant.
- SN Regularly occurring, usually migratory and typically non-breeding species in Alabama. A rank of S2N or S5N indicated a rare breeder but a common winter resident.
- SX Apparently extirpated from Alabama.

STATE - Alabama Department of Conservation and Natural Resources

- SP State Protected. Species with a state protected status are protected by the Nongame Species Regulation (Section 220-2.92, page 73-75) of the Alabama Regulations for 1999- 2000 on Game, Fish, and Fur Bearing Animals. Copies of these regulations may be obtained from the Division of Wildlife & Freshwater Fisheries, Alabama Department of Conservation & Natural Resources, 64 North Union Street, Montgomery, AL 36104.

The newly eligible R&D HD at MSFC includes four individually eligible buildings that also contribute to the district:

- Materials and Processes Lab (Building 4612)
- Structures, Dynamics and Thermal Vacuum Lab (Building 4619)
- Multi-purpose HB Facility and Neutral Buoyancy Simulator (NBS) (Building 4705)
- National Center for Advanced Manufacturing (Building 4707)

Two buildings contribute to the R&D HD–the Hydrogen TF (Building 4628 and the Shop and Calibration Lab (Building 4650) (Archaeological Consultants, Inc., 2007d).

Five buildings and structures at MSFC that are NHL-designated, but only one, the Multi-purpose HB Facility and NBS (Building 4705), contributes to the R&D HD. All NHL properties automatically are listed in the NRHP. The NHL designation recognizes properties that exemplify important trends in U.S. history (NASA, 2002a:11-23; NASA, 2007s; NASA, 2007x; NPS, 2007d). The MSFC was surveyed for archaeological resources in 2005 and the survey findings were published by Alexander and Alvey in 2006. It is customary not to publish the locations of potentially sensitive archaeological sites.

Exhibit 3-46 shows the properties eligible for listing and those listed in the NRHP at MSFC.

3.8.4 Hazardous and Toxic Materials and Waste

Resource Conservation and Recovery Act. NASA submitted a Part B RCRA permit application for post-closure operations at its former Industrial Waste Treatment Facility (IWTF) on August 1, 1991, to EPA and the Alabama Department of Environmental Management (ADEM).

3.8.4.1 Affected Environment

Storage and Handling. NASA does not have a large inventory of stockpiled chemicals at MSFC. There are numerous waste accumulation areas throughout the Center (NASA, 2007s).

To date, there are 15 registered USTs operating on MSFC. All 15 tanks are in compliance and have been upgraded to meet the standards for construction, monitoring, leak containment, and operational design. All of the tanks have single-wall steel construction, except for three fiberglass tanks (NASA, 2007s).

Waste Management. MSFC is an LQG of hazardous waste. The wastes are transported regularly from various accumulation areas within MSFC to a less-than-90-day storage facility and are properly manifested offsite for disposal (NASA, 2007s).

Contaminated Areas. RSA, under the DA, was placed on the NPL in 1994, thus requiring compliance with CERCLA. NASA's MSFC also was listed on the NPL with the DA's RSA. NASA submitted a Part B RCRA permit application for post-

3. AFFECTED ENVIRONMENT

This page intentionally left blank.

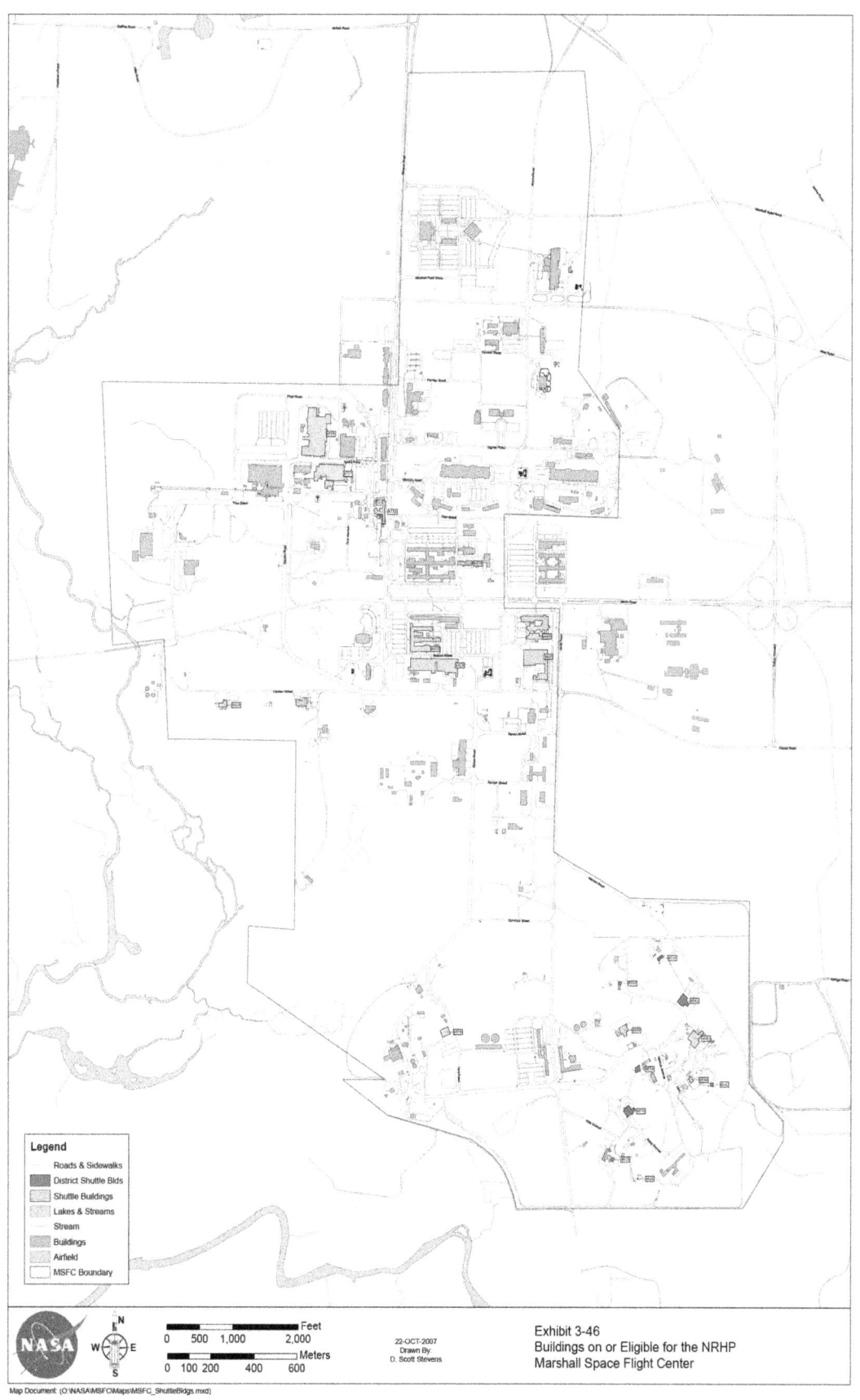

Exhibit 3-46
Buildings on or Eligible for the NRHP
Marshall Space Flight Center

This page intentionally left blank.

3.8.5 Health and Safety

3.8.5.1 Affected Environment

Hazardous materials are used for various operations at MSFC associated with propulsion testing in support of the SSP and other NASA programs. The following subsections outline MSFC's programs for protecting the health and safety of employees at MSFC and the public. Noise hazards at MSFC are outlined in Section 3.8.8.

Hazardous Materials. Hazardous materials, including propellants, are used at MSFC. The hazardous materials used and hazardous wastes generated are discussed in detail in Section 3.8.4. The degree of exposure to hazardous materials is minimized by the implementation of work practices and control technologies. The RSA Fire and Emergency Services Department (RSA FESD) serves as the first responder for all incidents involving hazardous materials.

Buildings at MSFC contain asbestos in the form of insulation and other building materials. MSFC complies with the 29 CFR 1910.1001 standard for the protection of employees from asbestos exposure. Buildings at MSFC also may contain LPB. All operations at MSFC that may disturb surfaces coated with LBP must be conducted in compliance with OSHA's construction standard for lead, 29 CFR 1926.62.

Hazardous Materials Transportation Safety. Hazardous materials such as fuels, chemicals, and hazardous wastes are transported in accordance with the DOT regulations for the interstate shipment of hazardous substances (49 CFR 100 through 199).

Explosions and Fire Hazards. Using certain hazardous materials, including propellants, in SSP operations presents a risk of explosions and fire hazards. Marshall Procedural Requirements (MPR) 1040.3 outlines the procedures for responding to explosions and fires at MSFC. The RSA FESD is responsible for fighting fires and performing rescue operations, as needed.

3.8.6 Hydrology and Water Quality

3.8.6.1 Affected Environment

Surface Waters. Surface waters at MSFC are part of the Indian Creek and Huntsville Spring Branch watersheds. Most drainage in MSFC is through manmade ditches to intermittent and perennial streams that flow west into tributaries of Indian Creek, or south and southeast into tributaries of Huntsville Spring Branch. Indian Creek is located to the west of MSFC and drains approximately 127 km^2 (49 square miles). Huntsville Spring Branch lies to the east and south of MSFC and drains approximately 109 km^2 (42 square miles). Huntsville Spring Branch occupies a mature floodplain largely inundated by Wheeler Lake. This stream joins Indian Creek in the backwaters of Wheeler Lake. Wheeler Lake overlaps the southwestern corner of MSFC and eventually discharges to the Tennessee River. At its closest point, MSFC is located 4.8 km (3 miles) north of the Tennessee River. Exhibit 3-47 shows the surface waters on MSFC.

Streams on MSFC are generally limited and of relatively low value. With the exception of surface waters in the southwestern portion of the site, most streams are intermittent or ephemeral ditches and swales. In addition to the streams and ditches, large, manmade reservoirs are located in the East and West Test Areas. These reservoirs are used for treating process water (NASA, 2002a).

Groundwater. Groundwater aquifers at MSFC are associated with three principal hydrogeological units–the residuum, the undifferentiated Tuscumbia Limestone, and the Fort Payne Chert (which comprises the Tuscumbia-Fort Payne Aquifer).

The residuum is the surficial geologic unit at MSFC. This unit consists of silty clay material with variable amounts of chert rubble and boulders that were formed from weathering of the underlying Tuscumbia Limestone. The thickness of the residuum generally ranges from about 0.3 to 24 m (10 to 80 ft). The residuum acts as a groundwater reservoir that stores large amounts of water and releases it slowly into the underlying bedrock aquifer. Groundwater recharge in the residuum is almost exclusively from precipitation.

Groundwater in the residuum under MSFC is believed to discharge to springs located to the south and west of MSFC. The flow rate for the discharge areas ranges from milliliters-per-minute for seeps to as much as 18.9 million L per day (5 million gallons per day [mgd]) for springs. Well-developed karst features in the limestone are believed to be the cause of larger spring discharges. The discharged water eventually flows to the Tennessee River via Indian Creek or Huntsville Spring Branch.

Beneath the residuum, the Tuscumbia Limestone and the Fort Payne Chert form the Tuscumbia-Fort Payne Aquifer. The Tuscumbia-Fort Payne is the primary aquifer in the region for water supply. This unit is composed of about 91 to 100 m (300 to 330 ft) of fossiliferous and dolomitic limestone with occasional interbedded chert. The Tuscumbia-Fort Payne is a karst aquifer, which means that groundwater occurs within solution-enlarged fractures, joints, and bedding planes in the formation. Water enters the aquifer from the land surface through sinkholes and disappearing and losing streams. Because of this connection with the land surface, water levels in the aquifer respond quickly to rainfall.

The Tuscumbia-Fort Payne Aquifer is the primary aquifer in the region for water supply. Municipal and private water supplies are obtained from this aquifer from both wells and springs. Wells completed in the Tuscumbia-Fort Payne Aquifer are reported to yield from about 151 to 11,350 L per minute (40 to more than 3,000 gallons per minute [gpm]). This variability in well yield depends on the

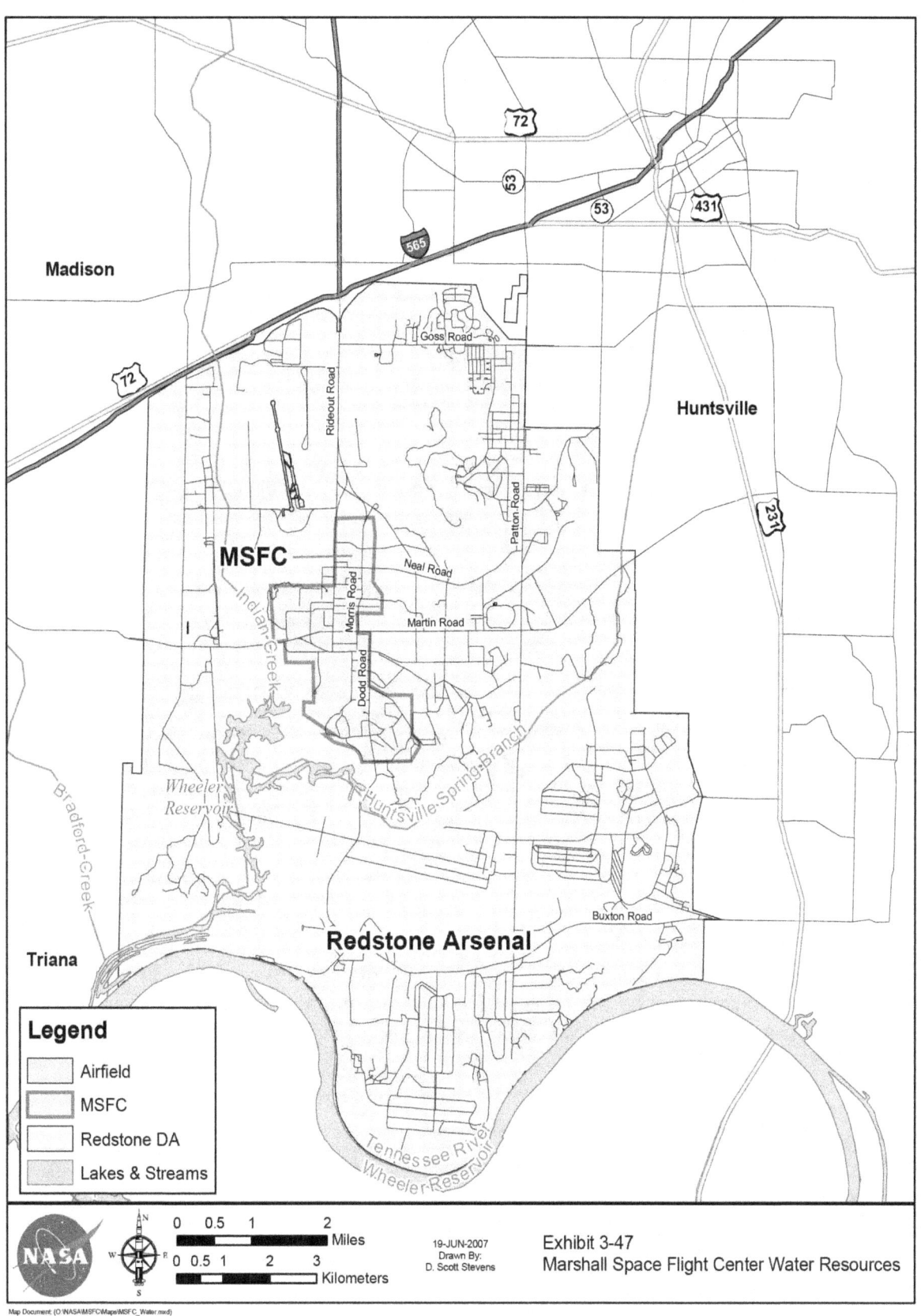

Exhibit 3-47
Marshall Space Flight Center Water Resources

intersections of the wells with solution-enlarged openings in the bedrock (NASA, 2002a).

Water Quality. Specific water quality criteria are governed by Alabama water laws. Surface waters are designated according to eight classifications, based on their potential use and value, as follows:

- Public Water Supply
- Swimming and Other Whole Body Water-Contact Sports
- Shellfish Harvesting
- Fish and Wildlife
- Agricultural and Industrial Water Supply
- Industrial Operations
- Navigation
- Outstanding Alabama Water

Indian Creek and Huntsville Spring Branch both have Fish and Wildlife-designated uses. Wheeler Reservoir is designated as a Public Water Supply and Fish and Wildlife Use.

Indian Creek and Huntsville Spring Branch are both included on Alabama's draft 2006 303d list. The causes of impairment are listed as priority organics for Indian Creek and metals and priority organics for Huntsville Spring Branch (ADEM, 2006).

Monitoring of surface water outside of NPDES compliance monitoring is infrequent.

Regulated Water Discharges and Withdrawals. MSFC holds NPDES Permit #AL0000221, which specifies discharge limitations and monitoring requirements for 27 outfalls. Effluent is monitored for pH, TSS, oil and grease, copper, TCE, and residual chlorine (NASA, 2007s). Indian Creek is the receiving water for 11 of the 27 outfalls (001, 016, 017, 018, 019, 020, 021, 022, 023, 027, and 028), Huntsville Spring Branch is the receiving water for 11 of the 27 outfalls (008, 009, 010, 011, 012, 013, 014, 015, 031, 032, and 033), and Wheeler Lake is the receiving water for the remainder of the outfalls (024, 025, 026, 029, and 030).

The majority of the outfalls at MSFC discharge storm water. As a requirement of federal and state law for facilities needing a storm water permit under the Clean Water Act (CWA), an SWP3 was developed as part of MSFC's *Consolidated Environmental Response Plan*.

As part of the NPDES program, facilities discharging to POTWs are required to have a State Indirect Discharge (SID) permit. MSFC also has received SID Permit IU 08 45 00027 to discharge industrial wastes resulting from metal finishing to PDR Properties' RSA Central STP.

MSFC's potable water is supplied by RSA via two intakes and two surface WTPs along the Tennessee River. No potable or non-potable water supply wells exist at

MSFC. MSFC's average water use in 2001 was 3,218,000 L per day (0.85 mgd) for potable water and 6,587,000 L per day (1.74 mgd) for industrial water use.

Currently, much of the water used at MSFC is discharged into the industrial sewer and the storm water drainage system. Treatment of wastewater occurs from the industrial sewer (Outfall 001) and Outfalls 012, 016, 018, 019, and 022. Sedimentation in naturally occurring ponds occurs at Outfalls 012, 016, 018, 019, and 022 before discharge. Dechlorination treatment is provided at the remaining outfall. MSFC currently is removing many discharges from the storm drainage system and all discharges from the industrial sewer, and is rerouting some of them to the sanitary sewer. Sources that will be rerouted to the storm sewer include non-contact cooling water, non-contact industrial water, groundwater sump discharges, and HVAC discharges.

The majority of the industrial wastewater at MSFC is sent to the IWTF at Building 4761. Discharge sources include the following:

- Cyanide plating wastewater
- Metal finishing wastewater
- Paint booth wastewater
- Photo processing wastewater

The IWTF has a capacity of treating 190,000 L per day (50,000 gpd) of wastewater under three different treatment schemes, but on average treats 76,000 L per day (20,000 gpd) (NASA, 2002a).

3.8.7 Land Use

3.8.7.1 Affected Environment

MSFC Facilities. MSFC uses its facilities and technical equipment for research, testing, and development activities in support of assigned programs. The laboratories, test stands, and HB facilities at MSFC can accommodate space system components through all stages of development and flight readiness testing. MSFC facilities make up a significant portion of the land use at the site, with approximately 281,450 square meters (m^2) (3,029,506 ft^2) of facility space.

Land Use Planning. The current, approved master plan for MSFC, completed in 2003, contains pertinent information necessary to achieve effective correlation of land areas and structures; to provide for adequate utilities, transportation, safety, and quantity distance zones, flow of material, and personnel; and to plan for future requirements. It contains the essential data and information to integrate current missions and programs with projections of future growth and long-range management decisions relative to facilities and land uses. The proposed land use described in the 2003 Master Plan was based on current uses of land at the time and also on expected future requirements for land based on the objectives and goals of NASA for MSFC in 2003 (NASA, 2002a).

Special Zones and Clearances–Test Areas. The test areas are the predominant land use of MSFC, and some of the test activities require large buffer zones for the safety of personnel and facilities. Test areas generally are located in the southern part of MSFC and have restricted access from Dodd Road.

Land Use Permitted Areas. The areas at MSFC described below are subject to planning restrictions.

Airfield and Flight Paths. MSFC air facilities are located at the Redstone Army Airfield. The airfield is located in the northwestern portion of RSA and is controlled by the Army Missile Command. The airfield consists of a north-south hard surface runway that is 2,190 m (7,300 ft) long and 45 m (150 ft) wide. Clear zones around the Redstone Airfield extend 900 m (3,000 ft) directly beyond the runway ends. Accident Potential Zones (APZs) I and II also are located at the southern and northern ends of the runway, respectively. APZ I south of the runway contains 1.4 km^2 (344 acres) that extend 1,500 m (5,000 ft) directly beyond the clear zone area and extend over a small portion of MSFC. This portion of the APZ is used for NASA research facilities. APZ II extends 2,100 m (7,000 ft) directly beyond APZ I and contains 1.9 km^2 (482 acres).

Specific Easements and Rights-of-Way. The ROW Plan, located in the Master Plan, indicates and reserves the necessary ROW and building setback (BSB) requirements for each basic type of road facility (arterial, collector, and local). The ROWs for the roads have been established to provide ample space for future widening without infringing on BSB space. For detailed information regarding the ROWs, refer to the Master Plan. The Master Plan also describes existing utilities such as water supply, sewer service, electrical supply, and telecommunications. The plan further discusses proposed plans and needs for increased supply and service. The plan covers topics such as existing vehicular parking and pedestrian walkways, landscape planting and conservation, security, and emergency services (NASA, 2002a).

3.8.8   Noise

3.8.8.1   Region of Influence

The ROI for noise generated by SSP-related activities are those areas that could be exposed to SPLs equal to or greater than 70 dBA. NASA MSFC has determined, based on experience gained from previous testing programs, that 70 dBA is the level of significance within the community, as determined by noise-related complaints (NASA, 2002a).

3.8.8.2   Affected Environment

The major noise at MSFC is generated by rocket motor and engine testing (NASA, 2002a). MSFC is located in the center of RSA, which provides a buffer zone between noise-generating activities and the nearest civilian population centers, including the

City of Huntsville and the City of Madison. The following sensitive receptors are located within a 2-mile radius of the RSA fence line:

- 36 schools
- 1 hospital
- 76 churches
- 28 shopping centers
- 20 parks
- 44 subdivisions

The primary source of noise at MSFC is rocket testing.

3.8.9 Site Infrastructure

The DA, under the direction of the RSA Support Agency (RASA), is responsible for supplying steam, electricity, water, and wastewater treatment services to MSFC (NASA, 2002a).

3.8.9.1 Region of Influence

The ROI is defined as the SSP-related facilities and the energy sources for these facilities at MSFC.

3.8.9.2 Affected Environment

Potable Water Supply. See "Hydrology and Water Quality," Section 3.8.6.

Wastewater System. See "Hydrology and Water Quality," Section 3.8.6.

Storm Water System. See "Hydrology and Water Quality," Section 3.8.6.

Energy Sources. RSA's electrical power system is made up of three subsystems–a transmission, a subtransmission, and a distribution system. The primary supply is obtained from the 161-kV, 3-phase transmission systems of the Tennessee Valley Authority (TVA). The part of TVA's transmission system to which RSA is connected is supplied by the following three separate 161-kV generating stations (NASA, 2002a):

- Wheeler Dam Station (including the Browns Ferry Nuclear Plant)
- Guntersville Dam Stations
- Widow's Creek Steam Generating Plant

Power normally is supplied to RSA by the Wheeler and Guntersville Dam Stations. The 161-kV transmission lines are transformed to a 44-kV, 3-phase subtransmission level by three government-owed primary substations at three different locations on RSA (NASA, 2002a).

NASA MSFC also has approximately an 1,800-kilovolt ampere (kVA) total capacity through several emergency generators for critical or special electrical circuits.

Finally, RSRA bills NASA for all electrical power consumed at MSFC (NASA, 2002a).

Steam is provided by 3 boiler plants and 22 modular boilers located in MSFC buildings. One of these boilers has been turned over to RSA. There are also 2 rental boilers and 1 portable boiler. The boiler plants, located in the test areas, are used exclusively for heat generation and to power processes in the test areas. RSA's main steam plant is the City of Huntsville Plant, Ogden Martin Systems. The plant is owned by RSA. MSFC is supplied with steam from RSA's steam supply. The steam lines operate at a pressure of 100 to 200 pounds per square inch (psi), depending on the area, and the condensate return lines have an operating pressure of approximately 40 psi. Some condensate is collected and returned to the boiler plants for further steam generating (NASA, 2002a).

Steam for the East Test Area is generated by three diesel-fired boilers in one of the boiler plants in Building 4567, with a combined capacity of 36 million British thermal units per hour (MMBtu/hr) operating at a pressure of 124 psi. Steam for the West Test Area is provided by two boiler plants in Buildings 4660 and 4675. Three diesel-fired boilers with a combined capacity of 37.8 MMBtu/hr are located in Building 4660. Building 4675 also contains three diesel-fired boilers with a combined capacity of 16.2 MMBtu/hr and that operate at a pressure of 125 psi each (NASA, 2002a).

Modular boilers are used in many buildings at MSFC. Currently, the modular boilers are kept for emergency use only (NASA, 2002a).

Buildings 4466, 4514, 4515, 4520, 4564, 4549, 4572, 4553, 4641, 4642, 4640, 4646, 4638, 4639, 4645, and 4648 and the RF Test Area (4100 Area) are the only facilities at MSFC not heated by steam (NASA, 2002a).

RSA receives its natural gas supply from the City of Huntsville. Natural gas is routed through MSFC in a 12-inch pipeline at a nominal pressure of 45 psi. The 12-inch pipe is tapped at three locations to serve MSFC buildings. Separate branch lines serve modular boilers located in Buildings 4487 and 4491. A 4-inch metered main is extended from the 12-inch main to serve Buildings 4707, 4708, 4718, 4752, 4755, and 4776 (NASA, 2002a).

The boilers at MSCF are listed on the facility's Title V Permit #0108900014 (March 2001) (NASA, 2007s).

3.8.10 Socioeconomics

3.8.10.1 Region of Influence

The economic ROI for MSFC is defined as Madison, Morgan, Limestone, Marshall, Lauderdale, Cullman, and Jackson counties in Alabama and Lincoln County, Tennessee (Exhibit 3-48), where approximately 98 percent of all MSFC civil service

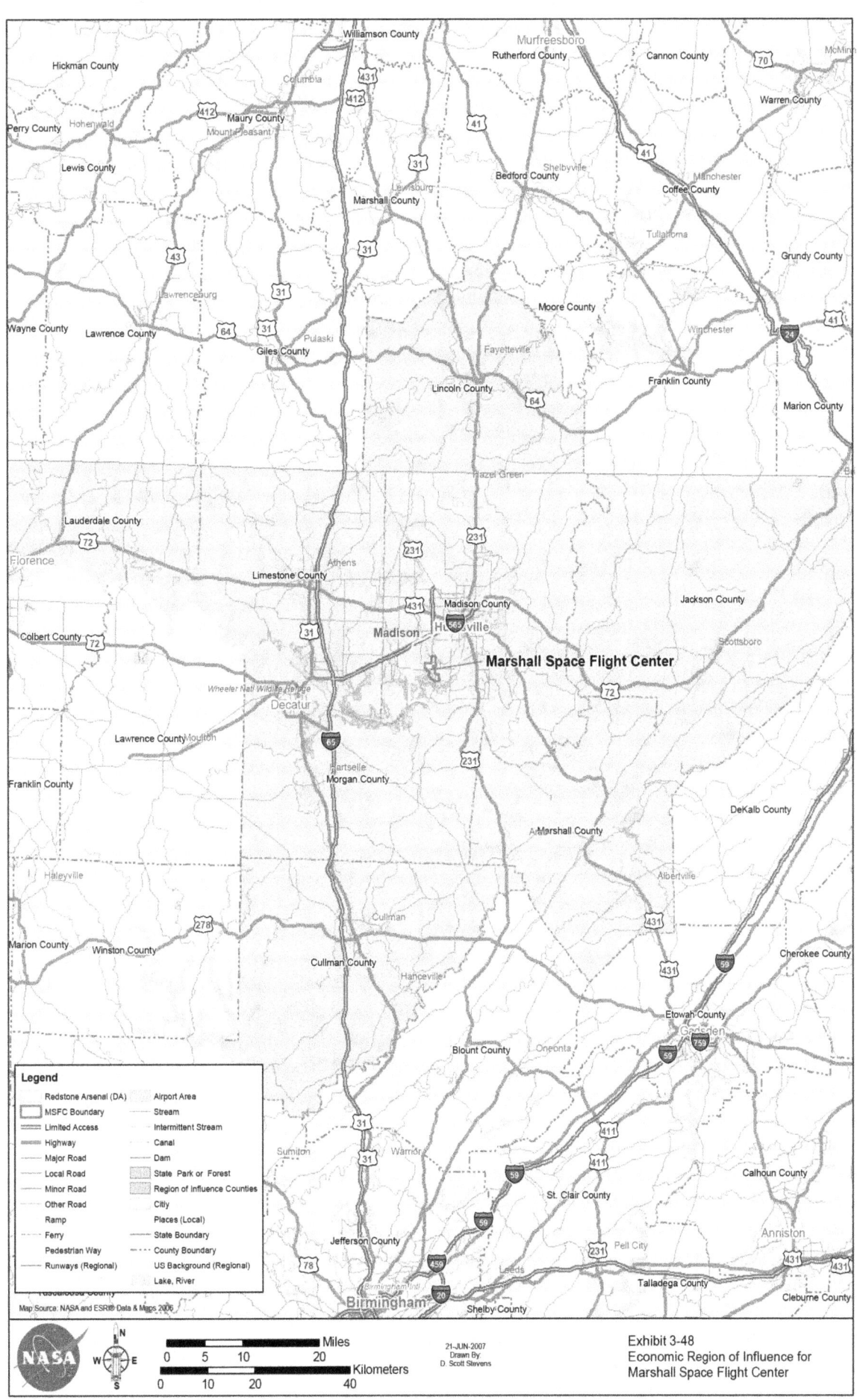

Exhibit 3-48
Economic Region of Influence for
Marshall Space Flight Center

This page intentionally left blank.

and prime contractor employees live. The majority (73 percent) of MSFC employees live in Madison County, Alabama (NASA, 2007a). Madison and Limestone counties are designated as the Huntsville MSA by the OMB (OMB, 2006). The Huntsville MSA is one of the fastest-growing science and technology centers in the nation. Huntsville is known as "Rocket City" in honor of the Saturn V rocket, which put man on the moon (NASA, 2002a).

3.8.10.2 Affected Environment

Population. In 2006, more than 821,000 people lived in the 8 counties of the ROI, an increase of 4.5 percent from the 2000 Census. Limestone and Madison counties (the Huntsville MSA) showed the largest increases at 7.3 percent and 7.7 percent, respectively. Between the 1990 and 2000 Censuses, the population in the Huntsville MSA grew by 18 percent. By 2010, the total population of the ROI is projected to grow by 6.6 percent (U.S. Census Bureau, 2006c; Center for Business and Economic Research [CBER], 2001).

Regional Employment and Economic Activity. During the past 45 years, the regional economy has expanded from primarily agriculture and space-related industries into a robust, diversified mix of manufacturing, testing, development, research, and support services, although agriculture remains important (NASA, 2002a).

NASA and defense agencies are a cornerstone of the regional economy, both as major employers and through the procurement of goods and services. In FY 2006, MSFC had an operating budget of $2.26 billion and contributed $302 million in payroll expenditures. Tourist dollars also are brought into the region through the U.S. Space and Rocket Center, which features a large space history museum, a space camp for students, and the NASA Educator Resource Center (NASA, 2007t; NASA, 2002a; NASA, 2007n).

In 2006, the total labor force of the ROI was almost 427,000 people, with an overall unemployment rate of 3.3 percent, which was similar to the state (3.6 percent) and below the national unemployment rate (4.6 percent) (BLS, 2006).

NASA at MSFC is one of the region's largest employers. In FY 2006, NASA employed more than 2,500 civil servants and nearly 5,000 contractors at MSFC. Other leading employers include the U.S. Army Aviation and Missile Command (AMCOM) and other defense agencies at RSA (14,600 employees), Huntsville Hospital System (5,100 employees), Huntsville City Schools (3,000 employees), and the Boeing Company (3,000 employees) (Chamber of Commerce of Huntsville/ Madison County Website, 2007).

Economic Contribution of the Space Shuttle Program. In 2006, the SSP accounted for only about 9 percent of total MSFC employment, with approximately 500 civil service FTEs and 200 prime contractor FTEs. This number only includes time directly charged to the SSP budget; it excludes NASA MSFC base operations and administrative personnel, time spent supporting other programs at MSFC, and jobs at offsite suppliers and subcontractors within and outside of the region (NASA, 2007a).

In FY 2006, the SSP directly contributed more than $170 million to the regional economy, including civil service and prime contractor salaries and non-payroll procurements to subcontractors and suppliers. Those expenditures generated additional economic output, jobs and income in supporting industries within the ROI. The total (direct plus indirect and induced[10]) effect of the SSP on economic output was approximately $500 million (less than 1 percent of the more than $35 billion[11] in overall economic activity in the 8-county region), $241 million in earnings, and 4,500 jobs (NASA, 2007cc).

3.8.11 Solid Waste

3.8.11.1 Affected Environment

The majority of solid waste generated at MSFC is transported to the City of Huntsville's Refuse-to-Steam Plant. This facility receives approximately 150 tons of solid waste each month from MSFC. Certain wastes are excluded from disposal at the facility, such as hazardous or radioactive waste, paint products, fuels, explosives, and construction debris. The construction waste, rubble, vegetation, and asbestos generated at MSFC are disposed at the RSA inert landfill.

3.8.12 Traffic and Transportation

3.8.12.1 Region of Influence

The transportation ROI for MSFC is defined as Madison, Morgan, Limestone, Marshall, Lauderdale, Cullman, and Jackson counties in Alabama and Lincoln County, Tennessee. At least 50 MSFC employees live in each of these 8 counties, with the majority living in Madison County. The counties together account for approximately 98 percent of all MSFC employees, based on zip code data for civil servants and prime contractors (NASA, 2007a). Madison and Limestone counties make up the Huntsville, Alabama, MSA (OMB, 2006).

3.8.12.2 Affected Environment

Transportation. The modes of transportation serving NASA at MSFC are roads, railroads, waterways, and air. The existing system forms an interrelated transportation system that provides two primary functions–the means by which people and goods move into MSFC, and the means for internal circulation within MSFC.

---

[10] Based on the "multiplier effect" using economic multipliers for the 8-county MSFC region from the U.S. Bureau of Economic Analysis.

[11] U.S. Bureau of the Census, 2002 Economic Census–Total sales, shipments, receipts, or revenue for all establishments (2-digit NAICS codes) in the ROI for MSFC.

Access Roads to MSFC. Per the surface treatment categories, the road system for MSFC consists of primary, secondary, and tertiary roads (Exhibit 3-49). The principal system configuration consists of north-south and east-west alignments.

The major north-south roads are Rideout Road, Toftoy Thruway, and Dodd Road. Major east-west roads are Martin Road, Fowler Road, and Neal Road. The majority of the bridges are single-lane, two-directional, a 4.5 to 4.8 m (15- to 16-foot) clearance above stream beds, and a 36-ton load limit. Currently, all traffic to and from MSFC and RSA is routed through six gates. The Main Gate (Gate 1) is on Martin Road on the eastern side of RSA. Gate 3 (Redstone Road) is also on the eastern side of RSA. Gate 7 (Martin Road) is on the western side of RSA. Gate 8 (Goss Road), Gate 9 (Rideout Road), and Gate 10 (Patton Road) provide access to RSA from the north. Six of these gates (3, 7, 8, 9, 10, and the Main Gate) are open to traffic during daylight hours. The Main Gate and Gates 8 and 9 are open 24 hours per day.

Railroads. The use of rail facilities on RSA was largely discontinued in 1973. Most of the track has been removed, and only a small section of rail remains in RSA. The use of planes and trucks for shipping purposes has decreased the demand for rail transportation. A railhead located near the northern boundary has been retained to serve NASA at MSFC as the need arises.

Airports. MSFC's air facilities are located at the Huntsville International Airport. The airfield has two hard surface runways with concrete approaches. The runways are designated as 18 (left and right) and 36 (left and right). The airport is located at a latitude of 34° 38' North, longitude 86° 46' West, and at elevation 629 ft above msl. The runway is 3,000 m long by 45 m wide (10,000 ft long by 150 ft) wide for the 18 (left) and 36 (right); and 2,400 m long by 45 m wide (8,000 ft long by 150 ft) wide for the 18 (right) and 36 (left). Precision approaches (Instrument Landing System [ILSs]) are available for each runway, in addition to non-precision approaches. A tower is located in the center of the field between the two runways. The runway, taxi, and apron lights are U.S. Standard (A). The runway can accommodate any aircraft in the U.S. The airport has an average of 5,000 arrivals and departures of aircraft per month. Of these, less than 35 percent are NASA or NASA-related flights.

3.9 White Sands Test Facility

NASA's WSTF is an aerospace testing facility located in southern New Mexico near Las Cruces (Exhibit 3-50). WSTF is a Directorate-level component of NASA's JSC. The majority of the testing and research conducted at WSTF is directly related to NASA's manned spaceflight programs.

WSTF's primary mission is to provide the expertise and infrastructure to test and evaluate spacecraft materials, components, and rocket propulsion systems to enable the safe human exploration and use of space. NASA also evaluates materials and

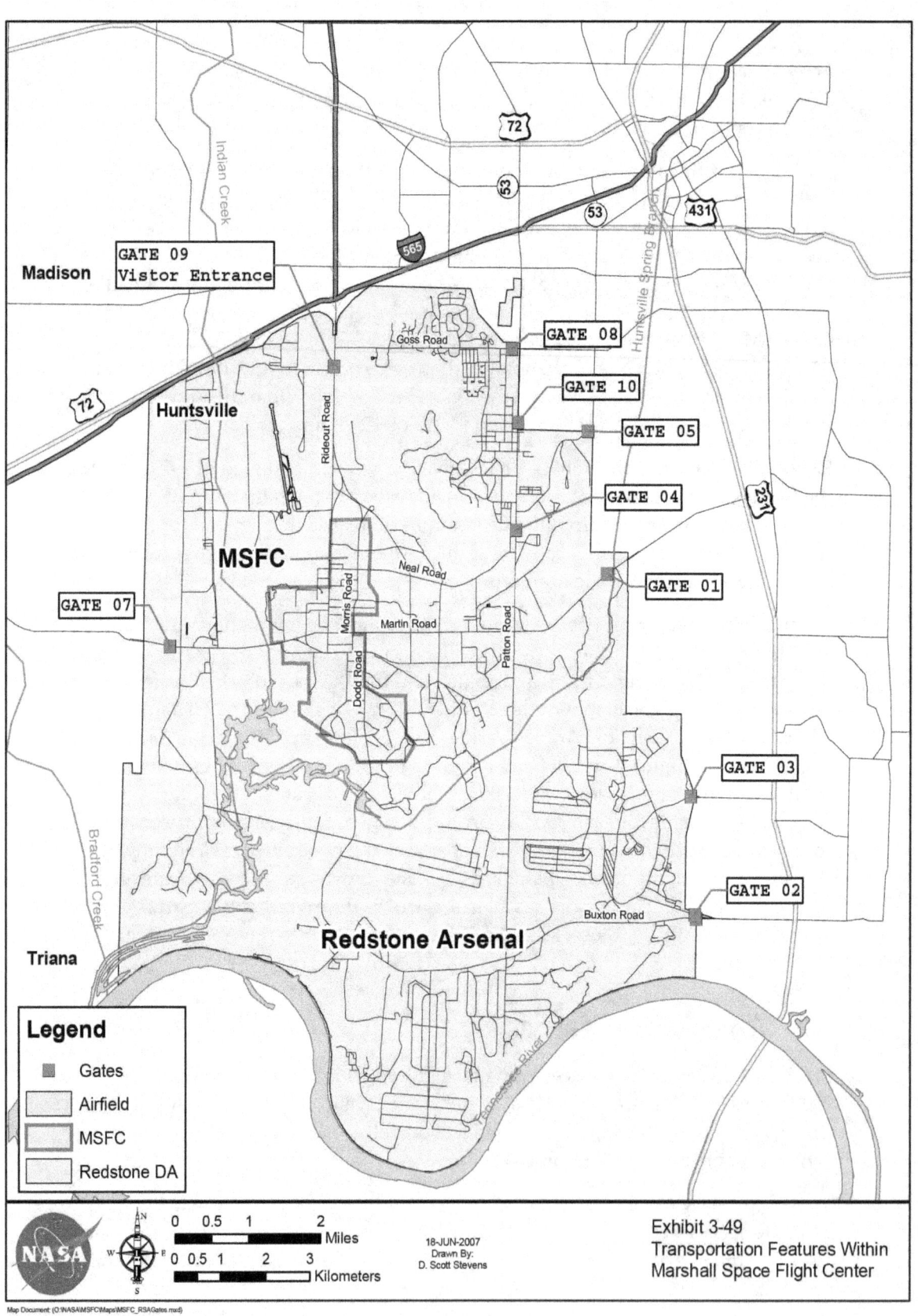

Exhibit 3-49
Transportation Features Within
Marshall Space Flight Center

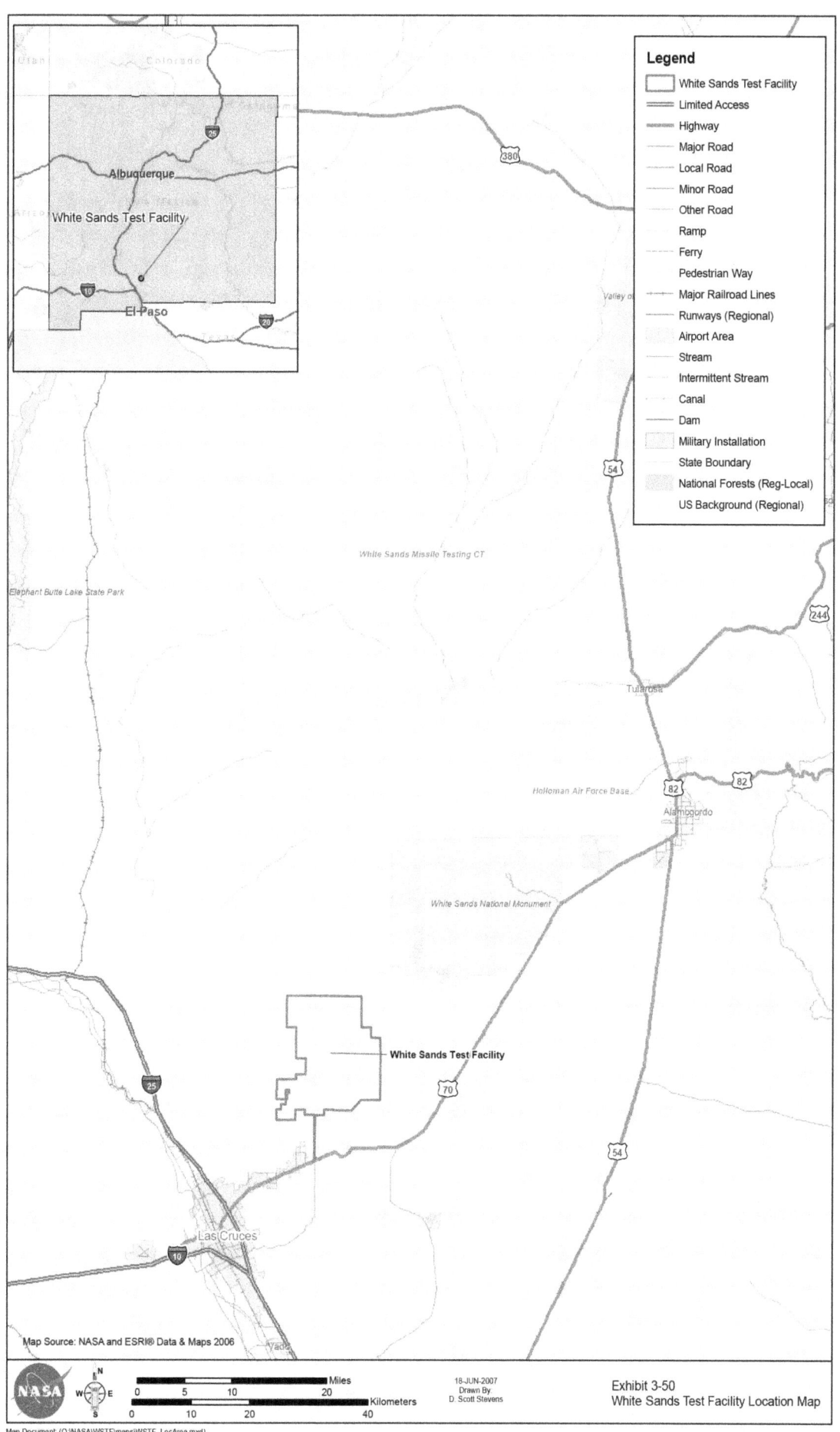

Exhibit 3-50
White Sands Test Facility Location Map

This page intentionally left blank.

components at WSTF for use in propulsion, power generation, and life-support systems, crew cabin equipment, payloads, and experiments carried aboard the Orbiter and the ISS. WSSH is the Orbiter approach and landing training facility. It also is an alternate landing site for the Orbiter if the conditions at KSC or DFRC are not favorable.

NASA has equipment to support the Orbiter in the event of a landing; however, there are no major SSP assets located at WSSH.

3.9.1 Air Quality

3.9.1.1 Affected Environment

Regional Climate. WSTF is situated in an area that has a predominantly high steppe and desert climate. The climate is characterized by abundant sunshine, relatively low humidity, slight rainfall, moderate winds, and a wide range in daily temperature variations. The mountainous terrain in the area influences the climate by blocking the incursion of moisture-laden maritime air masses. Cold air drainage down slopes causes a wide variation in the minimum temperatures experienced in the area. Precipitation, greatest in July and August, averages 10 inches annually. The growing season is about 200 days per year (NASA, 2001a).

The average annual temperature is 60°F. The warmest portion of the year is from the beginning of July to the end of August, when the temperature frequently rises to higher than 90°F. Day-to-night temperature variations ranging from 30°F to 35°F are common (NASA, 2001a).

The coldest periods are from November to mid-March, with the lowest temperatures occurring in December and January. Although freezing nighttime temperatures are normal for this period, average highs near 60°F prevail in even the coldest months (NASA, 2001a).

The greatest annual amount of recorded precipitation is 19.60 inches and the least is 3.62 inches. Most of the rainfall occurs in the summer. The average rainfall by season is as follows: winter 2.70 inches; spring 0.79 inch; summer 3.80 inches; and, fall 2.47 inches. Intense thunderstorms frequently release heavy rainfall within a short time span over a restricted geographical area (NASA, 2001a).

Although snow flurries typically occur in light amounts at WSTF, the records show that heavy snowfalls are not uncommon. Sleet seldom occurs, but skim ice has been found on the sewage lagoons in the early hours of the colder winter days. The average yearly snowfall is about 2.5 inches. Monthly, mean humidities for the area remain below 50 percent for every month of the year. Seasonal averages are as follows: winter 43 percent; spring 28 percent; summer 36 percent; and fall, 40 percent (NASA, 2001a).

Seasonal wind variations in the area are significant, with the strongest sustained winds occurring in late winter and spring months, primarily because of entrainment of the surface winds, with the strong westerly winds aloft, and the nature of the terrain (NASA, 2001a).

Thunderstorm activity, with the season of maximum occurrence from June through September, develops over the mountains, depending on the winds aloft (NASA, 2001a).

Emission Sources. NASA WSTF operates as a minor source of air emissions and has four construction permits (NASA, 2006a).

WSTF is located in attainment areas for CO, nitrogen dioxides, 1-hour ozone, 8-hour ozone, sulfur oxides, $PM_{2.5}$, PM_{10}, and lead (NASA, 2006a). WSTF is also in attainment with the New Mexico state-specific Ambient Air Quality Standards (New Mexico Environmental Department [NMED], 2001). WSTF has the following emission sources, with corresponding construction permits for each (NASA, 2007s):

- 300 Area Altitude Simulation System covers the following equipment:
    - Rocket test chamber
    - Two water spray condensers
    - Two cooling towers
    - One cooling pond
    - One 31.32-MMBtu per hour Cleaver-Brooks steam boiler
- 400 Area Altitude Simulation System the following equipment:
    - Three 800-horsepower (hp) Cleaver-Brooks steam boilers, which burn low-sulfur diesel fuel and are subject to the National Stationary Performance Standards (NSPS) for small steam-generating units
    - Three 195-hp Model 6030C Detroit diesel engines
    - Four 645-hp Model 7163-7000 Detroit diesel engines
    - One 68,000-brake horse power (bhp) Pratt & Whitney steam generator
- The High-energy Blast Facility (HEBF) is used to characterize explosive blasts of solid, cryogenic, and hypergolic propellants, and consists of two test pads and a nearby control center. No specific equipment is covered by this permit.
- 800 Area Test Cell 844. Test Cell 844 is set up for Shuttle Auxiliary Power Unit (APU) testing. The APU uses the exhaust gases from hydrazine decomposed over a catalyst to spin the power-generating turbine. The equipment covered by this permit consists of the APU, systems testing equipment, and various support equipment.

3.9.2 Biological Resources

3.9.2.1 Affected Environment

Vegetation. WSTF lies within the Chihuahuan desert shrub biotic community. As indicated by the soil associations, three primary vegetation types occur at WSTF, including rock outcrops, semi-desert, and grassland (NASA, 2001a). Approximately 45 percent of WSTF is rock outcrops and rock-dominated soil units (Exhibit 3-51). These areas, primarily along the eastern side of WSTF and also including the overlapping lands of the San Andres National Wildlife Refuge (SANWR) and part of the Joranada Experimental Range (JER), have limited usage and plant growth (NASA, 2001a). About 40 percent of WSTF is limey, gravelly, sandy to clayey soils. Most of the facilities on WSTF are constructed on this soil type. Common semi-desert vegetation found on these soils includes creosote bush (*Larrea tridentata*), mesquite (*Prosopis sp*), tarbush (*Flourensia cernua*), snakeweed (*Gutierrezia* sp.), burro grass (*Scleropogon brevifolius*), and some grama grasses (*Bouteloua* spp.) (NASA, 2001a).

The remaining 15 percent of WSTF consists of clay, sand, and foot-slope grassland vegetation types, including mesquite, yucca (*Yucca*), sagebrush (*Artemisia*), burro grass, and assorted grama grasses (NASA, 2001a).

Wildlife. WSTF and the area around WSTF support songbirds, small mammals, carnivores, and ungulates. Landscaping at WSTF consists of low-maintenance desert vegetation and, as a result, WSTF still supports abundant wildlife. The ERD (NASA, 2001a) lists the wildlife species found on and in the vicinity of WSTF.

Three regions of WSTF have been identified as high-sensitivity habitat, including the mesic woodland and arroyo vegetation associated with the Love Ranch area, the upper reaches of the Bear Canyon drainage, and the mesic woodland habitat associated with the northeast foothills of Quartzite Mountain and the San Andres Mountain Range. These areas are rich in topographic relief, biodiversity of both plant and animal species, and natural water catchments and cover for wildlife (NASA, 2001a).

Protected Species and Habitats. WSTF does not contain critical habitat for any federally listed threatened or endangered plant or wildlife species (NASA, 2001a).

Plants. The only exotic, threatened, or endangered plant species that has been observed on WSTF is the night-blooming cereus (*Peniocereus greggii*) (NASA, 2001a).

The only exotic, threatened, or endangered wildlife species that has been observed on WSTF is the desert bighorn sheep (*Ovis canadensis mexicana*) (NASA, 2001a).

3.9.3 Cultural Resources

3.9.3.1 Affected Environment

Archaeological Resources. The WSTF ERD lists 85 identified archaeological resources within the boundaries of the facility. Three of the sites have prehistoric and historic elements, while 14 are historic and 75 are pre-historic. The Love Ranch facility is the only one of these identified sites eligible for listing in the NRHP. The site shows the remains of a 2-acre, early 20th century ranch complex. To protect the resources from

3. AFFECTED ENVIRONMENT

This page intentionally left blank.

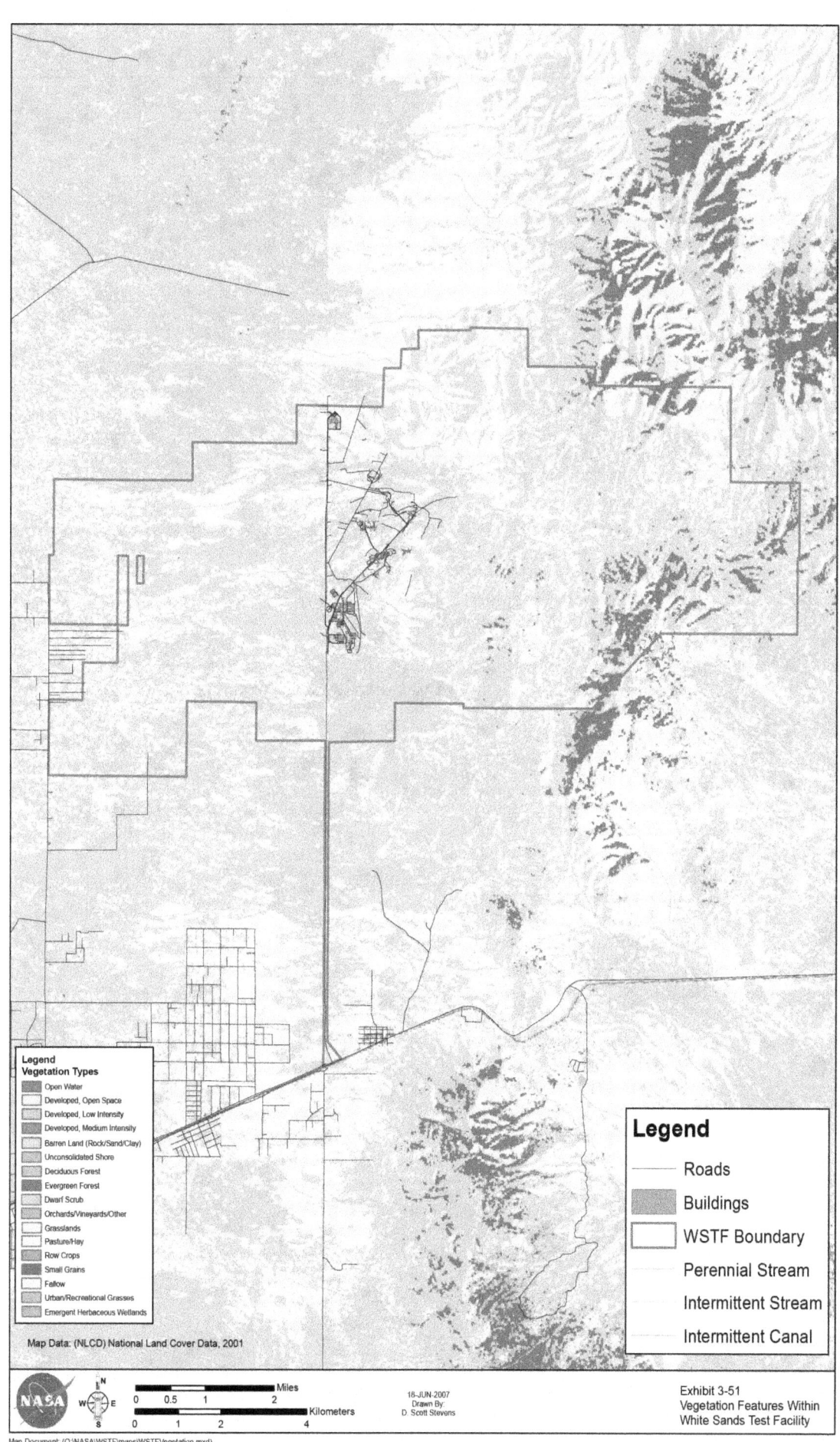

Exhibit 3-51
Vegetation Features Within
White Sands Test Facility

3. AFFECTED ENVIRONMENT

This page intentionally left blank.

vandalism, it is customary not to publish the exact locations of the sites (NASA, 2001a:115-23).

Historic Resources. A survey of 14 properties was conducted to determine their significance to the SSP. The results of the survey determined that the only facility eligible for its association with the SSP is the SLF (Archaeological Consultants, Inc., 2007e). There are NHL properties at WSMR, but they are located outside the boundaries of the NASA WSTF. The SLF is shown in Exhibit 3-52.

3.9.4 Hazardous and Toxic Materials and Waste

3.9.4.1 Affected Environment

Storage and Handling. NASA WSTF is classified an LQG waste generator. Hazardous waste generation activities at WSTF are regulated by RCRA Permit #NM8800019434-1. The RCRA Operating Permit covers the Evaporation Tank Unit (ETU) in the 200 Area and the Fuel (hydrazine) Treatment Unit (FTU) in the 500 Area. The permit was issued in 1993 and currently is pending renewal by the NMED Hazardous Waste Bureau (NASA, 2007s).

There are no major stockpiles of chemicals at WSTF or WSSH. Wastes are removed periodically and disposed per WSTF's RCRA permit. SSP primarily is responsible for two synthetically lined evaporation ponds, the FTU and the 400 Area ponds. However, it is likely that future NASA programs will use these three facilities when the SSP is transitioned (NASA, 2007s).

NASA currently operates and pays fees on two petroleum USTs. NMED has evaluated WSTF's tanks and determined them to be in compliance (NASA, 2007s).

Waste Management. NASA manages five RCRA-closed Hazardous Waste Management Units under the provisions of Post-closure Care Permit #NM8800019434-2. The RCRA closures include holding ponds and mixing tanks in both the 300 and 400 Areas, two underground chemical waste storage tanks in the 200 Area laboratories, and a dual pond system in the 600 Area that received some wastes from around WSTF, but primarily received wastes from the laboratory areas. These closures are monitored actively with upgradient and downgradient groundwater monitoring systems; the groundwater and hydrogeological data are reported annually to the NMED Hazardous Waste Bureau. The Post-closure Care permit has been in place for more than 10 years and is pending renewal (NASA, 2007s).

Contaminated Areas. NASA is regulated by an RCRA Part B permit. NASA entered into RCRA corrective action at WSTF and generated an RFI and Corrective Measures Study (CMS). The original requirements for the RFI and CMS work were delineated in an RCRA §3008(h) Administrative Order on Consent issued by EPA Region 6 (NASA, 2007s). As part of the RCRA corrective action at WSTF, NASA actively is investigating the extent of the effects that historic releases of chemical wastes have

This page intentionally left blank.

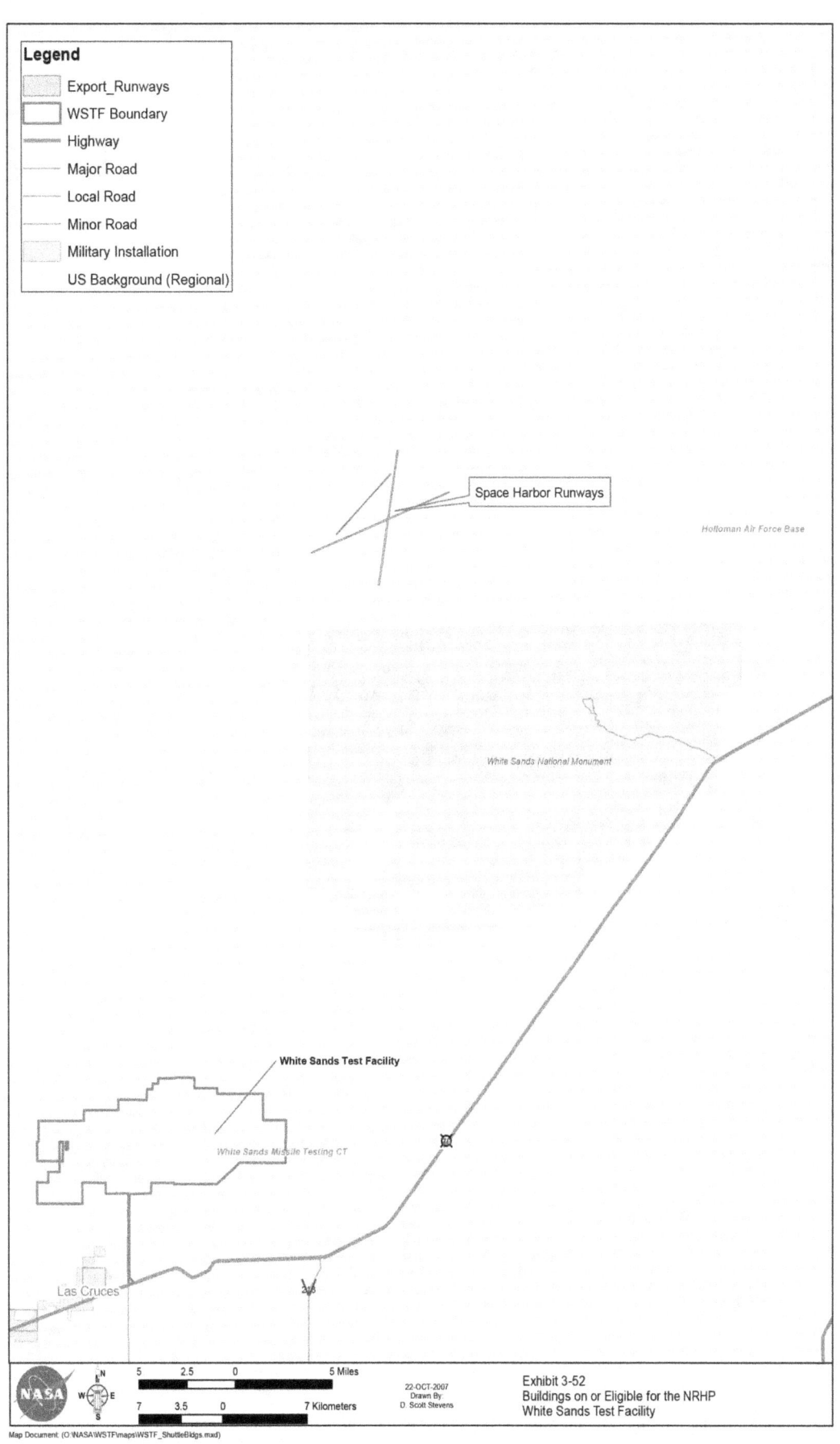

This page intentionally left blank.

had on soil and groundwater at WSTF, along with required actions and remediation technologies.

Toxic Substances.
Emergency Planning and Community Right-to-Know Act. NASA submits annual Tier II and Form R reports for chemical releases at WSTF. A Form R is reported for the release of methyl hydrazine and lead. The lead release is due to lead accumulated at the Security Department's small arms firing range (firearm certification and training) (NASA, 2007s).

Toxic Substances Control Act. WSTF has controlled PCB use in transformers, capacitors, oils, and light ballasts. Equipment that contains PBCs is replaced due to attrition, and PCB-containing wastes are disposed offsite (NASA, 2007s).

In 1988, NASA JSC issued an *Asbestos Control Manual* to provide the information, guidance, standards, and procedures necessary to implement its policy relating to asbestos-related activities. WSTF is in compliance with this manual and the applicable regulations (NASA, 2007s).

3.9.5 Health and Safety

3.9.5.1 Affected Environment

The following subsections outline WSTF's programs for protecting the health and safety of employees at WSTF and the public. Noise hazards at WSTF are outlined in Section 3.9.8.

Hazardous Materials. WSTF operates several laboratories for conducting analytical materials and hardware tests for SSP. These laboratories use hazardous materials that may include hypergols, oxygen gas, nitrogen gas, and other materials. WSTF also operates propulsion test areas, including four test stands, a control blockhouse, equipment and support buildings, instrumentation bunkers, and small office buildings. Rocket fuels and oxidizers are stored, pressurized, and transferred within these areas. The hazardous materials used and hazardous wastes generated are discussed in detail in Section 3.9.4. The degree of exposure to hazardous materials is minimized by implementing work practices and control technologies. Hazardous material spills or releases may be handled by the WSTF Emergency Services personnel (NASA, 2007s).

Buildings at WSTF contain asbestos in the form of insulation and other building material. WSTF complies with 29 CFR 1910.1001, OSHA's standard for the protection of employees from asbestos exposure.

WSTF maintains a health and safety program for all employees, and ensures that visitors are advised of potential hazards present at the facility. A staff of health and safety professionals works with the WSTF community to assist with the application of occupational and system safety requirements, to identify potential health and safety hazards, and to develop controls to protect employees and facility assets. All of this is done to ensure that personnel are fit and able to perform their assigned duties (NASA, 2007s).

Hazardous Materials Transportation Safety. Hazardous materials such as fuels, chemicals, and hazardous wastes are transported in accordance with the DOT regulations for interstate shipment of hazardous substances (49 CFR 100 through 199).

Explosions and Fire Hazards. Using certain hazardous materials, including fuels and propellants, for R&D operations presents a risk of explosions and fire hazards. There are several levels of fire protection provided to personnel, facilities, and the surrounding environment at WSTF, including the following:

- Fire-resistant construction and wide spacing of the buildings and test facilities
- Automatic fire detection and alarm systems
- Automatic suppression system in selected locations
- Level 3 hazardous material response team
- Emergency medical services
- Fire extinguishers and fire hose racks
- A 24-hour Fire Department (NASA, 2001a)

White Sands Space Harbor Operations. WSSH is used as a training area for astronauts for practicing landing and is used as a back-up landing site for the Orbiter if the conditions at KSC or EAFB are not favorable. WSSH is ideal for these training and development operations for the following reasons:

- Restricted, controlled airspace
- Year-round flying weather
- Wide-open area for various landing configurations, such as parachute or hydrofoil type equipment
- Area available for use is 160 km by 160 km (10 miles by 10 miles), with additional land available immediately adjacent (NASA, 2007s)

WSSH's remoteness decreases the potential for accidents to occur that would affect the public.

Aircraft Safety. All aircraft-related operations at EF are conducted in accordance with FAA regulations to protect the health and safety of the crew, EF employees, and the public during taxis, takeoffs, flights, and landings.

3.9.6 Hydrology and Water Quality

3.9.6.1 Affected Environment

Surface Water. Surface water is not present at WSTF. Runoff from snowmelt in the nearby mountains or rainfall generally flows for short distances across permeable alluvial fans before the water percolates downward or evaporates.

Groundwater. The Rio Grande aquifer system is the principal aquifer under WSTF. The aquifer system consists of a network of hydraulically interconnected aquifers in basin-fill deposits located along the Rio Grande Valley and nearby valleys (USGS, 2007). The area is characterized by mountain-sized tilted blocks of layered sedimentary rock. Where blocks tilted and were depressed, the depressions were filled with alluvial deposits eroded from adjacent mountains. Near the WSTF, the ground is partially underlain by Quaternary-aged alluvial deposits. The alluvial deposits generally consist of poorly sorted gravelly, silty sand and sandy silt (USACE, 1987). At WSTF, alluvial deposits are located to the western side, while impermeable limestone, andesite, and rhyolite underlie the rest of the facility (NASA, 2007s). The thickness of the alluvial basin fill is unknown in most areas, but is estimated at about 6,000 m (20,000 ft) near Albuquerque, New Mexico, and at about 600 m (2,000 ft) near El Paso, Texas (USGS, 2007).

Regulated Water and Wastewater. Domestic and industrial wastewater is discharged into clay and synthetically lined holding ponds for disposal by evaporation at WSTF. One discharge, treated water from the NASA WSTF Plume, is treated and reinjected in the ground. The NMED Groundwater Quality and New Mexico Energy, Minerals and Natural Resources Department–Oil Conservation Division operate the state's Underground Injection Control (UIC) Program under state regulations (20.6.2 New Mexico Administrative Code [NMAC]) established under the Safe Drinking Water Act (SDWA). WSTF has approved discharge plans for five wastewater systems (NMED, 2007). Injection information is listed in Exhibit 3-53.

EXHIBIT 3-53
WSTF Discharge Plans

Permit Name	Discharge Plan	Type of Discharge	Gallons per Day	Activity Number
NASA 100, 200, and 600 Areas	DP-392	Domestic	33,360	PRD20020003
NASA WSTF STGT	DP-584	Domestic	8000	PRD20020006
NASA 300 Area SASS Discharge System	DP-697	Industrial	25,000	PRD20020007
NASA WSTF 400 Area SASS	DP-1170	Industrial	16,805	PRD20030001
NASA WSTF Plume	DP-1255	Industrial	1,872,000	PRD20020005

Notes:
NASA = National Aeronautics and Space Administration
SASS = Small Altitude Simulation System
STGT = Satellite Station Ground Terminal
WSTF = White Sands Test Facility

Potable water is supplied by a groundwater well designed to serve 1,500 people. The water system identification number for the system is NM3590607 (NASA, 2007s). Routine sampling of WSTF's drinking water supply system is conducted to ensure compliance with state and federal safe drinking water regulations (NASA, 2006f).

WSTF maintains an SWP3 to document the site environmental conditions, location of waste management units, material inventory, staff training, and material-handling procedures. The SWP3 is reviewed annually, along with the individual management units, to determine the appropriate best management practices (BMPs) over time. The incorporation of the plan included a surface water site assessment for each waste management unit (NASA, 2007s).

3.9.7 Land Use

3.9.7.1 Region of Influence

The ROI for the NASA facility includes all lands controlled and operated by WSTF.

3.9.7.2 Affected Environment

NASA Land Use. The active WSTF land use is confined to six sections. Of these six, two sections (Sections 35 and 36, T20S, R3E) serve both hazardous testing and administrative and technical support purposes. Another section (Section 25, T20S, R3E) is designated for hazardous testing and three sections (Sections 1, 2, and 11, T21S, R3E) are set aside for administration and technical support. The remaining WSTF land use sections, the bulk of which are to the north and east, are to provide a safety buffer. Additionally, WSTF has the use of 5.6 km^2 (1,409 acres) of land at the water supply wells and limited use of approximately 18.8 km^2 (4,707 acres) of Bureau of Land Management (BLM) land along its southwestern border.

The current Master Plan (1994) satisfies all of WSTF's foreseeable major functional requirements and relationships. For example, it protects offsite adjacent land usage from objectionable or hazardous influences and incorporates the flexibility to accommodate current long-range planning goals and objectives.

WSTF is divided into seven major land use areas. These areas serve as a basis for the Facility Numbering System used by the installation. Generally, these areas are planned and confined to the uses indicated in the Land Use Plan and described below:

- 100 Project Control Area is set aside for installation support functions. It contains office facilities for administrative, management, drafting, procurement, environmental, program assurance, and engineering activities. It also contains special facilities for food service, vehicle and facility maintenance, emergency medical, fire fighting, and warehousing functions.

- 200 Laboratory Area provides general laboratory, data reduction and analysis facilities; and modification, checkout, and preparation facilities for propulsion system testing.
- 300 Propulsion Test Area is set aside for the facilities and systems necessary to accommodate cold flow and hot firing static testing of propulsion systems. Facilities in the area include one atmospheric, down-firing static test stand; one altitude simulation, down-firing test stand; two belowgrade structures for instrumentation and control signal conditioning equipment; a test control center; a remote command building; two shelters for equipment storage; and structures for test area support.
- 400 Propulsion Test Area also is set aside for the facilities and systems necessary for the performance of cold flow and hot firing static testing of propulsion systems. Facilities in this area include two vertical down-firing altitude simulation and one vertical down-firing atmospheric static test stands; two test stand support buildings; a test control building; and miscellaneous support facilities.
- 500 Storage Area is divided into two separate land areas. One is set aside for the facilities and systems necessary for the storage and transfer of propellants: fuels (hydrazine fuels, alcohol, and LH2) and oxidizers. The other is reserved for the storage of cryogenics and inert gases and includes storage dewars, high-pressure vessels, vaporization equipment, and various systems for the LOX, nitrogen, and gaseous nitrogen required for testing and site support.
- 700 Test Area is reserved for the performance of hazardous testing of explosives and potentially explosive materials, and for storage of solid rocket propellants.
- 800 Material Test Area contains facilities for performing tests on a variety of materials for ignition and combustion under various temperatures and pressures and in various liquid and gaseous atmospheres. WSTF's land usage has a direct effect on about 20.7 km^2 (5,190 acres) of the approximately 242 km^2 (60,500 acres) of land. This is the acreage developed or allocated for current testing, support, and administration. It does not include the buffer zones or water supply areas, because they are neither developed nor actively used for testing.

Easements and Rights-of-Way. In addition to these major land use areas, there are certain outgrants (easements) to external concerns shown in the Land Use Plan and discussed earlier in this subsection. Also, ROW easements have been established for an access road from the remote location of the developed portion of the site to the nearest public highway; and a road from the developed area of the site to the remote location of the WSTF under groundwater supply sources.

3.9.8 Noise

3.9.8.1 Affected Environment

Major sources of noise at WSTF include test operations, vehicles, heavy equipment, air handlers, aircraft, and miscellaneous activities (NASA, 2001a).

3. AFFECTED ENVIRONMENT

The loudest, and also the most infrequent, noise sources include the associated steam generation system at the 400 Area. The most common, regular, and repetitive sources of noise at WSTF include vehicular traffic, air handlers, and construction equipment. These sources of noises fall below harmful levels within 30 to 60 m (100 to 200 ft) of the source, and personnel working closer to the sources use protective hearing equipment. In addition, because of WSTF's location, layout design, and operation methodology, even the highest levels of noise from WSTF'S activities have only a minor impact on the personnel and no impact on neighboring residents. Exhibit 3-54 outlines the major sources of noise at WSTF (NASA, 2001a). The majority of the land surrounding WSTF is undeveloped, and as a result, there are no sensitive receptors. There is a 7.2-km (4.5-mile) buffer zone between WSTF's industrial area and the nearest sensitive receptor, which is a private home.

EXHIBIT 3-54
Major Sources of Noise at WSTF

Major Source	Approximate Levels Generated	Controls
Steam Generator (300 Area)	140 dBA at ejector	Distance buffer, hearing protection, and building design
Steam Generator (400 Area)	94 dBA at 200 Area	Distance buffer
Six Diesel Pumps	110 dBA at idle	Signs, distance buffer, ear protection
Vibration Laboratory Building 203	10,000 force pounds	Vibration isolation and attenuation facilities
Vehicular Traffic Noise	100-dBA maximum per vehicle	Restricted site access and speed limit onsite
Construction Noise	110 dBA	Personnel training, hearing protection, and annual hearing testing.
Air Handlers	80 dBA	Location above and away from personnel.
Miscellaneous (Open Detonation, etc.)	120 dBA	Safety procedures include immediate and protective hearing equipment.

Notes:
dBA = Decibel A-rated
Source: NASA, 2001a

These major sources of noise can be attributed to the following six principle activities:

- Test operations
- Vehicular traffic
- Heavy equipment and construction
- Building air handlers

- Aircraft movements
- Miscellaneous activities

A number of permanent and temporary measures are taken to reduce noise levels at WSTF. WSTF noise abatement measures for testing and general operation include the following:

- Property acquisition for use as a buffer zone
- Noise insulation of buildings
- Permanent noise barrier
- Proper scheduling of a specified activity to eliminate or alleviate noise impacts during critical periods

The effect of noise generated by the test stands has been minimized through engineering design and ongoing health and training programs. WSTF has a health and safety program that includes personnel training for ear plugs and muffs, the availability of protective hearing equipment, and, if necessary to protect personnel, the provision for audiometric testing and engineering changes. OSHA has outlined permissible noise exposures to ensure the protection of employees' hearing (NASA, 2001a).

3.9.9 Site Infrastructure

3.9.9.1 Region of Influence

The ROI is defined as the SSP-related facilities and the energy sources for these facilities at WSTF.

3.9.9.2 Affected Environment

Potable Water Supply. See "Hydrology and Water Quality," Section 3.9.6.

Wastewater System. See "Hydrology and Water Quality," Section 3.9.6.

Storm Water System. See "Hydrology and Water Quality," Section 3.9.6.

Energy Sources. WSTF is supplied with electricity and natural gas by WSMR. WSTF also has several electricity-generating units that support the SSP. The 300 Area Altitude Simulation System has one 31.32-MMBtu/hr Cleaver-Brooks steam boiler. The 400 Area Altitude Simulation System has three 800-hp Cleaver-Brooks steam boilers, which burn low-sulfur diesel fuel and are subject to the NSPS for small steam-generating units; three 195-hp Model 6030C Detroit diesel engines; four 645-hp Model 7163-7000 Detroit diesel engines; and one 68,000-bhp Pratt & Whitney steam generator (NASA, 2007s). The electricity-generating units onsite are covered by Air Permits #629-M-3 and #400-M-1 (NMED, 1993; NMED, 1997).

3.9.10 Socioeconomics

3.9.10.1 Region of Influence

The economic ROI for WSTF is defined as the three counties in which 99 percent of WSTF's employees live: Doña Ana, New Mexico; Otero, New Mexico; and El Paso, Texas (Exhibit 3-55). The majority (93 percent) of WSTF employees live in Doña Ana County. In FY 2006, approximately 59 percent of WSTF's expenditures were made within the 3-county region (NASA 2007a; Personal Communication, 2007f).

The City of Las Cruces is the primary population center of Doña Ana County. The demographic and economic stability of the region as a whole is firmly based on the presence of several major government installations–WSMR, Fort Bliss (including the cantonment area near El Paso and the McGregor and Doña Ana Ranges in New Mexico), and Holloman AFB–as well as New Mexico State University in Las Cruces and the University of Texas in El Paso (NASA, 2001a; El Paso Community Development, 2007).

3.9.10.2 Affected Environment

Population. In 2005, almost 975,000 people lived in the ROI, an increase of 6.3 percent from the 2000 Census. By 2010, the population in the ROI is projected to grow by nearly 14 percent overall, with the greatest change (26 percent) expected in Doña Ana County, far exceeding the projected 10-percent growth rate for the state (U.S. Census Bureau, 2006c; Doña Ana County Website, 2007).

Regional Employment and Economic Activity. Historically, the regional economy was based on ranching and agriculture, with major contributions from military and space-related work beginning in the 1950s. WSMR and Holloman AFB together provide an annual payroll of more than $255 million (military and civilian) and an economic impact of more than $485 million to the local economy (Sites Southwest & Bohannon Huston, Inc., 2005). Including salaries and local contracting, WSMR directly commits approximately $350 million per year into the economy of the region (NASA, 2007t).

In 2006, the total labor force of the ROI was more than 407,000 people, with an overall unemployment rate of 6.2 percent, higher than unemployment in the State of New Mexico (4.2 percent) and the nation (4.6 percent).

The major employers in the region are the U.S. Army at Fort Bliss (16,000 military and 7,500 civilian workers in 2007), WSMR (6,200 civil service, military, and contractor personnel in 2002), New Mexico State University (7,500 employees in 2001), and Holloman AFB (6,000 employees). WSMR and New Mexico State University together provide more than one third of the jobs in Doña Ana County (NASA, 2001a; NASA, 2007a; Sites Southwest & Bohannon Huston, Inc., 2005; NASA, 2007t).

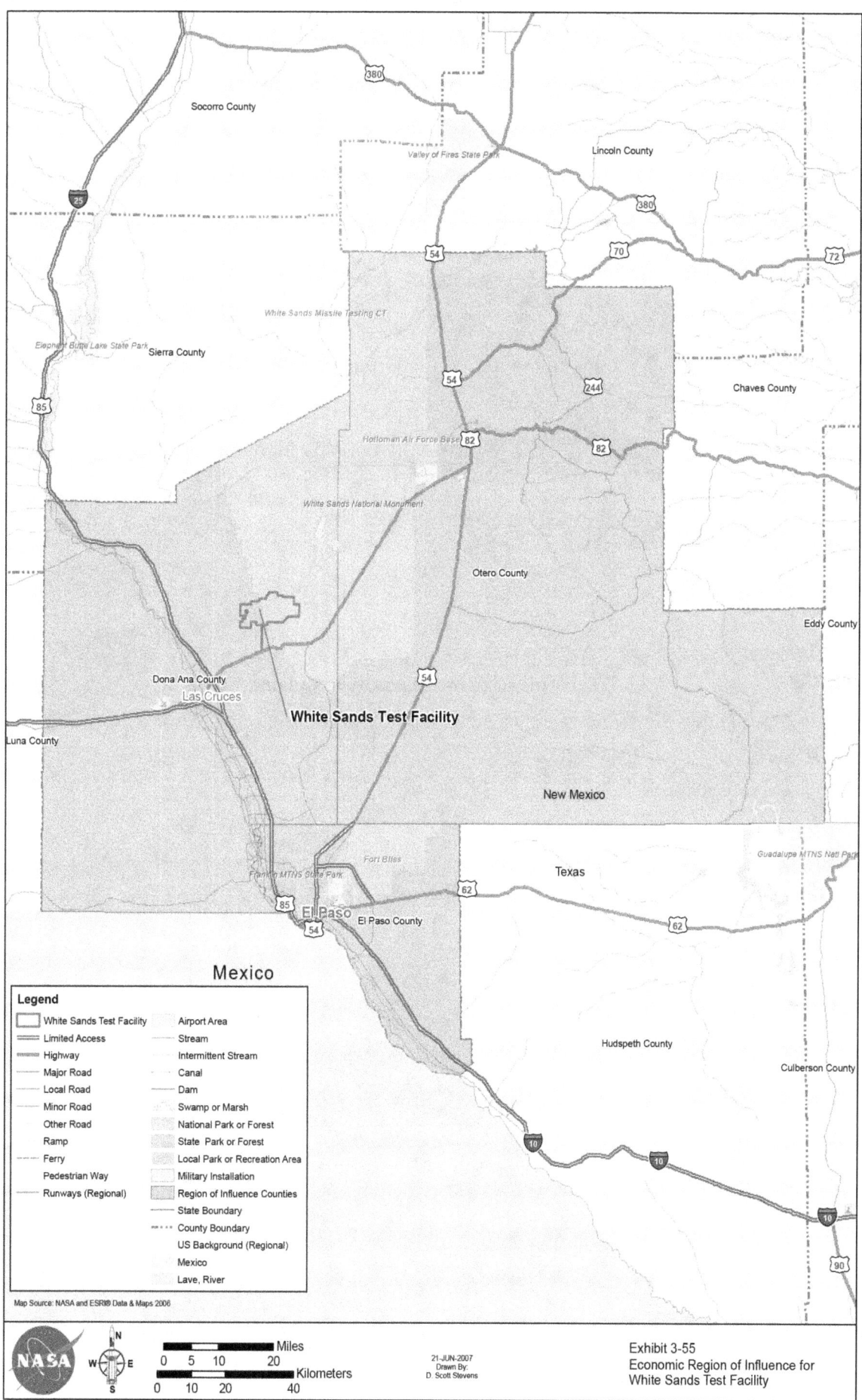

This page intentionally left blank.

Although a positive presence, WSTF is not a key contributor to the regional economy when compared to WSMR, Fort Bliss, and Holloman AFB. In 2007, WSTF employed nearly 700 civil servants and contractors, or about 11 percent of the total employment at WSMR. More than 600 WSTF employees are contractors, 350 of whom are employed by the primary test and evaluation contractor, with the rest involved in facilities operations and other support functions (Personal Communication, 2007f).

Economic Contribution of the Space Shuttle Program. Only 150 of the contractors and 20 of the civil servants at WSTF (24 percent of the workforce) are engaged in SSP-related work. In FY 2006, the SSP put nearly $18.2 million into the regional economy, including civil service and prime contractor salaries and non-payroll procurements to subcontractors and suppliers. Those expenditures generate additional economic output, jobs, and income in supporting industries within the 3-county ROI. The total (direct plus indirect and induced[12]) effect of the SSP on economic output was approximately $54 million (less than 1 percent of the $28 billion[13] in overall economic activity in the 3-county region), $23 million in earnings, and 440 jobs (NASA, 2007cc).

3.9.11 Solid Waste

3.9.11.1 Affected Environment

NASA operated a solid waste landfill at WSTF from 1965 to 1998. The total waste volume is estimated to be 78,000 cubic yards. Solid wastes consisting primarily of cardboard; office, shop, and nonhazardous laboratory wastes; construction and demolition debris; and waste generated by a central cafeteria were disposed at the landfill. An NOI to close the landfill was submitted to NMED in February 1998. Upon the completion of extensive closure activities and NMED's inspection, the 30-year closure period began on August 14, 1998. Closure activities included the installation of a geosynthetic clay liner and cover, grading for proper drainage, and natural seeding for revegetation of the site. NASA currently monitors groundwater and methane at the site to evaluate whether releases to the environment have occurred. Inspections are conducted quarterly for cover integrity, erosion, vegetative cover, and fence integrity (NASA, 2007s). The solid waste is disposed at Corralitos Landfill west of Las Cruces, New Mexico.

3.9.12 Traffic and Transportation

WSTF is located on the DA's WSMR.

---

[12] Based on the "multiplier effect" using economic multipliers for the WSTF ROI from the U.S. Bureau of Economic Analysis.

[13] U.S. Bureau of the Census, 2002 Economic Census—Total sales, shipments, receipts, or revenue for all establishments (2-digit NAICS codes) in the 3-county WSTF ROI.

3.9.12.1 Region of Influence

WSTF is located on WSMR. The transportation ROI for WSTF is defined as the 3 counties in which 99 percent of WSTF's employees live–Doña Ana, New Mexico; Otero, New Mexico; and El Paso, Texas. The majority (93 percent) live in Doña Ana County (NASA, 2007a). Each of these 3 counties is also an MSA, made up of a central city and the surrounding commuting area: the Las Cruces, New Mexico, MSA; the Alamagordo, New Mexico, MSA; and the El Paso, Texas, MSA (OMB, 2006).

3.9.12.2 Affected Environment

Transportation. I-10, I-40, and I-25 provide interstate access to WSTF. I-10 passes approximately 80 km (50 miles) south of the Main Post, with exits to WSMR at El Paso, Texas, and Las Cruces, New Mexico. I-40 passes east–west through the northern half of New Mexico and intersects with I-25 at Albuquerque, approximately 160 km (99 miles) north of WSMR. I-25 provides a north–south interstate connection to WSMR, with local exits at San Antonio (approximately 27 km [17 miles] from the Stallion Gate) and Las Cruces, New Mexico (approximately 36 km [22 miles] from the Las Cruces Gate).

Other major highways serving WSMR are U.S. 380, 70, and 54. U.S. 380 passes along the northern boundary of WSMR between San Antonio and Carrizozo and connects with I-25 at San Antonio. U.S. 70 crosses the southern half of WSMR between Las Cruces and Alamogordo with an exit 8 km (5 miles) north of the Main Post. U.S. 54 runs a parallel course along the entire eastern boundary of WSMR between Carrizozo and El Paso.

Access Roads to WSMR. There are seven primary entry points onto WSTF. Access to the entry points is provided by U.S. 380 at the Stallion Gate, U.S. 54 at the Tularosa and Oro Grande gates, U.S. 70 at the Las Cruces and Small Missile Range gates, Rural Route (RR) 10 at the Hollomon Gate, and RR 1 at the El Paso Gate. The entry points at the Stallion, Las Cruces, and El Paso gates are manned by security personnel and opened to approved traffic 24 hours a day (Exhibit 3-56). The Holloman and Oro Grande gates currently are unmanned and access is controlled by combination-type locks.

Railroads. The Southern Pacific rail line runs north-south along the entire length of WSMR's eastern boundary, with a railhead at Orogrande Range Center, Fort Bliss. RR 2 provides road access from the railhead to the Main Post and other parts of WSMR. The railhead serves as the primary delivery point for tanks and other heavy equipment to and from WSMR.

Airports. EPIA, Las Cruces International, and Alamogordo-White Sands Regional airports provide private, corporate, and/or commercial air facilities within 60 km (37 miles) of the Main Post. EPIA serves approximately 400 private aircraft and supports 160 daily arrivals and departures by commercial airlines. White Sands

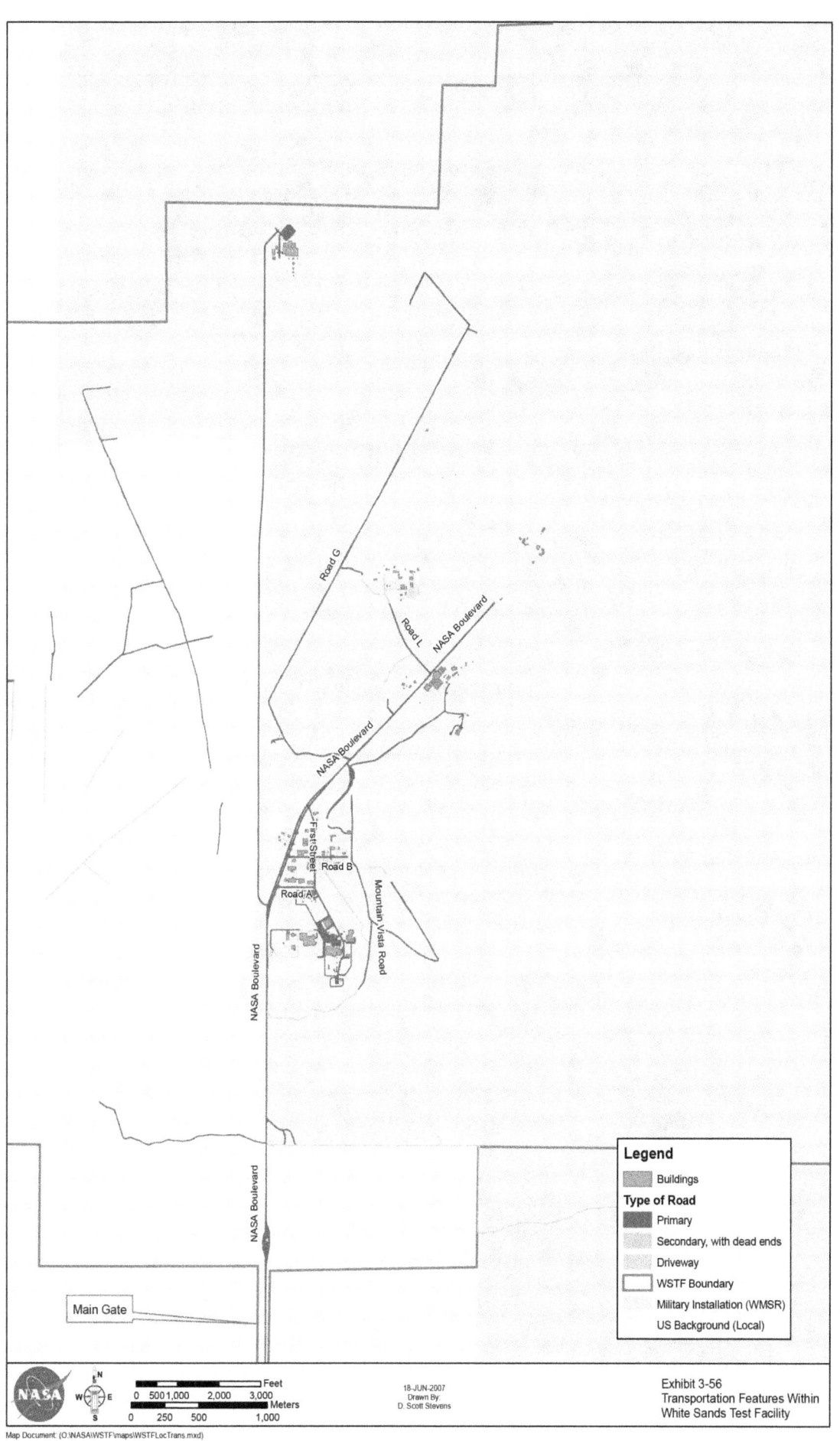

3. AFFECTED ENVIRONMENT

This page intentionally left blank.

Regional Airport is used primarily for private aviation, but is served by Mesa Airlines with daily flights to Albuquerque. Municipal airports that have general aviation facilities are available at Truth or Consequences and Socorro.

3.10 Dryden Flight Research Center

DFRC is a tenant organization on EAFB, which is located in the Antelope Valley region of the western Mojave Desert, approximately 105 km (65 miles) north north-east of Los Angeles, 108 km (67 miles) north north-west of San Bernardino, and 113 km (70 miles) east-southeast of Bakersfield, California (Exhibit 3-57). DFRC leases three locations with an area of approximately 339 ha (838 acres) in Kern County, on the shore of Rogers Dry Lake Bed, which currently is used as an emergency landing area.

DFRC is an aeronautical research facility developing new technologies to improve aircraft flight control components and systems and to transfer new concepts to the U.S. aerospace industry for commercial and military applications. DFRC's mission is to provide world-leading flight research in the following areas:

- Aerospace flight research and technology integration to revolutionize aviation and pioneer aerospace technology
- Development and operations of the remaining Space Shuttle missions, ISS, and future space vehicles
- Airborne remote sensing and science missions, flight operations, and development of piloted and uninhabited aircraft test-beds for research and science missions
- Validation of space exploration concepts

DFRC is the alternative landing site for the Space Shuttle when weather conditions at KSC prohibit landing. The Shuttle landing support includes preparations for the Shuttle's return to KSC.

3.10.1 Air Quality

3.10.1.1 Affected Environment

Regional Climate. The facility is in the northwestern portion of the Mojave Desert Air Basin (MDAB). The MDAB, one of the largest air basins in the state, encompasses the desert portions of San Bernardino and Riverside counties, the eastern parts of Kern and Los Angeles counties, and all of Imperial County. The Mojave Desert is sheltered from maritime weather influences of the Pacific Ocean by mountain barriers extending from north to south. The climate in this area is considered to be a continental desert regime (NASA, 2003c).

A review of the historical meteorological data from EAFB shows the prevailing wind direction to be from the west-southwest throughout the year, with an average wind

This page intentionally left blank.

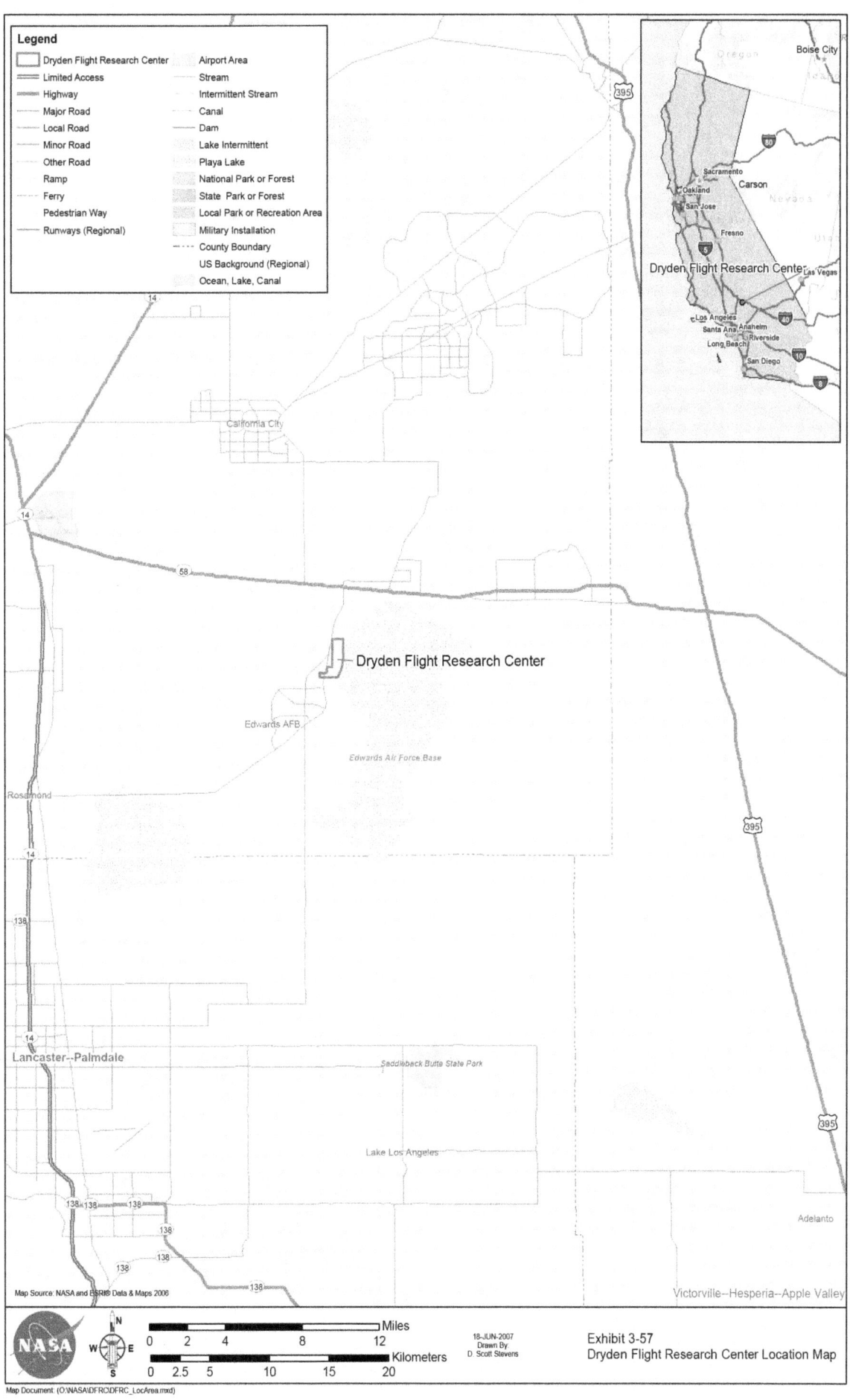

This page intentionally left blank.

speed of 8 mph. Atmospheric stability, the measure of the vertical dispersion of pollutants, is high at EAFB. Stable conditions, which are an indication of weak pollutant dispersion, exist about 57 percent of the time, indicating that the pollution potential in the area is relatively high (NASA, 2003c).

Temperature fluctuations have an important influence on basin wind flow, dispersion along mountain ridges, vertical mixing, and photochemistry. Precipitation is highly variable seasonally and summers are often completely dry. In the winter, an occasional storm from the high latitudes sweeps across the coast, bringing rain or snow (NASA, 2003c).

Average extreme monthly maximum temperatures range from 113°F in July to 4°F in January. The annual mean temperature is 62°F, with monthly average temperatures ranging from 82°F in July to 44°F in December and January (NASA, 2003c).

Precipitation is light, averaging about 5 inches annually. Approximately 96 percent of the annual rainfall occurs during the 6-month period from November through April. Thunderstorms infrequently occur, averaging only 10 days per year (NASA, 2003c).

Emission Sources. DFRC operates air emissions sources under a synthetic minor permit. DFRC is located in an ozone non-attainment area. Therefore, the facility falls under the NSR program and the General Conformity Rule. DFRC is in an attainment area for all other criteria pollutants (NASA, 2007i). DFRC also is located in a nonattainment area under the state ozone, $PM_{2.5}$, and PM_{10} standards (California Air Resources Board [CARB], 2007).

The following permitted sources at DFRC are used by the SSP: 16 electric generators; 16 emergency-use piston engines; 6 aircraft air conditioning units; 1 hydrazine vapor scrubber; 1 nitrogen vapor scrubber; and 1 degreasing operation (NASA, 2007i).

3.10.2 Cultural Resources

3.10.2.1 Affected Environment

Archaeological Resources. As a result of a cultural resources survey completed in 1996, four archaeological sites were discovered on DFRC's Buckhorn Ridge (NASA, 2003c:95). The unevaluated sites range from likely prehistoric flint knapping activities to a cabin site related to mining. None of these sites have been evaluated to determine NRHP eligibility. To protect the resources from vandalism, it is customary not to publish the exact locations of the sites. There are currently no archaeological sites listed, nor eligible for listing, in the NRHP within the boundaries of DFRC.

Historic Resources. Building 4802 at DFRC has been nominated for inclusion in the NRHP because of the aircraft design and testing activities that occurred in the structure. Eleven resources were surveyed to determine their significance to the SSP. Of those, only Building 4860, the Mate-Demate Device, was determined to be eligible for NRHP listing for its contributions to the SSP (Archaeological Consultants, Inc., 2007f). The DFRC ERD also lists the X-1E aircraft as a "historic landmark," but this could not be verified as an NHL property (NASA, 2003c:95-97; NASA, 2007s:28).

Exhibit 3-58 shows the properties listed and eligible for listing on the NRHP at DFRC.

3.10.3 Hazardous and Toxic Materials and Waste

3.10.3.1 Affected Environment

Storage and Handling.

Hazardous Materials. Five ASTs located at DFRC are used to store jet fuel (Jet Propellant [JP]-8 and Jet Propulsion Thermally Stable [JPTS]), unleaded gasoline, and diesel fuel. The ASTs are equipped with secondary spill protection for the tanks, valves, and pumping system. The central Chemical Crib is used to distribute other materials and chemicals directly to the workplaces where they are used (NASA, 2003c).

The Hazardous Materials Management System is used to collect and manage chemical usage data. EPA requires reporting of chemical usage that exceeds certain thresholds through the TRI reporting program. According to the chemical usage data collected, DFRC's use of each chemical is below the 4,500-kg (10,000-pound) annual use threshold for TRI reporting (NASA, 2003c).

All unused hazardous materials are returned to the Chemical Crib when finished. Additionally, a reduction in the quantity of hazardous material, or replacement of hazardous materials with nonhazardous materials, is required for the organizations at DFRC (NASA, 2003c).

Hazardous Waste. NASA is an LQG of hazardous waste at DFRC. EPA's Environmental and Compliance History Online (ECHO) website indicates that no compliance violations have been associated with this permit from 2003 to date. These wastes are stored at a 90-day accumulation point before disposal. After hazardous waste is characterized, it temporarily is stored in the Hazardous Waste Shipping Building for less than 90 days before disposal. Hazardous waste is then hauled by a certified transporter to a permitted disposal facility under a Uniform Hazardous Waste Manifest. Wastes generated by SSP activities include personal protective equipment (PPE) and neutralized fuel (NASA, 2003c).

Exhibit 3-58
Buildings on or Eligible for the NRHP
Dryden Flight Research Center

Waste Management.

Hazardous Materials. DFRC is responsible for managing the hazardous materials and waste associated with the flight research operations.

The DFRC flight research operations and industrial operations involve the use and storage of hazardous materials associated with aircraft repair and maintenance, ground equipment repair and maintenance, and vehicle maintenance and painting. The most commonly used hazardous materials at DFRC include jet fuels, gasoline and diesel motor vehicle fuels, lubricating oil, paints, thinners, cleaners, solvents, strippers, antifreeze, and refrigerants (NASA, 2003c).

Hazardous Waste. DFRC implements a Chemical Management Plan to maintain compliance with the RCRA requirements. The plan establishes policies, responsibilities, and procedures for hazardous waste management (NASA, 2003c).

DFRC's Safety, Health and Environmental Office (Code SH) performs all waste collection, identification, and disposal. On a weekly basis, the DFRC collects waste from five designated locations, each with containers for waste oils, contaminated materials, empty containers, and empty aerosol cans. In addition, Code SH provides work-site pick-up service for hazardous materials within 3 days after receiving a request (NASA, 2003c).

The disposal services for ordnance and explosives are provided for DFRC by EAFB. The general types of material include smoke grenades, signal kits, rocket launchers, and various ejection cartridges and initiators. EAFB treats explosive ordnance in the open burn thermal treatment facility, and another facility is an open detonation thermal facility. The wastes from these treatments are then characterized to establish the appropriate disposal method (NASA, 2003c).

Contaminated Areas. EAFB, and consequently DFRC, was listed on the NPL in 1990. The Environmental Restoration Program (ERP) at EAFB serves to identify, investigate, assess, and clean up hazardous waste in compliance with CERCLA. The former disposal and release sites at DFRC are being addressed under this ERP. As part of the ERP, a Preliminary Assessment was performed that located potential AOCs resulting from past activities at DFRC (NASA, 2003c).

A total of 471 ERP sites and AOCs that have potential contamination have been identified at EAFB. These ERP sites are grouped into 10 operable units (OUs) that generally are based on geographic locations. Contamination at DFRC associated with OU-6 consists of TCE in the groundwater; also, benzene, toluene, ethylbenzene, and xylenes (BTEX) contamination was identified at the location of the former gas station. The contamination is not a result of SSP operations and is not present at the Shuttle operations area (NASA, 2007s).

Asbestos and Lead-based Paint. Asbestos or LBP potentially are present in SSP buildings at DFRC.

3.10.4 Health and Safety

3.10.4.1 Affected Environment

DFRC is an alternate landing site for the Space Shuttle in instances when weather does not permit a landing at KSC. SSP operations at DFRC include the operation of GSE and maintenance of hangars for storing the Shuttle should it have to land at DFRC. The following subsections outline DFRC's programs for protecting the health and safety of employees at DFRC and the public. Noise hazards at DFRC are outlined in Section 3.10.7.

Hazardous Materials. Hazardous materials are used to maintain the aircraft used for training at DFRC. The hazardous materials used and hazardous wastes generated are as discussed in detail in Section 3.10.3. The degree of exposure to hazardous materials is minimized by the implementation of work practices and control technologies. The DFRC S&MA Directorate is responsible for training employees to handle the hazardous chemicals kept at their worksites (hazard communications) and for implementing appropriate spill response protocols in case of emergencies. Hazardous material spills or releases may be handled by a DFRC team and/or a team from EAFB, depending on the location and severity of the incident (NASA, 2003a). All activities associated with the protection of workers who use hazardous materials are conducted in accordance with the OSHA Hazardous Communication Regulations. Dryden Policy Directive (DPD)-8700.1 outlines the organizational and individual safety responsibilities for minimizing mishaps and close-call events at DFRC (NASA, 2005a). Emergency response operations associated with the release of hazardous substances are conducted in accordance with OSHA's Hazardous Waste Operations and Emergency Response Standards (HAZWOPER) under 29 CFR 1910.120.

Buildings at DFRC contain asbestos in the form of insulation and other building materials. DFRC complies with the 29 CFR 1910.1001 standard for the protection of employees from asbestos exposure.

Hazardous Materials Transportation Safety. Hazardous materials such as fuels, chemicals, and hazardous waste are transported in accordance with the DOT regulations for the interstate shipment of hazardous substances (49 CFR 100 to 199).

Explosions and Fire Hazards. Using certain hazardous materials, including fuels, in aircraft training operations presents a risk of explosions and fire hazards. DFRC maintains the Aircraft Maintenance and Safety Manual (AM&SM), a Center-wide procedure that includes protocols for fire prevention and handling fires. The hangars at DFRC are equipped with automatic fire detection systems that have sprinkler and foam systems (NASA, 2005d). EAFB provides fire protection services to DFRC under an alliance formed between the USAF and NASA (NASA, 2003c).

DFRC also maintains a policy to implement the use of electric-powered units equipped with explosion-proof motors, whenever possible, in preference to gasoline- or diesel-powered equipment to prevent explosions and fire hazards (NASA, 2005d).

Range Safety. The DFRC Range Safety Program addresses health and safety issues associated with the risk of a piece of a vehicle coming off the aircraft during flight (NASA, 2005e). DFRC maintains a Range Safety Program, as specified in DPD-8715.1. DPD-8715 applies to all unmanned aerial vehicles (UAVs) and Experimental Aerospace Vehicle programs and projects for which DFRC is responsible for Range Safety. It also applies to all programs and projects for which there is a planned release of an unmanned object in flight or for which there is an accepted risk that a piece of the vehicle or test article could come off an aircraft during flight (NASA, 2005e).

Aircraft Safety. All aircraft-related operations at DFRC are conducted in accordance with FAA regulations to protect the health and safety of the crew, DFRC employees, and the public during taxis, takeoffs, flights, and landings.

3.10.5 Hydrology and Water Quality

3.10.5.1 Region of Influence

The ROI for hydrology and water quality is the boundaries of DFRC.

3.10.5.2 Affected Environment

Surface Waters. There are no perennial streams at DFRC. Nonjurisdictional surface water resources include storm water drainages, flood-prone areas, imported surface water, artificial ponds, dry lakes, and ephemeral streams.

Storm water is conveyed through a system of drainage ditches, with Rogers Dry Lake serving as the terminus for the storm water runoff. There is no outlet from Rogers Dry Lake; water either evaporates or infiltrates into the ground. Rogers Dry Lake floods most winters. Once flooded, the lakebed tends to remain inundated for the winter because of the low permeability of the lakebed soils. Most development at DFRC is above the estimated flood of record level of 683 m (2,277.4 ft). However, a small portion of the NASA ramp is located below that elevation (NASA, 2007s).

Groundwater. Basin-fill sediments in the Antelope Valley constitute a vast groundwater basin. The basin is a single, undrained, closed system divided into three aquifers. The aquifers include a shallow unconfined aquifer (the upper aquifer), which is thin and generally unproductive; a deeper and thicker confined aquifer (the middle aquifer), which contains the majority of the groundwater; and the deepest confined aquifer (the lower aquifer), which is thinner and produces less water than the middle aquifer. The upper aquifer is termed the principal aquifer and the middle and lower aquifers collectively are known as the deep aquifer.

3. AFFECTED ENVIRONMENT

These aquifers consist of poorly consolidated, variably sorted beds of clay, silt, sand, and gravel separated by layers of fine-grained material and thick lacustrine deposits (NASA, 2007s).

Regulated Water Discharges and Withdrawals. Water at DFRC is supplied by EAFB. EAFB sources include groundwater, surface water, and reclaimed wastewater. Groundwater use at EAFB varies from 3.1 to 5.3 mgd. Groundwater is supplied by 15 groundwater wells located on base; of these, 10 provide drinking water. The 10 drinking water wells have a combined capacity of 18.24 mgd.

EAFB purchases surface water from the Antelope Valley-East Kern Water Agency (AVEK). An AVEK supply line runs north of the Base. AVEK requires that EAFB purchase a minimum of 7.6 million cubic kilometers per day (km^3)/d (2 mgd) and limits the base's maximum allotment to 4 mgd. However, there are no guarantees on this supply of water. Exports from the source, the Sacramento-San Joaquin Delta, may be reduced in the future (NASA, 2007s).

Wastewater at DFRC is treated by EAFB. The California Water Quality Control Board, Lahontan Region, regulates wastewater treatment and discharge at EAFB. The Main Base WWTP has an average daily flow capacity of 9.5 million km^3/d (2.5 mgd), with a peak daily flow of 1.5 million km^3/d (4 mgd). There is no direct discharge of effluent. The discharge from the plant is transferred to evaporation ponds during the non-irrigation season. During the irrigation season, effluent is transferred to a reclaimed water system, with excess flows going to the evaporation ponds. Treated wastewater effluent is used for some urban landscape irrigation and as a water source for some artificial ponds. The largest user of reclaimed water is the golf course (NASA, 2007s).

Storm water is handled at DFRC via a general permit under EAFB. The storm water runoff collection system comprises drainage ditches and some storm drains that generally flow eastward into Rogers Dry Lake or a storm water retention pond located on its western edge. The EAFB Bioenvironmental Engineering Office samples storm water runoff at 11 sampling points under the terms of the general permit (EAFB, 2003).

3.10.6 Land Use

3.10.6.1 Affected Environment

Dryden Flight Research Center Land Use. DFRC facilities are located in areas designated by the 1994 EAFB Comprehensive Plan as Engineering Test. This land use category is unique to EAFB; permissible uses include specialized facilities for the ongoing evaluation of flying machines and their auxiliary equipment (NASA, 2003c). In 1996, DFRC created a Facilities Master Plan that set out land use plans and policies to ensure that DFRC's land use is aligned with EAFB's land use designations. The planning objectives listed in the 1996 Facilities Master Plan are as follows (NASA, 2003c):

- Ensure that DFRC's long-term mission is supported by the provision of required facilities and land use areas
- Use real estate efficiently
- Ensure compatible land use relationships
- Avoid sprawl of new facilities and land development
- Increase operational efficiency by collocating and/or consolidating functionally related facilities to lessen the burden on maintenance and infrastructure; develop optimum land use that respects manmade and natural constraints

3.10.7 Noise

3.10.7.1 Affected Environment

The major noise source of noise at DFRC is aircraft operations, including fixed and rotary wing air traffic. Noise at ground level around the runways used by DFRC has been measured between 65 and 85 dBA during flight operations. The ground crew at DFRC is required to wear hearing protection. Aircraft noise levels are lower at DFRC than at EAFB, where noise levels can range from 84 dBA to 125 dBA. The EAFB flight line has been designated as a hazardous noise area. The closest sensitive receptor to DFRC is the West Housing area of EAFB, approximately 4.8 km (3 miles) away. The housing area is shielded from noise levels by the Bissell Hills land feature, and noise levels resulting from aircraft operations at DFRC are not expected to exceed the background noise levels (NASA, 2003c).

3.10.8 Site Infrastructure

3.10.8.1 Affected Environment

Potable Water Supply. Potable water at DFRC meets EPA's drinking water standards. In addition to metering USAF water deliveries, DFRC's water usage also is metered (NASA, 2003c). EAFB supplies potable water to DFRC, derived from 10 groundwater wells and surface water from AVEK. There are no guarantees on the supply of surface water because exports from the Sacramento–San Joaquin delta may be reduced in the future. In addition, population growth projected for other communities that are supplied may increase demands and reduce the amount of imported water available to the Antelope Valley, including DFRC (NASA, 2003c).

Wastewater System. See "Hydrology and Water Quality," Section 3.10.5.

Storm Water System. See "Hydrology and Water Quality," Section 3.10.5.

Energy Sources. The EAFB electrical distribution system has more than 800 km (500 miles) of aboveground primary lines and more than 480 km (300 miles) of underground primary lines. The present system has a maximum demand capacity of 79 megawatts (MW), nearly double the 44-MW maximum demand registered for

EAFB. The Western Area Power Authority sends a standard allotment of power to EAFB. Requirements beyond that allotment require purchase from Southern California Edison.

There are 51 real property installed equipment emergency power generators of varying output (with a total capacity of 11,545 kW) located throughout the installation. Powered by diesel fuel, natural gas, or propane, these units start, transfer power, and stop automatically during power outages. In addition, there are 29 mobile electric power units, ranging from 5 to 200 kW (with a total capacity of 930 kW), stored and maintained by EAFB's Civil Engineer (NASA, 2003c).

Pacific Gas and Electric delivers gas to EAFB through 30-inch and 6-inch high-pressure gas lines. Industrial use consumes approximately 45 percent of the demand, with commercial and residential uses consuming the remainder (NASA, 2003c).

3.10.9 Solid Waste

Solid waste disposal in California is regulated by California State Regulation, Title 27, Division 2. CR 27, Division 2, specifies the requirements for the treatment, storage, processing, and disposal of solid waste in California.

3.10.9.1 Affected Environment

Non-RCRA wastes (wastes not considered hazardous under RCRA) generated at DFRC include scrap metal, used motor oil, drained used oil filters, and spent fluorescent light tubes. These non-RCRA wastes are disposed offsite by recyclers and reclaimers. Other municipal-type solid wastes and nonhazardous wastes generated on EAFB are disposed in the EAFB landfill (NASA, 2003c).

The Main Base Landfill at EAFB operates a nonhazardous waste (municipal solid waste) landfill. This permitted Class III landfill is owned and operated by the USAF and is not open to the public. No other operational nonhazardous solid waste landfills are present at EAFB. In 1999, DFRC generated approximately 11,000 to 14,000 kg (25,000 to 30,000 pounds) of municipal solid waste; 4,500 to 6,800 kg (10,000 to 15,000 pounds) were recycled. In 2001, the Main Base Landfill accepted approximately 8,327 metric tons (8,195 tons) of solid waste. The landfill is expected to reach its permitted capacity in the year 2028 (NASA, 2003d).

Because of the volume of construction and demolition waste generated on EAFB, most current construction contracts require the contractor to dispose of such wastes at an approved off-Base landfill to reduce the impacts to the Main Base Landfill (Air Force Flight Test Center [AFFTC], 2002).

3.10.10 Traffic and Transportation

3.10.10.1 Region of Influence

DFRC is located on EAFB in Kern County, California. The transportation ROI for DFRC is defined as the counties from which 98 percent of the people employed in Kern County commute–Kern County, Los Angeles County, and Tulare County, California (U.S. Census Bureau, 2000b).

3.10.10.2 Affected Environment

Transportation. U.S. Highway 395, SR 58, and SR 14 connect EAFB to the local communities and the interstate highway system. U.S. Highway 395 parallels the eastern boundary and leads to I-15, 64 km (40 miles) to the south near Victorville. SR-58 parallels the northern boundary and provides a connection 80 km (50 miles) to the east to I-15. To the west, it connects to I-5, 123 km (77 miles) to the west through Mojave, Tehachapi, and Bakersfield to I-5. SR-14 parallels the western boundary intersecting SR-58 at Mojave at the northwestern corner of the installation. From there, it leads south through Lancaster and Palmdale to I-5, 85 km (53 miles) to the south.

Access Roads to DFRC. Because of the sparse population and undeveloped desert character of the DFRC area, few improved roadways occur immediately adjacent to DFRC other than those supporting DFRC or EAFB (Exhibit 3-59). Most of the area's county and local roads are aligned with the land survey system section lines, with named avenues running east and west and numbered streets running north and south along the section lines. Many of the local roads adjacent to the base boundary are simply tracks in the desert, subject to infrequent use and little, if any, maintenance.

The principal paved local roadways adjacent to the base are two-lane rural asphalt roads and include Sierra Highway, parallel to and just west of the base western boundary; East Avenue E, along the southern boundary; and 120th Street East, 140th Street East, and 200th Street East. 120th Street becomes Lancaster Boulevard on the base, 140th Street intersects Mercury Boulevard to provide access to the southeastern portion of the base, and 200th Street parallels a portion of the southeastern base boundary.

Railroads. Freight service is provided to the installation by the Burlington Northern Santa Fe (BNSF) Railroad. A rail spur connects the BNSF main line to the government-owned trackage servicing the Main Base. The West Gate is located on Rosamond Boulevard approximately 14 km (9 miles) from the western boundary.

This gate handles approximately 42 percent of all Base traffic. The South Gate is located on Lancaster Boulevard approximately 3.2 km (2 miles) from the southern boundary. This gate handles approximately 38 percent of all Base traffic. The North

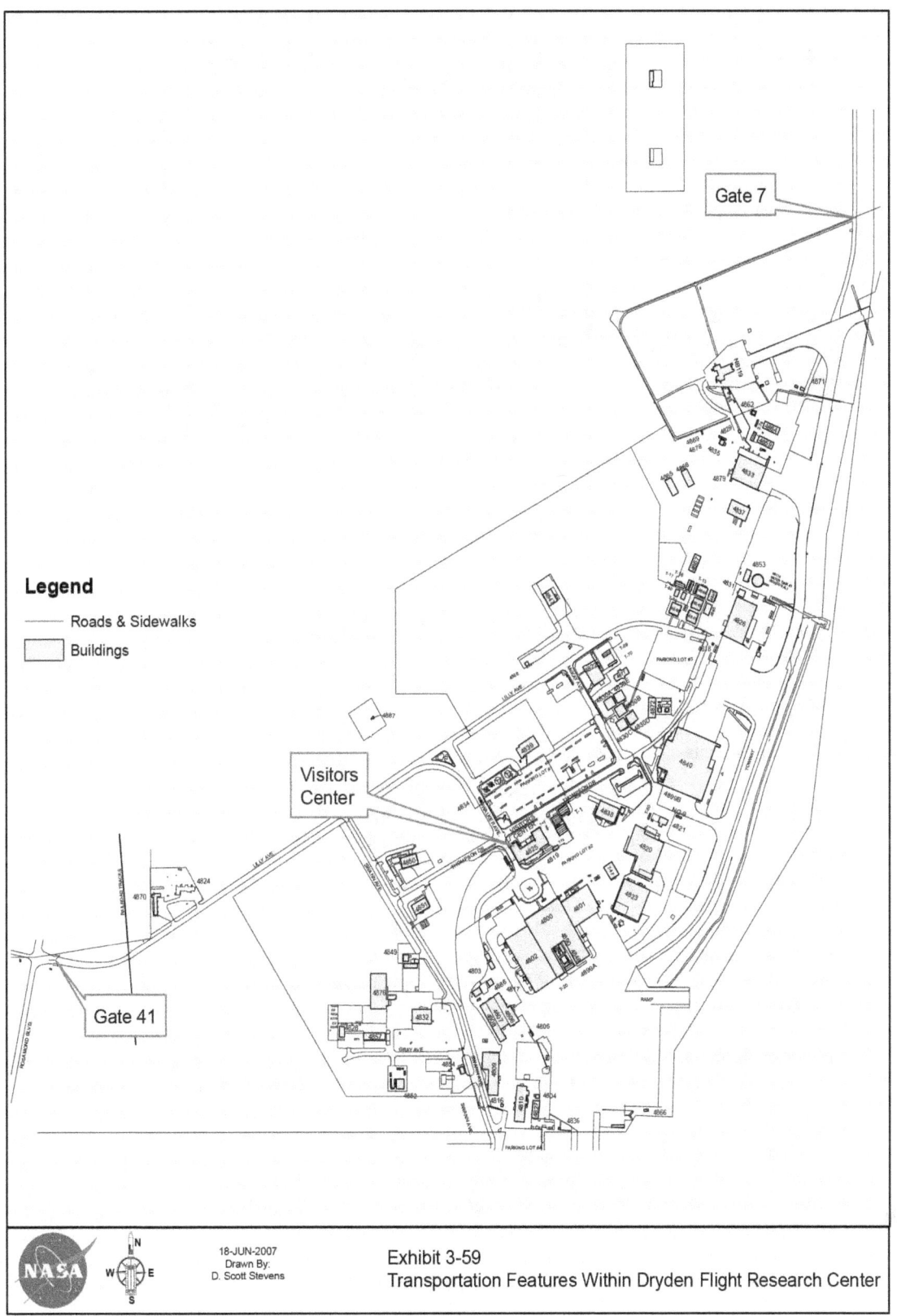

Exhibit 3-59
Transportation Features Within Dryden Flight Research Center

Gate is located on Rosamond Boulevard at the northern boundary. This gate handles approximately 20 percent of all Base traffic.

Airports. Palmdale AFP NR 42 Airport is located approximately 40 km (25 miles) to the south of the AFP 42.

Transit. No direct bus service to DFRC is available.

3.11 Palmdale

Palmdale (also known as AFP 42 Site 1), is a tenant of Wright-Patterson AFB, is owned by the USAF, leased by NASA, and operated by Boeing Company. Palmdale is located on 0.15 km² (36.5 acres), with 0.1 km² (25 acres) fenced (Exhibit 3-60). Palmdale consists of 4 permanent buildings, 14 modular buildings, and 14 minor buildings, and has been dedicated to support the SSP since 1972. NASA has access to runways located on the base.

NASA owns Buildings 154, 151C, 163, 164, 165, 171, 172, and 173. The other buildings on the site are owned either by USAF or Boeing. USAF is responsible for the land at Palmdale. Activities at Palmdale include aerospace vehicle testing and assembly, thermal protection fabrication and installation, plaster pattern shop and foam shop activities, TCS blanket fabrication, electrical wire harness fabrication, new generation tile fabrication, phantom works, and R&D.

3.11.1 Air Quality

3.11.1.1 Affected Environment

Regional Climate. The average warmest month at Palmdale is July, with an average high temperature of 97°F and an average low temperature of 66°F. The highest recorded temperature was 113°F in 1972. The average coolest month is December, with an average high temperature of 59°F and an average low temperature of 33°F.

The lowest recorded temperature was 6°F in 1963 (TWC, 2007c). The maximum average precipitation occurs in February, with a monthly average of 1.69 inches. The minimum average precipitation occurs in June and July, which each has a monthly average of 0.06 inch (TWC, 2007c).

Emission Sources. Boeing holds no Title V Air Permits for its operations at Palmdale, because the EAFB Title V Air Permit covers all base operations. Palmdale is not a major source of HAPs regulated by NESHAPs. The compliance certification is returned to the USAF for its files because the USAF has a certification from EPA that covers all base operations (NASA, 2007s). Palmdale is classified as non-attainment for 8-hour ozone, CO, PM_{10}, and $PM_{2.5}$ (EPA, 2007a). Therefore, Palmdale follows the NSR program and also must evaluate all new projects under the General Conformity rule. Palmdale also is classified as non-attainment under the state ozone, $PM_{2.5}$, and PM_{10} standards (CARB, 2007).

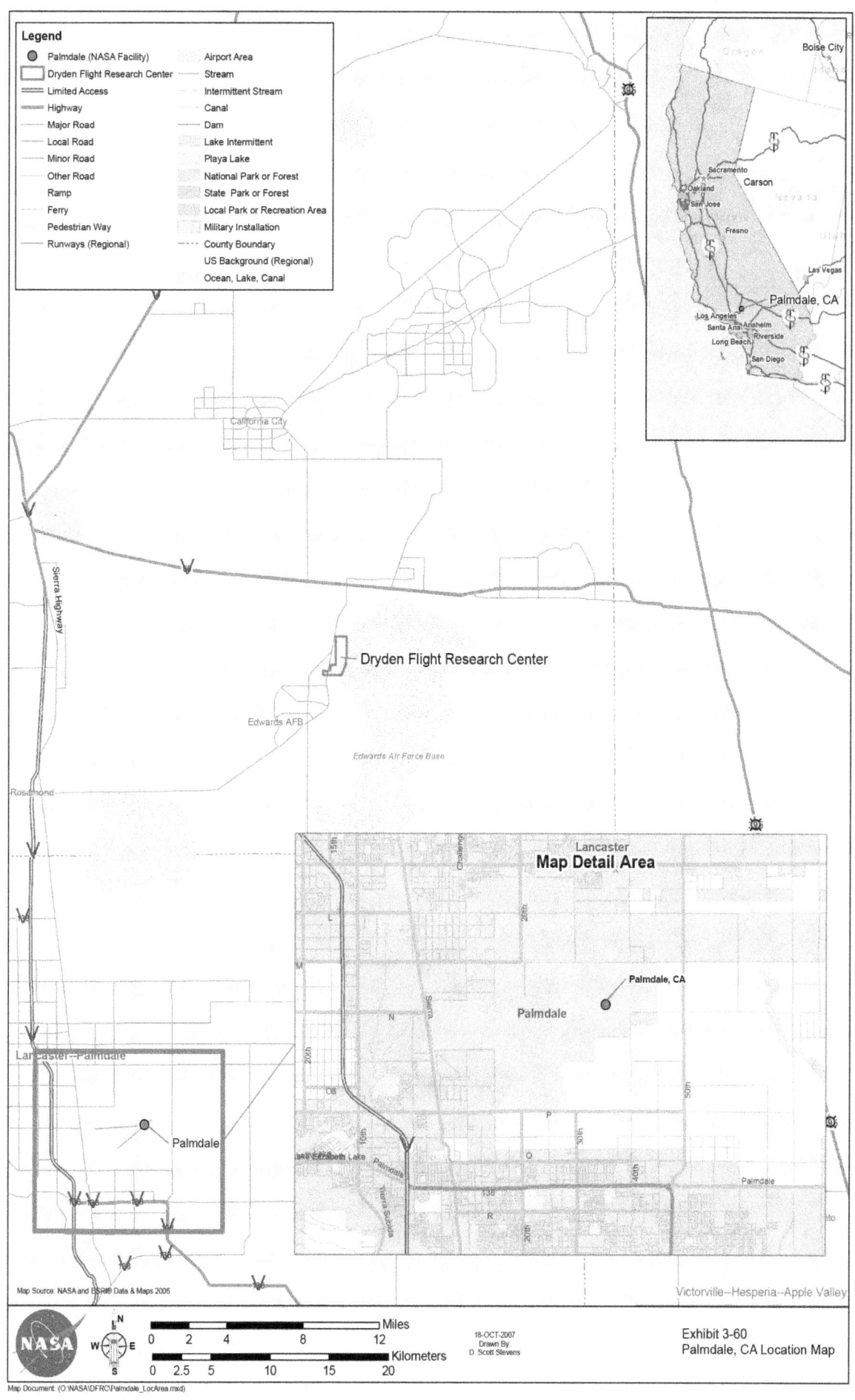

This page intentionally left blank.

Palmdale has 22 permitted devices (15 at Site 1 North and 7 at Site 1 South), including paint booths, open air spray guns, emergency internal combustion equipment, and scrubbers (NASA, 2007s).

3.11.2 Cultural Resources

3.11.2.1 Affected Environment

Archaeological Resources. Currently, no archaeological sites are listed, or eligible for listing, in the NRHP within the boundaries of Palmdale.

Historic Resources. Five properties were surveyed to determine their significance to the SSP. The results of the survey found that two properties were NRHP-eligible for their contributions to the SSP: the Assembly Building (Building 150) and the Orbiter Lifting Frame (NASA, 2003c:95-97; NASA, 2007s:28; Archaeological Consultants, Inc., 2007h).

3.11.3 Hazardous and Toxic Materials and Waste

3.11.3.1 Affected Environment

Storage and Handling. There are large amounts of Shuttle-related hazardous materials in the Tooling Storage Area, located in the northeastern corner of Palmdale (NASA, 2007s).

Hazardous material storage areas are located throughout Palmdale at the locations where the materials will be used (NASA, 2007s).

Waste Management. AFP 42 is an LQG. AFP 42 generated approximately 20,100 pounds of hazardous wastes during 2005. The largest waste streams onsite are oil and water, contaminated debris, and chemical processing.

Hazardous waste satellite accumulation areas also are located throughout Palmdale at various generation points (NASA, 2007s).

E-waste is composed of environmentally sensitive electronics (ESE) and is housed at the North Waste Yard. California regulates the handling and disposal of E-waste. Boeing's requirements cover all non-reusable ESE (including scrap electronics that have resale value), which must be sent to a Boeing-approved, California certified, E-waste recycling facility (NASA, 2007s).

Contaminated Areas. Palmdale is not on the federal NPL for cleanup. Contaminated soil exists at two areas at AFP 42 in Palmdale. The soil is being addressed as part of an interim removal action under the USAF's Installation Restoration Program (IRP). Identified as Sites 4 and 6, these IRP sites are located in the southeastern portion of AFP 42, known as the Common Area. The soil removal plan, called the Draft Removal Action Work (RAW) Plan, was prepared by the USAF and submitted to the California EPA (Cal-EPA) Department of Toxic Substances Control (DTSC) (NASA, 2007s).

Asbestos and Lead-based Paint. There are no identified asbestos- or LBP-containing structures at Palmdale.

3.11.4 Health and Safety

3.11.4.1 Affected Environment

SSP-related operations at Palmdale include vehicle testing and assembly, thermal protection fabrication and installation, plaster pattern shop and foam shop activities, TCS blanket fabrication, electrical wire harness fabrication, new generation tile fabrication, phantom works, and R&D (NASA, 2007s). All aerospace vehicle testing operations are conducted in compliance with EAFB's regulations to ensure the protection of employees and the public. The following subsections outline Palmdale's programs for protecting the health and safety of employees at Palmdale and the public. Noise hazards at Palmdale are outlined in Section 3.11.5.

Hazardous Materials. Hazardous materials are used to conduct the SSP manufacturing and assembly operations at Palmdale. The hazardous materials used and hazardous wastes generated are as discussed in detail in Section 3.11.3. The degree of exposure to hazardous materials is minimized by the implementation of work practices and control technologies.

Hazardous Materials Transportation Safety. Hazardous materials such as fuels, chemicals, and hazardous wastes are transported in accordance with the DOT regulations for the interstate shipment of hazardous substances (49 CFR 100 through 199).

Explosions and Fire Hazards. Using certain hazardous materials in SSP manufacturing and assembly operations presents a risk of explosions and fire hazards. Palmdale is located at EAFB, which provides fire protection services to Palmdale.

3.11.5 Noise

3.11.5.1 Affected Environment

The noise levels generated at Palmdale are low in comparison to the noise levels generated by aircraft activity at EAFB and do not extend past the EAFB fence line. Employees at Palmdale who are exposed to 8-hour TWA SPLs of 85 dBA and 90 dBA are monitored and provided with hearing protection, respectively.

3.11.6 Site Infrastructure

3.11.6.1 Affected Environment

Potable Water Supply. The Water Quality Program covers the non-transient and non-community water system at the facility and is regulated under the California Department of Public Health's Drinking Water Program (NASA, 2007s). Potable water at Palmdale is supplied by two onsite wells. The facility has two well sites–North and South. The two functioning wells are designated Well 01 and Well 03.

The North well site was abandoned in place. The two functioning wells are located at the South well site. Sampling at both the North and South well sites has indicated TCE contamination at concentrations of 2.2 micrograms per liter (µg/L) or parts per billion (ppb). The USAF currently is investigating the source of the TCE contamination.

Wastewater System. Boeing holds the NPDES Permit at Palmdale. The permit requires semiannual self-monitoring reports (SMRs) by Boeing (NASA, 2007s). The Los Angeles County Sanitation District regulates site industrial wastewater under a Los Angeles County Sanitation District Permit. Wastewater treatment is performed onsite at the cold-plate line, operated by Boeing, before discharge to the county WWTP from one regulated discharge point (NASA, 2007s).

Storm Water System. Storm water from the site is regulated under an NPDES General Permit held by AFP 42 (NASA, 2007s).

All storm water systems at Palmdale are owned and operated by AFP 42.

Energy Sources. NASA is supplied with electricity and natural gas by EAFB. EAFB owns the distribution systems (NASA, 2007s). NASA uses the EAFB electricity and natural gas supply at Palmdale. NASA holds no permits for energy sources at the facility (NASA, 2007s).

3.11.7 Solid Waste

3.11.7.1 Affected Environment

No onsite disposal occurs at Palmdale. Solid wastes generated at Palmdale are disposed at an offsite landfill by a solid waste disposal contractor. E-waste generated at Palmdale is housed at the North Waste Yard. ESE generated at Palmdale is sent to a Boeing-approved, California certified, E-waste recycling facility (NASA, 2007s).

3.11.8 Traffic and Transportation

3.11.8.1 Region of Influence

Palmdale AFP 42 is located at a detachment of EAFB in the city of Palmdale, Los Angeles County, California. The transportation ROI for Palmdale is defined as the counties from which 98 percent of people employed in Los Angeles County commute–Los Angeles, Orange, San Bernardino, and Ventura counties, California (U.S. Census Bureau, 2000).

3.11.8.2 Affected Environment

Transportation. SR-14 is located on the west of DFRC. One major arterial, Sierra Highway, connects the project site to the local communities and the interstate highway system. SR-138 and SR-18 connect DFRC to I-15, which runs north-south, located at the eastern side of the project site. I-5, which is located on the western side of DFRC, is accessed through SR-14.

Access Roads to Palmdale. The principal roadways, East Avenue M parallel to and just north of the Site 1 boundary and East Avenue M-12, connect the project site to the Sierra Highway. The Sierra Highway is a major arterial that runs north-south and connects the project site to SR-14. Another local road that connects DFRC to East Avenue M is Site 1 Road (Exhibit 3-60).

Railroads. Southern Pacific Railroad runs west of the project site. This railroad runs north-south, parallel to the Sierra Highway.

Airports. Palmdale AFP NR 42 Airport is adjacent to and south of the project site. Agua Dulce Airpark is approximately 25.6 km (16 miles) southwest of DFRC. EAFB is located on the northern portion of the site.

4. Environmental Consequences

This section describes the potential environmental consequences of the two alternative actions, Proposed Action and No Action, by comparing these activities with the potentially affected environmental components. Section 4.1 provides an overview of cultural resources and socioeconomics. Sections 4.2 through 4.11 provide discussions of the potential environmental consequences of the activities. The amount of detail presented in each subsection is proportional to the potential for impacts. Sections 4.12 and 4.13 discuss environmental justice and cumulative impacts, respectively.

Potential impacts to resources resulting from the implementation of the two alternatives were identified and placed into one of the following pre-determined classifications (NASA, 2007h):

- No Impact–No impacts expected
- Minimal–Impacts are not expected to be measurable, or are measurable but are too small to cause any change in the environment
- Minor–Impacts are measurable but are within the capacity of the affected system to absorb the change, or the impacts can be compensated for with little effort and few resources so that the impact is not substantial
- Moderate–Impacts are measurable but are within the capacity of the affected system to absorb the change, or the impacts can be compensated for with effort and resources so that the impact is not substantial
- Major–Environmental impacts that, individually or cumulatively, could be substantial

The following subsections describe the potential environmental impacts resulting from the implementation of the Proposed Action and the No Action alternative. NASA currently anticipates that much of the SSP property would be reused by future space flight programs on the basis of the ongoing planning phases for these programs. It is anticipated that the other options listed under the preferred alternative would not be used to a great extent. Therefore, the text concludes that none of the potential impacts, other than some cultural resources at various Centers (described below), are moderate. The text also concludes that other resource area impacts are "minimal to no" or "no" impact.

4.1 Overview of Cultural Resources and Socioeconomics

4.1.1 National Perspectives on Cultural Resources

Cultural resources are broadly understood as the physical remains of historic and prehistoric cultural systems. These resources are used to interpret, explain, and study all aspects of a culture. These tangible cultural remains help us to better understand our heritage, to appreciate architecture and engineering, and to learn about past accomplishments.

The goal of preserving historic properties as important reflections of our cultural heritage became national policy in the early twentieth century with the passage of the Antiquities Act of 1906, and was then furthered with the Historic Sites Act of 1935 and the NHPA of 1966, as amended. (Appendix D contains a list of the applicable federal laws and regulations.) Multiple presidential EOs have followed in the ensuing decades to refine the goals of historic preservation, including EO 13287, signed in 2003. This order, creating the "Preserve America" initiative, established a policy to provide leadership in the preservation of our cultural heritage by actively advancing the protection, enhancement, and contemporary use of historic resources owned by the federal government (Preserve America, 2007).

The NHPA process has produced documentation regarding the appearance and importance of districts, sites, buildings, structures, and objects significant in our history and prehistory. Thousands of properties around the nation have been documented that illustrate for the generations to come the broad patterns of local, state, and national experience throughout U.S. history (NPS, 1995).

The federal government recognizes the cultural and societal value of irreplaceable historic and prehistoric resources and is committed to protecting them from damage (NASA, 2007h). Conservation of cultural resources is a component of NASA's environmental management program, in accordance with Section 110 of NHPA. The NASA Environmental Management Division (EMD) includes NASA's Federal Preservation Officer (FPO), who coordinates with the NASA Senior Historian to preserve historically significant NASA properties.

One of NASA's property management goals is to "ensure that historic properties are managed in a manner that promotes the long-term preservation and use of those properties as federal assets and, where appropriate and consistent with NASA's mission, contributes to the local community and its economy" (NASA, 2005c:6). NASA's goal is to provide responsible stewardship of its historic assets to achieve the best possible value for the public's investment (NASA, 2005c:7).

The NASA EMD serves as the agency lead in assuring that NASA meets its federal stewardship responsibilities under NHPA, while at the same time carrying out its primary mission of understanding and protecting the planet, exploring the larger universe, and inspiring the next generation of explorers (NASA, 2007d).

4.1.2 National Perspective on Socioeconomic Impacts

As indicated in Section 1 of this Programmatic EA, President Bush has directed NASA to transition and retire the SSP in 2010, and Congress has endorsed that directive. The Presidential decision to discontinue the SSP has already been made; as a Presidential decision, it is not a topic for NEPA analysis. NASA is in the planning stages of T&R activities for the SSP that will address the efficient reuse of critical skills, human capital, and property. This Programmatic EA evaluates NASA's decision about how to disposition the SSP's real and personal property assets (whether to use the approach of NASA's Proposed Action or the No Action alternative). Therefore, the socioeconomic impact analysis in this Programmatic EA addresses only the impacts of NASA's discretionary actions regarding the disposition of the SSP's real and personal property and does not address the broader socioeconomic impacts of the President's decision to discontinue the SSP.

Nevertheless, to provide context for this EA's socioeconomic analysis, the following introductory discussion provides information regarding the current and projected socioeconomic influence of the SSP and other NASA programs. A focused report (*Baseline Socioeconomic Resources, Space Shuttle Program, Fiscal Year 2006*) (NASA, 2007bb) was prepared to assess in more detail the current socioeconomic "footprint" of the SSP in the regions where the major NASA Centers are located. Brief summaries of that information are provided for each of those NASA Centers in Section 3, to describe baseline socioeconomic resources.

Section 4.1.2.1 provides an overview of the current economic footprint of the SSP in the regional economies and the anticipated effect as the SSP T&R takes place.

Section 4.1.2.2 provides a general discussion of the President's Vision for Space Exploration, describes NASA's current plans for developing future space flight programs, and illustrates NASA's proposed budgets for the SSP and other space operations during the SSP T&R period.

Section 4.1.2.3 describes the overall potential for effects from the Proposed Action (NASA's planned T&R of SSP assets, including real and personal property) on socioeconomic resources. Section 4.1.2.4 describes the overall potential for effects from the No Action Alternative on socioeconomic resources.

4.1.2.1 Socioeconomic Effects of Federal Agency Actions

Socioeconomic resources can be affected adversely by substantial changes in employment and procurement by federal agencies. The SSP currently provides an important source of revenue for local firms through the procurement of goods and services, as well as civil service and prime contractor salaries. The economic "multiplier effect" means that changes in SSP expenditures would be felt both in the industries that provide supplies and services to NASA and also in the businesses that depend on employee spending. The subsections in Section 3 provide baseline data for SSP employment and expenditures at each of the major NASA Centers, including an estimate of the total "multiplier effect" of the SSP's direct expenditures on the economic output, employment, and income in the regional economies.

New NASA programs and projects will help fill the void left by the SSP T&R activities; however, localities that host NASA Centers that are heavily involved in the SSP would experience adverse socioeconomic impacts. Ripple effects on population and the associated demand for community services (such as housing, school enrollment, shopping, and police and fire protection) could occur if employment changes caused large numbers of employees to move into or out of an area.

Although SSP expenditures and employment make a positive contribution to the regional economies, it is relatively modest in proportion to the overall economic activity of the regions. At most of the Centers, the total direct and secondary effects of the SSP on economic output, earnings, and employment were less than 1 percent of regional levels in FY 2006, except in the KSC region, where the effects were less than 3 percent. (See Exhibit 4-1 and the Center-specific socioeconomic discussions in Section 3.)

However, it is important to note that the social and economic influence of NASA's Centers, especially at KSC and MAF (which primarily support SSP operations), extends well beyond the direct and secondary economic effects of Shuttle-related expenditures and salaries. NASA's operations and technical R&D programs have attracted other aerospace and related businesses to these areas, and thus, serve as an economic driver for the regional economies in a broader sense. NASA also supports higher education and research conducted by universities and non-profits. NASA and the State of Louisiana are collaborating to build on existing public-private commercial partnerships for technical R&D at MAF, which will contribute to New Orleans' economic recovery. In addition, NASA Visitor Centers attract considerable tourism dollars, especially in the KSC, JSC, and MSFC regions.

NASA will continue to invest in other space operations at existing Centers and will distribute the new work across NASA's existing Centers, as discussed below and in Section 3. However, a detailed analysis of changes in employment and expenditures at each Center is limited by the fact that the new Constellation Program is at an early stage of development, with major procurements not yet awarded, and would be subject to adjustments and changes as requirements become better defined (NASA, 2007t).

EXHIBIT 4-1
Contribution of the Space Shuttle Program to Regional Economies in FY 2006

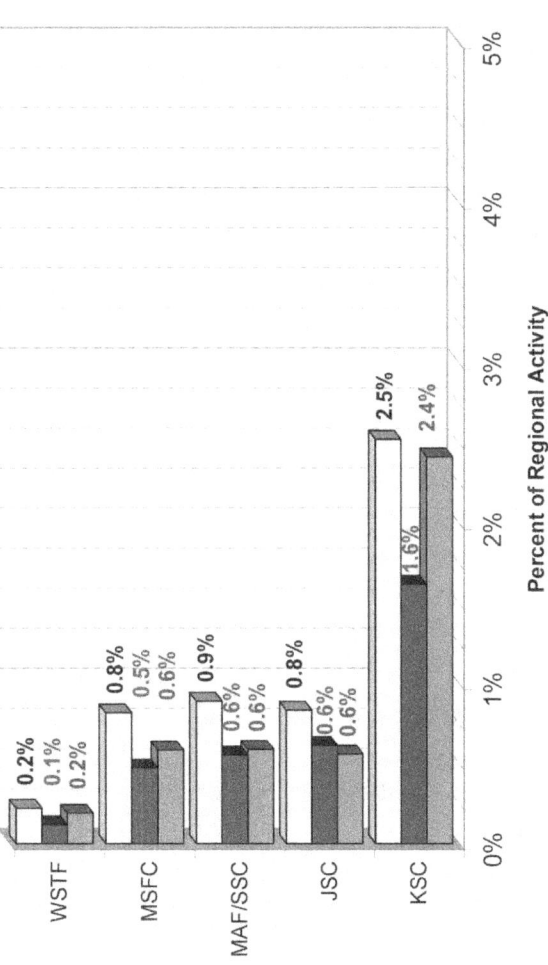

Output is compared to: U.S. Bureau of the Census, 2002 Economic Census – Total sales, shipments, receipts, or revenue for all establishments (2-digit NAICS codes).

Employment is compared to: U.S. Bureau of Economic Analysis, 2004 – Total wage and salary employment by place of work (jobs in the region).

Earnings are compared to: U.S. Bureau of Economic Analysis, 2004 – Total wage and salary disbursements by place of work.

Percentages should be considered only as illustrative.

4.1.2.2 NASA's Vision for Exploration Systems and Space Operations

The President's FY 2008 budget request for NASA shows a steadily increasing investment in exploration systems and space operations over the budget period of FY 2006 through FY 2012 (Exhibit 4-2).

This portion of NASA's budget covers the SSP, ISS, and Constellation Programs, as well as ongoing activities that support human space flight and advanced capabilities development. As the SSP transitions and retires, the Constellation Program plans to increase the pace of development and testing of the nation's new space vehicles, leading to an initial operating capability by 2015.

In addition, the SSP T&R will require some minimal level of spending after 2010 to retire the remaining SSP real and personal property.

NASA has assigned its Centers responsibility for developing and implementing the proposed Constellation Program. This distribution of work across NASA's Centers reflects NASA's intention to productively use personnel, facilities, and resources from across the Agency to accomplish the Vision for Space Exploration. Assignments align the work to be performed with the capabilities of the individual NASA Centers. In addition to primary work assignments, the Centers would support additional Constellation program and project activities. The primary work assignments for each Center are described in Section 3.

Additional information is available in the Final Cx PEIS (NASA, 2007t).

Although these work assignments would result in budget and personnel allocations at the Centers and component installations, detailed meaningful estimates of these allocations and the associated socioeconomic impacts would not be available until after prime contracts are awarded for all of the program's major projects and procurements. However, it is fair to say that NASA's plans for implementing Constellation would tend to minimize workforce dislocations, compared to other action alternatives that initially were considered for that program. Even with the new programs, there will be an approximate 4-year gap between the termination of the SSP and the operation of the new vehicles, during which employment and expenditures would be affected.

NASA recognizes that a skilled NASA and contractor work force is an essential ingredient to successful implementation of the Constellation Program. NASA is examining a variety of personnel initiatives to effect a smooth transition to Constellation operations, and is committed to preserving the critical and unique capabilities provided by each NASA Center (NASA, 2007t).

EXHIBIT 4-2
NASA FY 2008 Budget Request for Exploration Systems and Space Operations

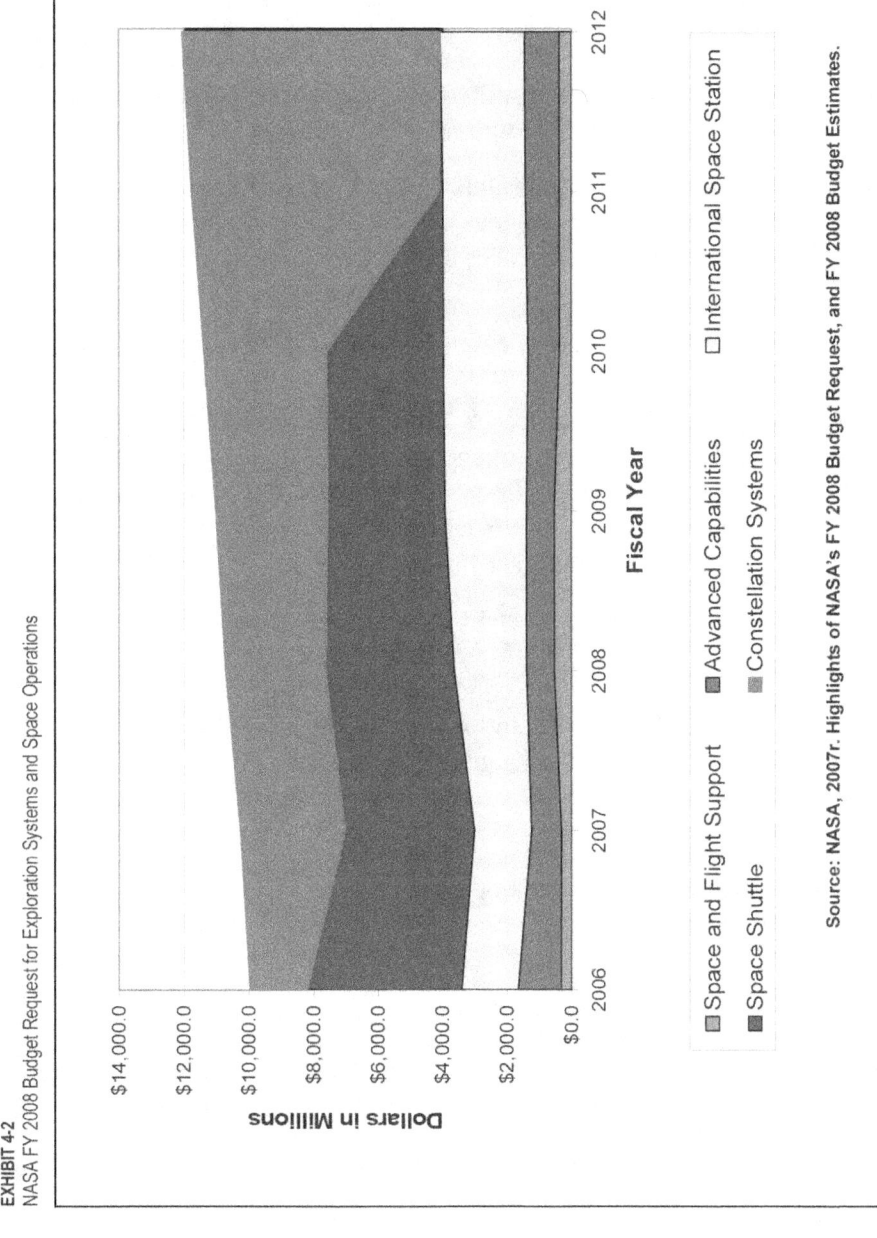

Source: NASA, 2007r. Highlights of NASA's FY 2008 Budget Request, and FY 2008 Budget Estimates.

4.1.2.3 Overall Effects of the Proposed Action

Under the Proposed Action alternative, NASA proposes to implement a centralized process for the disposition of the SSP real and personal property consisting of a coordinated series of actions. SSP real and personal property would be evaluated in accordance with NPR 8800.15, "Real Estate Management Program Implementation Manual," and NPR 4300.1, "NASA Personal Property Disposal Procedural Requirements," to select the best option for disposition.

Real Property. The major NASA Centers and GO/CO facilities will continue to operate under other programs besides the SSP. The disposition of selected buildings and smaller parcels of land within large and otherwise active facilities typically has minimal to no impact on socioeconomics outside the fence line.

The conveyance of real property to another NASA entity, or to new owners through a release to the GSA, whether the property is transferred to another federal agency, local government, or the private sector, would promote economic reuse of the property and generate employment and operational expenditures. If reuse were materially different from the existing use, additional NEPA documentation would be required. Mothballing the resource (that is, maintaining its functionality for reuse by NASA at a later time) would delay economic reuse.

Demolition temporarily would benefit the regional economy through the contracts for demolition and the hiring of the required workers. More importantly, demolition would allow another economically productive use of the land should another use be identified. If the land were to be transferred out of NASA's ownership, it could become available for conversion to a recreational or conservation use that could make a different type of social and economic contribution via tourism.

Personal Property. The disposition of personal property would have minimal to no discernable impact on the regional economies surrounding the NASA Centers where such property is located. One possible exception is for museums and visitor centers that receive Shuttle personal property, which would experience additional tourism, depending on the type and importance of the personal property newly available for display. Storage would delay or prevent this economic advantage, but is not likely to be of long duration for the Shuttle personal property most desirable to museums.

4.1.2.4 Overall Effects of No Action Alternative

Under the No Action alternative, NASA would not implement the proposed comprehensive and coordinated effort to disposition SSP property under a structured and centralized SSP process. The disposition of SSP property would instead occur on a Center-by-Center and item-by-item basis in the normal course of NASA's ongoing facility and program management. The No Action Alternative does not include continuing the SSP; it only pertains to the disposition of real and personal property. Just as the specific methods for the disposition of real and personal property are likely to have minimal to no impact on socioeconomics, the selection of the No Action Alternative would have minimal to no effects.

Real Property. The major NASA Centers and GO/CO facilities would continue to operate under other programs besides the SSP. The disposition of selected buildings and smaller parcels of land within large and otherwise active facilities typically has minimal to no impact on socioeconomics outside the fence line.

The conveyance of real property to another NASA entity, or to new owners through a release to the GSA, whether the property is transferred to another federal agency, local government, or the private sector, would promote economic reuse of the property and generate employment and operational expenditures. If reuse were materially different from the existing use, additional NEPA documentation would be required. Mothballing the resource (that is, maintaining its functionality for reuse by NASA at a later time) would delay economic reuse.

Demolition temporarily would benefit the regional economy through the contracts for demolition and the hiring of the required workers. More importantly, demolition would allow another economically productive use of the land. If the land were to be transferred out of NASA's ownership, it could become available for conversion to a recreational or conservation use that could make a different type of social and economic contribution via tourism.

The environmental impact would be expected to be similar to that of the Proposed Action Alternative. However, if a centralized process were not used to disposition assets (i.e., Proposed Action), the property disposal process could become overwhelmed with the volume of property to disposition. The volume of property that would be processed could result in schedule and cost impacts if a structured disposal process were not implemented. In addition, the amount of solid and hazardous waste that would require disposal could exceed landfill and less-than-90-day hazardous waste storage yard capacities at some Centers.

Personal Property. The disposition of personal property would have minimal to no discernable impact on the regional economies surrounding the NASA Centers where such property is located. One possible exception is for museums and visitor centers that receive Shuttle personal property, which would experience additional tourism, depending on the type and importance of the personal property newly available for display. Storage would delay or prevent that advantage, but is not likely to be of long duration for the Shuttle personal property most desirable to museums.

In addition, if a centralized process were not used to disposition assets (i.e., Proposed Action), the property disposal process could become overwhelmed with the volume of property to disposition. The volume of property that would be processed could result in schedule and cost impacts if a structured disposal process were not implemented. Also, artifacts may not be properly identified and made available to museums for display. In addition, the amount of solid and hazardous waste that would require disposal could exceed landfill and less-than-90-day hazardous waste storage yard capacities at some Centers.

4.2 Kennedy Space Center

Exhibit 4-3 outlines the major SSP real and personal property at KSC and the preliminary plans for their disposition.

EXHIBIT 4-3
Major SSP Real and Personal Property at KSC

SSP Asset/Facility	Description	Disposition
Vehicle Assembly Building (VAB) (Building K6-848)	The VAB is divided into three sections known as the transfer aisle, high-bay (HB), and low bay. The transfer aisle contains overhead cranes that are used to transfer Shuttle elements to the HBs. The VAB contains four HBs. Two of the HBs are equipped with extendable platforms used for Shuttle assembly and integration on the MLP. The other two HBs contain ET checkout cells. One of the HBs is used to safe a fully stacked vehicle in the event of a hurricane and one can accommodate Orbiter storage.	This facility would be used by the Constellation Program for vehicle processing.
Launch Control Center (Building LCC–K6-900)	The LCC is an automated Shuttle checkout and launch facility. The hardware and software used by the LCC is custom made for the SSP. The LCC uses three primary subsystems–the Shuttle Data Center; Checkout, Control, and Monitor Subsystem; and Record and Playback Subsystem.	This facility would be used by the Constellation Program for launch operations.
Orbiter Processing Facilities (OPF) HBs 1, 2, and 3, and SSME Facility (OPF HBs 1, 2, and 3) (Buildings K6-894, K6-696)	There are three OPFs at KSC that are responsible for Orbiter pre- and post-flight operations, as well as for routine maintenance activities for the TPS, SSME removal and installation, and hardware trouble shooting. NASA currently has three Orbiters that will need to be dispositioned that are maintained in the OPFs. The OPFs are equipped with HBs, as well as office annexes. The primary workload in the OPFs entails preparing the Orbiter for flight. OPF-3 houses the SSME Shop, where SSME maintenance activities are conducted. The OPFs have access platforms that surround the Orbiter and allow interior access. There are also zero-G counterweight devices for operating the Orbiter payload doors and a fixed crane system. A Launch Process System is used to check out the interface system between the Orbiter and the LCC. The payload operations conducted in the OPF entail down-mission payload removal, mission kit reconfiguration, and up-mission horizontal payload installation.	Constellation has identified the possibility of using one or more of the OPF HBs for processing of the Ares V upper stage.

EXHIBIT 4-3
Major SSP Real and Personal Property at KSC

SSP Asset/Facility	Description	Disposition
Thermal Protection System Facility (TPSF) (Building K6-794)	The TPSF houses offices, machine tools, processing equipment, and areas for storage. The operations conducted in the TPSF include producing Orbiter tiles from raw stock, thermal control system blankets, fibrous insulation blankets, and gap fillers and thermal barriers.	This property has not been identified for use by new programs.
Crawler Maintenance Facility (Building K6-743)	The Crawler Maintenance Facility is used to perform maintenance on the Crawler.	This facility would be used by the Constellation Program for maintenance of the Crawler-Transporters.
Launch Complex (LC) 39A and LC-39B (Buildings J8-1798 (A) and J7-337 (B))	The LC is a collection of facilities used for SSP launches.	This property would be used by the Constellation Program for launching space vehicles.
Hypergol Maintenance Facilities (HMF) (Buildings M7-1061, M7-961, and M7-1212), and HMF Support Building #2 (Building M7-1059)	The HMF consists of three buildings that process and store the hypergolic-fueled modules that make up the Orbiter's reaction control system, orbital maneuvering system, and auxiliary power units.	This property has not been identified for use by new programs.
Shuttle Landing Facility (SLF)	The SLF has a 15,000-foot-long runway that is equipped with navigational aids. The SLF also maintains and uses equipment to support Orbiter recovery, safing, processing, and towing operations.	Although NASA will no longer need this facility for Shuttle operations, it will continue to be used as an airfield to support cargo and equipment operations. The Mate-Demate Device will no longer be needed.
Operations Support Building (OSB) (Building K6-1096) Operations Support Building II (OSBII) (Building K6-1249)	The OSBs are office buildings that include a technical documentation center, library, and photograph analysis area.	These properties would be used by future programs as administrative space.
Component Refurbishment and Chemical Analysis (CRCA) (Building K6-1696)	The operational heart of the CRCA facility is a large clean-room area where instrumentation and pneumatic equipment of all types are serviced.	This property would be used by the Constellation Program for laboratory and cleaning operations.
Logistics Facility (Building K7-1547)	The Logistics Facility houses 190,000 SSP hardware parts and operates a state-of-the-art parts retrieval system, which includes automated handling equipment to find and retrieve specific SSP parts.	This property would be used by the Constellation Program for warehousing.
Rotation, Processing and Surge Facility (RPSF)	The RPSF consists of four buildings located north of the VAB. This facility is used to offload SRM segments from railcars and to build up the aft booster for the SRBs. The building contains 200-ton overhead bridge cranes, two surge buildings for the storage of processed SRM components, and a support building.	This property would be used by the Constellation Program for SRM handling.

EXHIBIT 4-3
Major SSP Real and Personal Property at KSC

SSP Asset/Facility	Description	Disposition
Hangar AF	Hangar AF is located on property owned by the USAF; however, NASA is responsible for the building and the associated processes. The operations conducted at Hangar AF are associated with recovery of the SRBs after a Shuttle launch. There are two ships with licensed crews and certified divers that recover the SRBs and associated hardware (Frustum and parachutes) from the ocean and perform an initial anomaly check. The SRBs and associated hardware are then towed to Hangar AF for disassembly operations. The SRBs are washed at Hangar AF and then disassembled by performing ordnance safing and removal operations, RSRM disassembly, forward and aft skirt disassembly, and TVC safing. SRB refurbishment activities also take place at Hangar AF, including TPS and substrate finish removal; and manual and robotic grit, hydro, and bead blasting operations. Once the SRB surfaces have been refurbished by blasting operations, an alodine and primer top coat application is applied to the parts, and they are sent to the ARF for reuse.	This property would be used by the Constellation Program for SRB recovery operations.
Hangar N	Hangar N is located on property owned by the USAF; however, NASA is responsible for the building and the associated processes. The ongoing processes at Hangar N include quality control tests primarily associated with checking SRB component welds, along with other Shuttle-related items. The hangar is equipped primarily with X-ray, infrared, and ultrasound equipment. The operations in the hangar also include dye penetrant, magnetic particle, eddy current, and thermography testing to evaluate fractures and welding anomalies in the SRB components. A bay located in the hangar is capable of performing tensile tests. There also is a robot that is used to scan SRB components to check for fractures in metals and flaws in welds. The facility has made great strides in moving to digital images rather than film images.	This property would be used by the Constellation Program for SRB checkout.

EXHIBIT 4-3
Major SSP Real and Personal Property at KSC

SSP Asset/Facility	Description	Disposition
Hangar S Annex	Hangar S is a Shuttle operations training facility.	This property has not been identified for use by new programs.
SSP SRB ARF	The assembly and maintenance operations for the SRB aft skirt, forward skirt, frustrum, and nose cap component assembly and TPS applications are performed in the ARF. In addition, the aft assembly acceptance checkout for the avionics and the aft skirt thrust vector control system hot fire testing are performed at the Aft Skirt Test Facility. The forward assembly acceptance checkout for avionics and range safety, as well as the ordinance installation and checkout, is conducted in the ARF. The parachutes that deploy, once the SRMs have spent their fuel upon launching the Shuttle, are packed in the ARF.	These properties would be used by the Constellation Program for SRB assembly and checkout.
Parachute Refurbishment Facility	The Parachute Refurbishment Facility washes, dries, and repacks the SRB main, drogue, and pilot chutes, as well as the Orbiter chutes. The water used to wash the parachutes is filtered and reused.	This property would be used by the Constellation Program for SRB parachute refurbishment.
Hangar M–Annex	Hangar M is located on USAF property, but NASA is responsible for the buildings and ongoing processes.	Hangar M would be used by the Constellation Program.
Mobile Launch Platforms (MLP) and Crawler-Transporter	There are three MLPs at KSC, which provide GSE for Shuttle checkout, servicing, and launch. They are two-story transportable launch bases for the Shuttle stack. The exterior of the MLPs provide for an SRB hold-down post, Orbiter tail service masts, and sound suppression water nozzles for deluge water. The MLPs are transported from the VAB to the launch pad by the Crawler-Transporter. The Crawler-Transporter weighs 6 million pounds	One MLP would be used by the Constellation Program for vehicle stacking, the remaining two MLP have no use identified. Both Crawler-Transporters would be used by the Constellation Program for transporting launch vehicles.
Transoceanic Abort Landing Sites (TALs)	NASA has various TAL sites and ELSs that are used in the case of an emergency during the Space Shuttle's accent into orbit. The TAL sites are located in Eastern Europe at Moron AFB; in Spain at Zaragoza AFB; and in Istres-le-Tube AFB, France.	This asset has not been identified for use by new programs.
Orbiters	The Orbiters are housed at KSC	The Orbiters will not be used by new programs.
Canister Rotation Facility (CRF)	The CRF was built in 1993 in the Industrial Area to handle the challenges of canister rotation. The 142-foot HB includes a 100-ton bridge crane and other specialized equipment required for lifting.	This asset has not been identified for use by new programs.

4. ENVIRONMENTAL CONSEQUENCES

EXHIBIT 4-3
Major SSP Real and Personal Property at KSC

SSP Asset/Facility	Description	Disposition

Notes:
AFB = Air Force Base
ARF = Assembly and Refurbishment Facility
CRCA = Component Refurbishment and Chemical Analysis
CRF = Canister Rotation Facility
ELS = Emergency Landing Site
ET = External tank
ft = Feet
ft^2 = Square foot
GSE = Ground support equipment
HB = High bay
HMF = Hypergol Maintenance Facilities
KSC = Kennedy Space Center
kW = Kilowatt
LC = Launch complex
LCC = Launch control center
LH2 = Liquid hydrogen
MAF = Michoud Assembly Facility
MLP = Mobile launch platform
MOA = Memorandum of Agreement
NASA = National Aeronautics and Space Administration
OPF = Orbiter Processing Facilities
OSB = Operations Support Building
OSBII = Operations Support Building II
RPSF = Rotation, Processing, and Surge Facility
SAF = U.S. Air Force
SLF = Shuttle Landing Facility
SRB = Solid rocket booster
SRM = Solid rocket motor
SSME = Space Shuttle Main Engine
SSP = Space Shuttle Program
TAL = Transoceanic Abort Landing
TPS = Thermal protection system
TPSF = Thermal Protection System Facility
TVC = Thrust Vector Control
VAB = Vehicle Assembly Building

4.2.1 Environmental Consequences for KSC

The environmental resources that were evaluated and subsequently determined to have no potential for environmental impacts are provided in Exhibit 1-2. The environmental consequences for the resource areas present at KSC are summarized in Exhibit 4-4.

EXHIBIT 4-4
Summary of Environmental Consequences for KSC

Resource Area	Alternative	Overall Effects of Alternative	Impact
Air Quality	No Action and Proposed Action	The demolition or disposition of property would have the potential to temporarily increase emissions at KSC during the demolition or disposition operations.	Minimal to No Impact

EXHIBIT 4-4
Summary of Environmental Consequences for KSC

Resource Area	Alternative	Overall Effects of Alternative	Impact
Biological Resources	No Action and Proposed Action	**Vegetation.** Most of NASA's operational areas at KSC are on developed landscapes and are devoid of natural vegetation. Natural vegetation may spread into the developed areas once the property has been disposed, thereby increasing the distribution of natural vegetation in the area. NASA operational areas that currently support natural vegetation would remain undisturbed with the disposition of NASA property. The demolition of NASA property on KSC would have a minimal impact on vegetation and could have the potential to increase natural vegetation on the installation.	Minimal Impact
		Wildlife. Increased human activity and noise due to the disposition and demolition of property temporarily could increase the disturbance of wildlife. However, wildlife probably would return to the area after demolition was complete.	Minimal Impact
		Protected Species. Disposition of real property on KSC would have minimal to no impacts on protected species and habitats because NASA would continue to use protective measures for the habitat of these species.	Minimal Impact
Cultural Resources	No Action and Proposed Action	**Historic Resources.** As the SSP approaches the end of its mission, a variety of buildings and facilities at several NASA installations will no longer be of use to SSP. Once SSP identifies and reports to a host installation that it no longer needs a building or facility, NASA will initiate the standard process for addressing excess infrastructure [as described in Section 2.1]. Termination of SSP by NASA will not lead to a specific decision or action on the future of each infrastructure asset and the associated environmental impacts to that asset. NASA will conduct an appropriate level of federally mandated NEPA analysis before final decisions on the disposition of SSP infrastructure assets are made. If any such properties are listed in or eligible for listing in the NRHP, NASA will take no action that would affect any such property until the NHPA Section 106 process is complete.	Moderate Impact
Hazardous and Toxic Materials and Waste	Proposed Action	**Storage and Handling.** If the facilities were reutilized, mothballed, demolished, or released to the GSA, minimal impacts on the storage and handling of hazardous materials associated with real property would be expected because waste generation would be expected to remain at the same level.	Minimal to No Impact
		Waste Management. If the facilities were reutilized, minimal impacts on the waste management procedures would be expected because KSC would be reutilized by a similar NASA program. If the facilities were mothballed, demolished, or released to the GSA, it would be likely that the waste management procedures associated with the SSP would no longer be applicable, because no wastes would be generated.	Minimal to No Impact

EXHIBIT 4-4
Summary of Environmental Consequences for KSC

Resource Area	Alternative	Overall Effects of Alternative	Impact
		Contaminated Areas. If the facilities were mothballed, demolished, or released to the GSA, it would be possible that new contaminated areas could be identified during closure activities (such as the closure of ASTs or USTs). Newly identified contaminated areas would be addressed by the Center's restoration programs. However, major impacts would be unlikely because the Center has undergone investigation efforts under RCRA and CERCLA.	Minimal Impact
		Asbestos and Lead-based Paint. If real property were to be demolished, it probably would contribute to the generation of asbestos waste or LBP. Asbestos and LBP surveys would be conducted before demolition. If ACMs were determined to be present, they would be removed appropriately before demolition. Such wastes would need to be disposed according to the hazardous waste classification determined.	Minimal to No Impact
	No Action	**Storage and Handling**. The property disposal process may become overwhelmed with the volume of property to disposition. The volume of property that would be processed could result in schedule and cost impacts if a structured disposal process were not implemented. In addition, the amount of hazardous waste that would require disposal could exceed landfill and less–than-90-day hazardous waste storage yard capacities.	Minimal Impact
		Waste Management. The property disposal process may become overwhelmed with the volume of property to disposition. The volume of property that would be processed could result in schedule and cost impacts if a structured disposal process were not implemented. In addition, the amount of hazardous waste that would require disposal could exceed landfill capacities.	Minimal Impact
		Contaminated Areas. If the facilities were mothballed, demolished, or released to the GSA, it would be possible that new contaminated areas could be identified during closure activities (such as the closure of ASTs or USTs). Newly identified contaminated areas would be addressed by the Center's restoration programs. However, major impacts would be unlikely because the Center has undergone investigation efforts under RCRA and CERCLA.	Minimal Impact
		Asbestos and Lead-based Paint. If real property were to be demolished, it probably would contribute to the generation of asbestos waste or LBP. Asbestos and LBP surveys would be conducted before demolition. If ACMs were determined to be present, they would be removed appropriately before demolition. Such wastes would need to be disposed according to the hazardous waste classification	Minimal Impact

EXHIBIT 4-4
Summary of Environmental Consequences for KSC

Resource Area	Alternative	Overall Effects of Alternative	Impact
		determined. The property disposal process could become overwhelmed with the volume of property to disposition. The volume of asbestos due to demolition operations could result in schedule and cost impacts if a structured disposal process were not implemented. In addition, the amount of ACM that would require disposal could exceed landfill capacities.	
Health and Safety	No Action and Proposed Action	Health and safety risks associated with real property at KSC could include contamination or damage resulting from major spills or accidents. Buildings at KSC could contain asbestos and LBP. Employees conducting renovation or demolition work must meet the safety standards outlined in 29 CFR 1926.1101, OSHA's construction standard for asbestos, or 29 CFR 1962.62, OSHA's construction standard for lead, to prevent exposure. The appropriate level of PPE must be worn depending on the level of the abatement, in accordance with OSHA's construction standard, to ensure worker safety.	Minimal Impact
Hydrology and Water Quality	No Action and Proposed Action	Disposition or removal of buildings or structures at KSC could result in minimal temporary soil disturbances, thus resulting in erosion during removal activities. The methods used for removal of any designated facilities would vary in relation to the type of structure, its location, the materials encountered in demolition, and the contractor's experience. Best engineering practices, codes, specifications, and standards would be followed to prevent or limit potential impacts. These would include the implementation of erosion and turbidity controls. Storm water permits might need to be obtained and soil stabilization measures might need to be implemented.	Minimal Impact
Land Use	No Action and Proposed Action	If the existing facilities were destroyed or released to GSA, new facilities potentially could be constructed in their place. It is anticipated that they would be compatible with the existing land use categories.	Minimal Impact
Noise	No Action and Proposed Action	The demolition or disposition of property would have the potential to temporarily increase noise levels at KSC during the associated demolition or disposition operations. Any demolition or disposition activities would comply with the OSHA hearing protection standards for employees and other individuals in the vicinity.	Minimal Impact
Site Infrastructure	No Action and Proposed Action	The existing utilities are sufficient to support the current activities at KSC. It is anticipated that a large percentage of the SSP property at KSC would be transferred to other NASA programs upon disposition of SSP real and personal property. In this case, it is assumed that the infrastructure would continue to operate at similar levels, because it is assumed that programs receiving SSP property would have similar infrastructure needs. If the property were not transferred but remained unused, a decreased load on	Minimal Impact

EXHIBIT 4-4
Summary of Environmental Consequences for KSC

Resource Area	Alternative	Overall Effects of Alternative	Impact
		the site infrastructure would result. Any impacts would result from decreased use.	
Socioeconomics	No Action and Proposed Action	The specific disposition methods selected for SSP real and personal property are likely to have minimal to no impact on the population, regional economy, and community services in the region surrounding KSC. KSC will continue to provide testing and launch services for other NASA programs. It is expected that most of the buildings at KSC that are used by the SSP would be reused for other NASA projects, with the same or similar functions. A few could be transferred, demolished, or reused. It is not anticipated that demolition or conveyance of individual buildings (and land) would affect the socioeconomic resources in the surrounding area appreciably.	Minimal to No Impact
Solid Waste	Proposed Action	If it were determined that the real property would be demolished, the overall impacts probably would include the generation of solid waste consisting of concrete, asphalt, glass, metals (conduit, piping, and wiring), lumber, asbestos, and LBP. Items and materials that could be reused would be salvaged to the extent possible for NASA's future use. Non-salvageable solid waste would be disposed in accordance with the applicable health and safety and environmental regulations, either at the Schwartz Road Class III Landfill or at an appropriate offsite, permitted disposal facility, depending on the waste classification.	Minimal Impact
	No Action	The property disposal process may become overwhelmed with the volume of property to disposition. The volume of property that would be processed could result in schedule and cost impacts if a structured disposal process were not implemented. In addition, the amount of solid waste that would require disposal could exceed landfill capacities.	Minimal Impact

EXHIBIT 4-4
Summary of Environmental Consequences for KSC

Resource Area	Alternative	Overall Effects of Alternative	Impact
Transportation	No Action and Proposed Action	Real property demolition could generate more destruction-related truck trips. It would increase the traffic on the surrounding streets in the study area.	Minimal Impact

Notes:
ACM = Asbestos-containing material
AST = Aboveground storage tank
CCSMP = Cape Canaveral Spaceport Master Plan
CERCLA = Comprehensive Environmental Response, Compensation, and Liability Act
CFR = *Code of Federal Regulations*
GSA = General Services Administration
KSC = Kennedy Space Center
LBP = Lead-based paint
MAF = Michoud Assembly Facility
NASA = National Aeronautics and Space Administration
NEPA = National Environmental Policy Act
OSHA = Occupational Safety and Health Administration
PPE = Personal protective equipment
PTE = Potential to emit
RCRA = Resource Conservation Recovery Act
SSP = Space Shuttle Program
UST = Underground storage tank

4.3 Johnson Space Center

Many of the operations conducted at JSC contribute to the SSP. Overviews of each directorate and its responsibilities are provided in Exhibit 4-5, along with descriptions of the key buildings that support the SSP and the preliminary plans for their disposition.

EXHIBIT 4-5
Major SSP Real and Personal Property at JSC

SSP Asset/Facility	Description	Disposition
Engineering Directorate–Government Furnished Equipment (GFE) Flight Systems Design and Development Laboratories	The GFE Flight Systems Design and Development Laboratories support the design, development, integration, test, and sustaining engineering of GFE hardware and software flight systems. The primary laboratories supported include the Wireless Instrumentation Development Laboratory and the Crew Health Care System Development Laboratory.	This property has not been identified for use by new programs.
Engineering Directorate–Flight Systems Integration and Test Facilities	The Flight Systems and Integration and Test Facilities support the integration of flight hardware and software systems. In addition, the functional and performance testing of flight systems is conducted in the GN&C Rapid Development Lab, Pyrotechnics Lab, Electrical Power Systems Lab, and the various pressure chambers.	The ISS program may continue to use SSP-developed space suits or transition to the Russian-developed suits in the future. NASA will evaluate future disposition options for this property.

EXHIBIT 4-5
Major SSP Real and Personal Property at JSC

SSP Asset/Facility	Description	Disposition
Engineering Directorate– End-to-End Integrated System Test Facilities	The End-to-End Integrated System Test Facilities provide high-fidelity, end-to-end, integrated hardware and software systems' testing and evaluation of critical functions and performance.	These facilities are reconfigurable and may include flight hardware and software systems, ground-based systems and facilities, and functional simulations. These facilities include the Electronic Systems Test Lab, Orion Avionics Integration Lab, and a Shuttle cabin with an airlock that is unique to the SSP. NASA will evaluate future disposition options for this property.
Engineering Directorate– Environmental Test Facilities (Building 32)	The Environmental Test Facilities provide high-fidelity, simulated fight environments for engineering unit testing and qualification and acceptance testing of flight hardware systems and spacecraft. The facilities include vacuum chambers, thermal and solar human-rated test facility, and a vibration and acoustic test facility.	The Constellation Program would use this for crewed thermal vacuum testing and altitude chambers.
Engineering Directorate– Arc Jet Test Facility (Building 222)	The Arc Jet Test Facility is used to simulate the conditions on a spacecraft during reentry.	Currently, it is anticipated that future space programs will use the Environmental Test Facilities. NASA will evaluate future disposition options for this property.
Engineering Directorate– High-fidelity Simulation and Analysis Facilities	The High-fidelity Simulation and Analysis Facilities provide high-fidelity, multi-system simulation facilities for engineering evaluations, operations procedures development, and crew training. These facilities include the following: • Aerosciences Laboratory • Systems Engineering Simulator • Six Degrees of Freedom Test System • Virtual Reality Laboratory	This property has not been identified for use by new programs.
Engineering Directorate– Long-duration, Integrated Simulation Facilities	The Long-duration, Integrated Simulation Facilities are high-fidelity, multi-system simulation facilities for long-duration testing of integrated systems, including advanced technology hardware and software systems, integrated real-time simulation systems, crew accommodations, and crew. These facilities include a 20-foot, human-rated chamber advanced life support and long-duration testing chambers.	These systems would be used to support lunar programs associated with long-duration missions and are expected to be used by other programs during the Shuttle transition activities. NASA will evaluate future disposition options for this property. The following locations in the facilities have been identified to support the Constellation Program: • 3rd floor – component and small unit bench top testing

EXHIBIT 4-5
Major SSP Real and Personal Property at JSC

SSP Asset/Facility	Description	Disposition
		• 8-ft chamber – unscrewed integrated EVA life support system operational vacuum testing. • 11-ft chamber – Crewed EVA system vacuum testing • Thermal Vacuum glovebox – thermal vacuum testing of gloves and small tools.
Engineering Directorate–Advanced Technology Development Laboratories	The Advanced Technology Development Laboratories support the design, development integration, and testing of advanced technology hardware and software systems. These test facilities include the following: • Advanced Portable Life Support System Development Laboratory • Regenerative Wastewater Processing Systems Development Laboratory • Wireless and Radio Frequency Identification Laboratory • Nanotube Development Laboratory	This property has not been identified for use by new programs.
Engineering Directorate–General Infrastructure Support Test Facilities	The General Infrastructure Support Test Facilities provide the infrastructure support services required by multiple Engineering Directorate core competencies and facilities, including manufacturing, integration and assembly, clean rooms, NDE, calibration and metrology, bonded storage, and gas and chemical analyses. These facilities include manufacturing processes, materials evaluation laboratories, and avionics development laboratories.	This property has not been identified for use by new programs.
Mission Operations Directorate–Mission Control Center and Integrated Planning System	The Mission Control Center and Integrated Planning System has the capability to provide ISS and SSP with real-time command and control operation to train and certify flight crews and controllers.	The cost to operate these facilities has been shared between ISS and SSP. Current plans are that the Constellation Program would provide the funding to operate these facilities, with no resulting gap due to the SSP retirement.

EXHIBIT 4-5
Major SSP Real and Personal Property at JSC

SSP Asset/Facility	Description	Disposition
Mission Operations Directorate–Training and Simulator Facilities	Three Shuttle simulators (Building 5), two fixed-based and one motion-based, are maintained by this directorate. Each simulator consists of numerous computers, workstations, special-purpose interface devices, visual image processing, cockpit mockups, and hydraulic motion systems. Fight operation trainers consist of single system trainers (three), flight controller trainers (three), and a payload trainer in Building 4 and dynamics skills trainers (seven) in Building 16. The operations that are conducted include the check out of special-purposed interface devices between the systems in the Space Shuttle. In addition, Mission Control Center workstations are developed and tested in this facility. Approximately 4,500 pieces of equipment were used to support these operations, along with about 13,000 spare parts.	The shuttle simulators would not be needed for future programs and would be dispositioned accordingly. The simulators have been identified as potential historical artifacts or landmarks. NASA would evaluate future disposition options for this property. Some of the equipment that supports Flight Operation Trainers would be used by the ISS operations, but most would be dispositioned upon the retirement of the SSP. There is also a simulator with a 40-foot dome used for astronaut training to dock the Shuttle with the ISS. NASA will evaluate future disposition options for this property. This directorate has developed an equipment replacement program that incorporates a phase-down leading up to transition. However, some of the spare parts stock is being increased so that new equipment would not have to be purchased for Shuttle fly-out, because replacement parts for existing equipment may not be available. Property that is shared between the SSP and ISS would become ISS property when the SSP retires. It is anticipated that Orion would use these facilities in the future.
Mission Operations Directorate–Space Vehicle Mockup Facilities.	This facility is located in Building 9 and has an inventory that is unique to the SSP. The equipment includes a full fuselage trainer, two crew compartment trainers, a crew escape system trainer, and TPS inspection and repair hardware. These trainers are all needed through SSP fly-out. Some of the platforms that support the trainers may be used by other programs. The Shuttle-specific portions of the trainers have been identified as potential historical artifacts.	The Shuttle-specific portions of the trainers have been identified as potential historical artifacts or landmarks. NASA would evaluate future disposition options for this property.

EXHIBIT 4-5
Major SSP Real and Personal Property at JSC

SSP Asset/Facility	Description	Disposition
Mission Operations Directorate–Software Production Facility	The Software Production Facility is a large computational facility consisting of an IBM Z900 mainframe and an IBM Shark data access storage device. The facility also contains tape silos and virtual tape systems. There are seven flight equipment interface device boxes that are unique to the SSP and various input and output and security devices.	These facilities would be needed through SSP fly-out, and none have been identified as being needed for future programs or as being of historical significance. NASA would evaluate future disposition options for this property.
Photographic Technology Laboratory (Building 8)	The Photographic Technology Laboratory handles approximately 100 rolls of film from each mission, along with about 1,000 digital images that are downloaded during the mission activities. This facility mixes its chemicals to develop photographs from the rolls of film and the digital images that will be stored in the archives.	The SSP has directly funded the operations in the laboratory. However, the Constellation Program, as well as the ISS, is beginning to fund the operations in the laboratory. The storage requirement for archiving the digital photographs is the largest issue facing the SSP transition activities from the operations in the laboratory. NASA will evaluate future disposition options for this property.
Technical Services Shop and Systems Integration Facility (Buildings 10 and 9S)	The Technical Services Shop (Building 10) has an extensive fabrication shop for metal, wood, and plastic to create up to full-scale spacecraft prototypes. The tooling is standard machine shop tooling that is not specific to the Shuttle. However, there is a Shuttle tile repair shop that has operations specific to the SSP. The Systems Integration Facility (Building 9S) houses technical and engineering personnel and provides for the construction of wood, plastic, and metal spacecraft hardware items. This building also has paint and model shops. There is a plating shop in this building used to perform plating operations for developing test models.	The tooling in the Technical Services Shop is standard machine shop tooling that is not specific to the Shuttle. However, there is a Shuttle tile repair shop that has operations specific to the SSP. NASA will evaluate future disposition options for this property. The Systems Integration Facility will have flight hardware associated with the SSP that will need to be excessed. However, the chemicals in use and the operations will not change, except that some "environmentally friendly" chemical replacements may be used in the future. It is anticipated that some of the work currently being implemented at Palmdale will be transferred to this area of JSC.

EXHIBIT 4-5
Major SSP Real and Personal Property at JSC

SSP Asset/Facility	Description	Disposition
Energy Systems Test Area (Building 357)	The Energy Systems Test Area performs tests on the electrical systems of spacecraft such as the fuel cells and solar cells, and is responsible for the pyrotechnic charges required for all aspects of spaceflight. This facility is capable of testing fuel cells in vacuum chambers under hot and cold conditions, along with the associated lithium batteries. The pyrotechnic testing includes age testing of the charges and space flight certification testing. Approximately 100 explosive devices are used for each flight. There are 35 explosive devices that are used in the event of an emergency; these devices can be used for other programs.	When the SSP retires, the contractors (USA) at KSC will be responsible for removing the explosive devices from the Orbiter as part of the safing process. The excess pyrotechnics are offered to the Harris County Bomb Squad for training purposes at its facilities. This facility was used to test hypergolics; however, because of encroachment from offsite development, this testing is now conducted at WSTF. This facility currently performs testing for the ISS, SSP, and other developmental projects. It is anticipated that future space flight programs would use this facility in the near future and that staff will be added to accommodate the increased workload. NASA will evaluate future disposition options for this property.

Notes:
°F = Degrees Fahrenheit
GFE = Government Furnished Equipment
GN&C = Guidance, Navigation, and Control
HCFC = Hydrochlorofluorocarbon
ISS = International Space Station
JSC = Johnson Space Center
NASA = National Aeronautics and Space Administration
NDE = Non-destructive evaluation
SSP = Space Shuttle Program
TPS = Thermal protection system
USA = United Space Alliance
WSTF = White Sands Test Facility

4.3.1 Environmental Consequences for JSC

The environmental resources that were evaluated and subsequently determined to have no potential for environmental impacts are provided in Exhibit 1-2. The environmental consequences for the resource areas present at JSC are summarized in Exhibit 4-6.

EXHIBIT 4-6
Summary of Environmental Consequences for JSC

Resource Area	Alternative	Overall Effects of Alternative	Impact
Air Quality	No Action and Proposed Action	The demolition or disposition of property would have the potential to temporarily increase emissions at JSC during the demolition or disposition operations.	Minimal to No Impact
Biological Resources	No Action and Proposed Action	**Wetlands.** No facilities on JSC are located in wetlands.	No Impact
		Floodplains. No facilities on JSC are located in floodplains.	No Impact
Cultural Resources	No Action and Proposed Action	**Archaeological Resources.** There would be minimal to no impact on archaeological resources under the Proposed Action and No Action Alternative if any sites are identified in the future.	No impact
		Historic Resources. As the SSP approaches the end of its mission, a variety of buildings and facilities at several NASA installations will no longer be of use to SSP. Once SSP identifies and reports to a host installation that it no longer needs a building or facility, NASA will initiate the standard process for addressing excess infrastructure [as described in Section 2.1]. Termination of SSP by NASA will not lead to a specific decision or action on the future of each infrastructure asset and the associated environmental impacts to that asset. NASA will conduct an appropriate level of federally mandated NEPA analysis before final decisions on the disposition of SSP infrastructure assets are made. If any such properties are listed in or eligible for listing in the NRHP, NASA will take no action that would affect any such property until the NHPA Section 106 process is complete.	Moderate Impact
Hazardous and Toxic Materials and Waste	Proposed Action	**Storage and Handling.** If the facilities were reutilized, mothballed, demolished, or released to the GSA, minimal impacts on the storage and handling of hazardous materials associated with real property would be expected because waste generation would be expected to remain at the same level.	Minimal to No Impact
		Waste Management. If the facilities were mothballed, demolished, or released to the GSA, it would be likely that the waste management procedures associated with the SSP would no longer be applicable, because no wastes would be generated.	Minimal to No Impact

EXHIBIT 4-6
Summary of Environmental Consequences for JSC

Resource Area	Alternative	Overall Effects of Alternative	Impact
		Contaminated Areas. If the facilities were mothballed, demolished, or released to the GSA, it would be possible that new contaminated areas could be identified during closure activities (such as the closure of ASTs or USTs). Newly identified contaminated areas would be addressed by the Center's restoration programs. However, major impacts would be unlikely because the Center has undergone investigation efforts under RCRA/CERCLA.	Minimal Impact
		Asbestos and Lead-based Paint. If real property were to be demolished, it probably would contribute to the generation of asbestos waste or LBP. Asbestos and LBP surveys would be conducted before demolition. If ACMs were determined to be present, they would be removed appropriately before demolition. Such wastes would need to be disposed according to the hazardous waste classification determined.	Minimal to No Impact
	No Action	**Storage and Handling**. The property disposal process could become overwhelmed with the volume of property to disposition. The volume of property that would be processed could result in schedule and cost impacts if a structured disposal process were not implemented. In addition, the amount of hazardous waste that would require disposal could exceed landfill and less-than-90-day hazardous waste storage yard capacities.	Minimal Impact
		Waste Management. The property disposal process could become overwhelmed with the volume of property to disposition. The volume of property that would be processed could result in schedule and cost impacts if a structured disposal process were not implemented. In addition, the amount of hazardous waste that would require disposal could exceed landfill capacities.	Minimal Impact
		Contaminated Areas. If the facilities were mothballed, demolished, or released to the GSA, it would be possible that new contaminated areas could be identified during closure activities (such as the closure of ASTs or USTs). Newly identified contaminated areas would be addressed by the Center's restoration programs. However, major impacts would be unlikely because the Center has undergone investigation efforts under RCRA/CERCLA.	Minimal Impact

4. ENVIRONMENTAL CONSEQUENCES

EXHIBIT 4-6
Summary of Environmental Consequences for JSC

Resource Area	Alternative	Overall Effects of Alternative	Impact
		Asbestos and Lead-based Paint. If real property were to be demolished, it probably would contribute to the generation of asbestos waste or LBP. Asbestos and LBP surveys would be conducted before demolition. If ACMs were determined to be present, they would be removed appropriately before demolition. Such wastes would need to be disposed according to the hazardous waste classification determined. The property disposal process may become overwhelmed with the volume of property to disposition. The volume of asbestos due to demolition operations could result in schedule and cost impacts if a structured disposal process were not implemented. In addition, the amount of ACM that would require disposal could exceed landfill capacities.	Minimal Impact
Health and Safety	No Action and Proposed Action	Health and safety risks associated with real property at JSC could include contamination or damage resulting from major spills or accidents. Buildings at JSC could contain asbestos and LBP. Employees conducting renovation or demolition work must meet the safety standards outlined in 29 CFR 1926.1101, OSHA's construction standard for asbestos, or 29 CFR 1962.62, OSHA's construction standard for lead, to prevent exposure. The appropriate level of PPE must be worn, depending on the level of the abatement, in accordance with OSHA's construction standard to ensure worker safety.	Minimal Impact
Hydrology and Water Quality	No Action and Proposed Action	Disposition or removal of buildings or structures at JSC could result in minimal temporary soil disturbances, thus resulting in erosion during removal activities. The methods used for removal of any designated facilities would vary in relation to the type of structure, its location, materials encountered in demolition, and the contractor's experience. Best engineering practices, codes, specifications, and standards would be followed to prevent or limit potential impacts. These would include the implementation of erosion and turbidity controls. Storm water permits might need to be obtained and soil stabilization measures might need to be implemented.	Minimal Impact
Land Use	No Action and Proposed Action	If the existing facilities were destroyed or released to GSA, new facilities potentially could be constructed in their place.	Minimal Impact
Noise	No Action and Proposed Action	The demolition or disposition of property would have the potential to temporarily increase noise levels at JSC during the associated demolition or disposition operations. Any demolition or disposition activities would comply with the OSHA hearing protection standards for employees and other individuals in the vicinity.	Minimal Impact

EXHIBIT 4-6
Summary of Environmental Consequences for JSC

Resource Area	Alternative	Overall Effects of Alternative	Impact
Site Infrastructure	No Action and Proposed Action	The existing utilities are sufficient to support current activities at JSC. It is anticipated that most of the SSP property at JSC would be transferred to other NASA programs upon the disposition of SSP real and personal property. In this case, it is assumed that the infrastructure would continue to operate at similar levels, because it is assumed that programs receiving SSP property would have similar infrastructure needs. If the property were not transferred but remained unused, a decreased load on the site infrastructure would result. Any impacts would result from decreased use.	Minimal Impact
Socioeconomics	No Action and Proposed Action	The disposition of selected buildings and smaller parcels of land in an active facility, such as JSC, typically have minimal to no impact on socioeconomics outside the fence line. It is likely that many of the buildings at JSC that have been used by the SSP would be reused for these projects, with similar functions. Those that are unique to the SSP could be transferred for a different use or be demolished and the land reused. Otherwise, it is anticipated that demolition or conveyance of individual buildings (and land) would have minimal to no impact on socioeconomic resources in the surrounding area. The disposition of personal property would have minimal to no discernable impact on the regional economy surrounding JSC. The transfer of historic artifacts indirectly could benefit the museums (outside the region), Space Center Houston at JSC, or other museums by attracting visitors.	Minimal to No Impact
Solid Waste	Proposed Action	If it were determined that the real property would be demolished, the overall impacts probably would include the generation of solid waste consisting of concrete, asphalt, glass, metals (conduit, piping, and wiring), lumber, asbestos, and LBP. Items and materials that could be reused would be salvaged to the extent possible for NASA's future use. Non-salvageable solid waste would be disposed in accordance with all applicable health and safety and environmental regulations at JSC, where nonhazardous refuse would be taken to roll-off boxes at the Central Waste Collection Facility and shipped to the City of Houston landfill.	Minimal Impact
	No Action	The property disposal process could become overwhelmed with the volume of property to disposition. The volume of property that would be processed could result in schedule and cost impacts if a structured disposal process were not implemented. In addition, the amount of solid waste that would require disposal could exceed landfill capacities.	Minimal Impact

EXHIBIT 4-6
Summary of Environmental Consequences for JSC

Resource Area	Alternative	Overall Effects of Alternative	Impact
Transportation	No Action and Proposed Action	Real property demolition could generate more destruction-related truck trips. It would increase the traffic on the surrounding streets in the study area. A traffic control plan could be required to control the movement of truck traffic during the demolition.	Minimal Impact

Notes:
ACM = Asbestos-containing material
AST = Aboveground storage tank
CERCLA = Comprehensive Environmental Response, Compensation, and Liability Act
CFR = *Code of Federal Regulations*
EPA = National Environmental Policy Act
GSA = General Services Administration
JSC = Johnson Space Center
LBP = Lead-based paint
NASA = National Aeronautics and Space Administration
OSHA = Occupational Safety and Health Administration
PPE = Personal protective equipment
PTE = Potential to emit
RCRA = Resource Conservation Recovery Act
SSP = Space Shuttle Program
UST = Underground storage tank

4.4 Ellington Field

Exhibit 4-7 outlines the SSP property at EF.

EXHIBIT 4-7
Ellington Field SSP Property

SSP Asset or Facility	Description	Disposition
Three Aircraft Maintenance Hangars	EF maintains the following facilities in support of aircraft maintenance operations: - Wash rack that also is used as a hangar for the Guppy - Aircraft simulator and test facility for avionics - Maintenance Shops - Engine Testing Facility - Paint Shop - Tire Shop - X-Ray Facility	This property has not been identified for use by new programs.
T-38s	These aircraft are for astronaut transport between NASA facilities and for training purposes.	This property has not been identified for use by new programs.

4. ENVIRONMENTAL CONSEQUENCES

EXHIBIT 4-7
Ellington Field SSP Property

SSP Asset or Facility	Description	Disposition
Gulfstream 2	The Gulfstream 2 is used as an STA and the left side of the cockpit has been modified to simulate the flight controls of the Space Shuttle. In addition, other modifications have been made to the aircraft to simulate the flight characteristics of the Space Shuttle.	This property has not been identified for use by new programs.
Gulfstream 3	The Gulfstream 3 aircraft supports the transport of management teams and is capable of flying overseas, if necessary, to transport astronauts.	This property has not been identified for use by new programs.
C-9	The C-9 aircraft is used to support microgravity experiments and training for the astronauts. The aircraft will reach about 60,000 ft and will fly a parabolic pattern to simulate zero gravity.	This property has not been identified for use by new programs.
Guppy	The Guppy primarily is used by the ISS project to transport modules and other large components between NASA Centers.	This property has not been identified for use by new programs.
B-57	The B-57 supports high-altitude research programs and is able to test optical equipment and to collect air samples.	This property has not been identified for use by new programs.

Notes:
EF = Ellington Field
ft = Feet
ISS = International Space Station
NASA = National Aeronautics and Space Administration
STA = Shuttle Training Aircraft

4.4.1 Environmental Consequences for Ellington Field

The environmental resources that were evaluated and subsequently determined to have no potential for environmental impacts are provided in Exhibit 1-2. The environmental consequences for the resource areas present at EF are summarized in Exhibit 4-8.

EXHIBIT 4-8
Summary of Environmental Consequences for EF

Resource Area	Alternative	Overall Effects of Alternative	Impact
Air Quality	No Action and Proposed Action	The demolition or disposition of property would have the potential to temporarily increase emissions at EF during the demolition or disposition operations.	Minimal to No Impact

EXHIBIT 4-8
Summary of Environmental Consequences for EF

Resource Area	Alternative	Overall Effects of Alternative	Impact
Hazardous and Toxic Materials and Waste	Proposed Action	**Storage and Handling.** If the facilities were reutilized, mothballed, demolished, or released to the GSA, minimal impacts on the storage and handling of hazardous materials associated with real property would be expected because waste generation would be expected to remain at the same level.	Minimal to No Impact
		Waste Management. If the facilities were mothballed, demolished, or released to the GSA, it would be likely that the waste management procedures associated with the SSP would no longer be applicable, because no wastes would be generated.	Minimal to No Impact
		Contaminated Areas. Minimal to no impacts would be expected because there are no reported contaminated SSP areas.	Minimal to No Impact
		Asbestos and Lead-based Paint. If real property were to be demolished, it probably would contribute to the generation of asbestos waste or LBP. Asbestos and LBP surveys would be conducted before demolition. If ACMs were determined to be present, they would be removed appropriately before demolition. Such wastes would need to be disposed according to the hazardous waste classification determined.	Minimal to No Impact
	No Action	**Storage and Handling.** The property disposal process could become overwhelmed with the volume of property to disposition. The volume of property that would be processed could result in schedule and cost impacts if a structured disposal process were not implemented. In addition, the amount of hazardous waste that would require disposal could exceed landfill and less–than–90-day hazardous waste storage yard capacities.	Minimal Impact
		Waste Management. The property disposal process could become overwhelmed with the volume of property to disposition. The volume of property that would be processed could result in schedule and cost impacts if a structured disposal process were not implemented. In addition, the amount of hazardous waste that would require disposal could exceed landfill capacities.	Minimal Impact
		Contaminated Areas. If the facilities were mothballed, demolished, or released to the GSA, it would be possible that new contaminated areas could be identified during closure activities (such as the closure of ASTs or USTs). Newly identified contaminated areas would be addressed by the Center's restoration programs. However, major impacts would be unlikely because the Center has undergone investigation efforts under RCRA/CERCLA.	Minimal Impact

EXHIBIT 4-8
Summary of Environmental Consequences for EF

Resource Area	Alternative	Overall Effects of Alternative	Impact
		Asbestos and Lead-based Paint. If real property were to be demolished, it probably would contribute to the generation of asbestos waste or LBP. Asbestos and LBP surveys would be conducted before demolition. If ACMs were determined to be present, they would be removed appropriately before demolition. Such wastes would need to be disposed according to the hazardous waste classification determined. The property disposal process may become overwhelmed with the volume of property to disposition. The volume of asbestos due to demolition operations could result in schedule and cost impacts if a structured disposal process were not implemented. In addition, the amount of ACM that would require disposal could exceed landfill capacities.	Minimal Impact
Health and Safety	No Action and Proposed Action	Health and safety risks associated with real property at EF could include contamination or damage resulting from major spills or accidents. Buildings at EF could contain asbestos and LBP. Employees conducting renovation or demolition work must meet the safety standards outlined in 29 CFR 1926.1101, OSHA's construction standard for asbestos, or 29 CFR 1962.62, OSHA's construction standard for lead, to prevent exposure. The appropriate level of PPE must be worn depending on the level of the abatement, in accordance with OSHA's construction standard to ensure worker safety.	Minimal Impact
Hydrology and Water Quality	No Action and Proposed Action	Disposition or removal of buildings or structures at EF could result in minimal temporary soil disturbances, thus resulting in erosion during removal activities. The methods used for removal of any designated facilities would vary in relation to the type of structure, its location, the materials encountered in demolition, and the contractor's experience. Best engineering practices, codes, specifications, and standards would be followed to prevent or limit potential impacts. These would include the implementation of erosion and turbidity controls. Storm water permits might need to be obtained and soil stabilization measures might need to be implemented.	Minimal Impact
Land Use	No Action and Proposed Action	If the existing facilities were destroyed or released to GSA, new facilities potentially could be constructed in their place. It is anticipated that they would be compatible with the existing land use categories.	Minimal Impact

EXHIBIT 4-8
Summary of Environmental Consequences for EF

Resource Area	Alternative	Overall Effects of Alternative	Impact
Noise	No Action and Proposed Action	The demolition or disposition of property would have the potential to temporarily increase noise levels at EF during the associated demolition or disposition operations. Any demolition or disposition activities would comply with the OSHA hearing protection standards for employees and other individuals in the vicinity.	Minimal Impact
Site Infrastructure	No Action and Proposed Action	The existing utilities are sufficient to support current activities at EF. It is anticipated that most of the SSP property at EF would be transferred to other NASA programs upon disposition of SSP real and personal property. In this case, it is assumed that the infrastructure would continue to operate at similar levels, because it is assumed that programs receiving SSP property would have similar infrastructure needs. If the property were not transferred but remained unused, a decreased load on the site infrastructure would result. Any impacts would result from decreased use.	Minimal Impact
Solid Waste	Proposed Action	If it were determined that the real property would be demolished, the overall impacts probably would include the generation of solid waste consisting of concrete, asphalt, glass, metals (conduit, piping, and wiring), lumber, asbestos, and LBP. Items and materials that could be reused would be salvaged to the extent possible for NASA's future use. Non-salvageable solid waste would be disposed in accordance with all applicable health and safety and environmental regulations at EF, where nonhazardous refuse would be taken to roll-off boxes at the Central Waste Collection Facility and shipped to the City of Houston landfill.	Minimal Impact
	No Action	The property disposal process could become overwhelmed with the volume of property to disposition. The volume of property that would be processed could result in schedule and cost impacts if a structured disposal process were not implemented. In addition, the amount of solid waste that would require disposal could exceed landfill capacities.	Minimal Impact

EXHIBIT 4-8
Summary of Environmental Consequences for EF

Resource Area	Alternative	Overall Effects of Alternative	Impact
Transportation	No Action and Proposed Action	Real property demolition could generate more destruction-related truck trips. It would increase traffic on the surrounding streets in the study area. A traffic control plan could be required to control the movement of truck traffic during demolition.	Minimal Impact

Notes:
ACM = Asbestos-containing material
AST = Aboveground storage tank
CCSMP = Cape Canaveral Spaceport Master Plan
CERCLA = Comprehensive Environmental Response, Compensation, and Liability Act
CFR = Code of Federal Regulations
EF = Ellington Field
GSA = General Services Administration
LBP = Lead-based paint
NASA = National Aeronautics and Space Administration
NEPA = National Environmental Policy Act
OSHA = Occupational Safety and Health Administration
PPE = Personal protective equipment
PTE = Potential to emit
RCRA = Resource Conservation Recovery Act
SSP = Space Shuttle Program
UST = Underground storage tank

4.5 El Paso Forward Operation Location

Exhibit 4-9 outlines the major SSP property at EPFOL.

EXHIBIT 4-9
Major SSP Property at EPFOL

SSP Asset/Facility	Description	Disposition
Hangar 1 (STA hangar)	Operations in Hangar 1 include the following: • Astronaut training: Providing aircraft to train astronauts for Shuttle missions • Aircraft turn-around: Providing oversight to refueling operations and performing flight checks • Unscheduled maintenance: Providing maintenance for any items identified during flight checks or inspections. Such maintenance could include engine or thrust reverser replacement. • Aircraft washing: Approximately two airplanes are washed each month.	This property has not been identified for use by new programs.

EXHIBIT 4-9
Major SSP Property at EPFOL

SSP Asset/Facility	Description	Disposition
Hangar 2 (T-38 Hangar)	Operations in Hangar 2 include the following: • Handling the corrosion prevention program: Providing corrosion prevention treatment to T-38 aircraft. This treatment involves physical grinding and the application of primers, paints, and sealants in a paint booth. • Performing structural maintenance: Providing structural maintenance on aircraft on a non-routine basis. • Performing avionics system upgrade operations.	This property has not been identified for use by new programs

Note:
STA = Shuttle Training Aircraft

4.5.1 Environmental Consequences for EPFOL

The environmental resources that were evaluated and subsequently determined to have no potential for environmental impacts are provided in Exhibit 1-2. The environmental consequences for the resource areas present at EPFOL are summarized in Exhibit 4-10.

EXHIBIT 4-10
Summary of Environmental Consequences for EPFOL

Resource Area	Alternative	Overall Effects of Alternative	Impact
Air Quality	No Action and Proposed Action	The demolition or disposition of property would have the potential to temporarily increase emissions at EPFOL during the demolition or disposition operations.	Minimal to No Impact
Hazardous and Toxic Materials and Waste	Proposed Action	**Storage and Handling.** If the facilities were reutilized, mothballed, demolished, or released to the GSA, minimal impacts on the storage and handling of hazardous materials associated with real property would be expected because waste generation would be expected to remain at the same level.	Minimal to No Impact
		Waste Management. If the facilities were mothballed, demolished, or released to the GSA, it would be likely that the waste management procedures associated with the SSP would no longer be applicable, because no wastes would be generated.	Minimal to No Impact

EXHIBIT 4-10
Summary of Environmental Consequences for EPFOL

Resource Area	Alternative	Overall Effects of Alternative	Impact
		Asbestos and Lead-based Paint. If real property were to be demolished, it probably would contribute to the generation of asbestos waste or LBP. Asbestos and LBP surveys would be conducted before demolition. If ACMs were determined to be present, they would be removed appropriately before demolition. Such wastes would need to be disposed according to the hazardous waste classification determined.	Minimal to No Impact
	No Action	**Storage and Handling.** The property disposal process could become overwhelmed with the volume of property to disposition. The volume of property that would be processed could result in schedule and cost impacts if a structured disposal process were not implemented. In addition, the amount of hazardous waste that would require disposal could exceed landfill and less–than–90-day hazardous waste storage yard capacities.	Minimal Impact
		Waste Management. The property disposal process could become overwhelmed with the volume of property to disposition. The volume of property that would be processed could result in schedule and cost impacts if a structured disposal process were not implemented. In addition, the amount of hazardous waste that would require disposal could exceed landfill capacities.	Minimal Impact
		Asbestos and Lead-based Paint. If real property were to be demolished, it probably would contribute to the generation of asbestos waste or LBP. Asbestos and LBP surveys would be conducted before demolition. If ACMs were determined to be present, they would be removed appropriately before demolition. Such wastes would need to be disposed according to the hazardous waste classification determined. The property disposal process could become overwhelmed with the volume of property to disposition. The volume of asbestos due to demolition operations could result in schedule and cost impacts if a structured disposal process were not implemented. In addition, the amount of ACM that would require disposal could exceed landfill capacities.	Minimal Impact

EXHIBIT 4-10
Summary of Environmental Consequences for EPFOL

Resource Area	Alternative	Overall Effects of Alternative	Impact
Health and Safety	No Action and Proposed Action	Health and safety risks associated with real property at EPFOL could include contamination or damage resulting from major spills or accidents. Buildings at JSC could contain asbestos and LBP. Employees conducting renovation or demolition work must meet the safety standards outlined in 29 CFR 1926.1101, OSHA's construction standard for asbestos, or 29 CFR 1962.62, OSHA's construction standard for lead, to prevent exposure. The appropriate level of PPE must be worn depending on the level of the abatement, in accordance with OSHA's construction standard to ensure worker safety.	Minimal Impact
Hydrology and Water Quality	No Action and Proposed Action	Disposition or removal of buildings or structures at EPFOL could result in minimal temporary soil disturbances, thus resulting in erosion during removal activities. The methods used for the removal of any designated facilities would vary in relation to the type of structure, its location, the materials encountered in demolition, and the contractor's experience. Best engineering practices, codes, specifications, and standards would be followed to prevent or limit potential impacts. These would include the implementation of erosion and turbidity controls. Storm water permits might need to be obtained and soil stabilization measures might need to be implemented.	Minimal Impact
Noise	No Action and Proposed Action	The demolition or disposition of property would have the potential to temporarily increase noise levels at EPFOL during the associated demolition or disposition operations. Any demolition or disposition activities would comply with the OSHA hearing protection standards for employees and other individuals in the vicinity.	Minimal Impact
Site Infrastructure	No Action and Proposed Action	The existing utilities are sufficient to support current activities at EPFOL. It is anticipated that most of the SSP property at EPFOL would be transferred to other NASA programs upon disposition of SSP real and personal property. In this case, it is assumed that the infrastructure would continue to operate at similar levels, because it is assumed that programs receiving SSP property would have similar infrastructure needs. If the property were not transferred but remained unused, a decreased load on the site infrastructure would result. Any impacts would result from decreased use.	Minimal Impact

EXHIBIT 4-10
Summary of Environmental Consequences for EPFOL

Resource Area	Alternative	Overall Effects of Alternative	Impact
Solid Waste	Proposed Action	If it were determined that the real property would be demolished, the overall impacts probably would include the generation of solid waste consisting of concrete, asphalt, glass, metals (conduit, piping, and wiring), lumber, asbestos, and LBP. Items and materials that could be reused would be salvaged to the extent possible for NASA's future use. Non-salvageable solid waste would be disposed in accordance with all applicable health and safety and environmental regulations.	Minimal Impact
	No Action	The property disposal process could become overwhelmed with the volume of property to disposition. The volume of property that would be processed could result in schedule and cost impacts if a structured disposal process were not implemented. In addition, the amount of solid waste that would require disposal could exceed landfill capacities.	Minimal Impact
Transportation	No Action and Proposed Action	Real property demolition could generate more destruction-related truck trips. It would increase traffic on the surrounding streets in the study area. A traffic control plan could be required to control the movement of truck traffic during the demolition.	Minimal Impact

Notes:
ACM = Asbestos-containing material
AST = Aboveground storage tank
CCSMP = Cape Canaveral Spaceport Master Plan
CERCLA = Comprehensive Environmental Response, Compensation, and Liability Act
CFR = Code of Federal Regulations
EPFOL = El Paso Forward Operation Location
GSA = General Services Administration
LBP = Lead-based paint
MAF = Michoud Assembly Facility
NASA = National Aeronautics and Space Administration
NEPA = National Environmental Policy Act
OSHA = Occupational Safety and Health Administration
PPE = Personal protective equipment
PTE = Potential to emit
RCRA = Resource Conservation Recovery Act
SSP = Space Shuttle Program
UST = Underground storage tank

4.6 Stennis Space Center

Exhibit 4-11 outlines the major SSP property at SSC.

EXHIBIT 4-11
Major SSP Property at SSC

SSP Asset/Facility	Description	Disposition
Test Stand A-1	Test stands A-1 and A-2 have been used since 1975 to test the SSME.	Test stand A-1 is scheduled to transfer to the Constellation Program.
Test Stand A-2	Test stand A-2 has been used for SSME testing.	Test Stand A-2 also is proposed for Constellation use after 2010.
B1/B2 Test Stand	SSMEs are not tested at the B1/B2 Test Stand.	The Constellation Program will use this asset.
E-Complex	The E-Complex has three test stands.	The Constellation Program will use this asset.
Test Control Centers	Supports operation of the test stands.	NASA will evaluate future disposition options for this property.
Data Acquisition Facilities	Supports operation of the test stands.	NASA will evaluate future disposition options for this property.
Cryogenic Propellant Facility	Supports operation of the test stands.	NASA will evaluate future disposition options for this property.
Electrical Power-Generating Plant	Supports operation of the test stands.	NASA will evaluate future disposition options for this property.
Navigation Canal and Locks	NASA maintains a 7-mile manmade navigation canal and locks system for the transfer of liquid gases in supports operation of the test stands. SSC has nine barges used to transfer liquid gasses. Three of the barges are used to transfer hydrogen and six are used to transfer oxygen.	Currently, it is anticipated that the nine barges would be transferred to the Constellation Program.
Water Storage Reservoir	NASA maintains a 66-million-gallon water storage reservoir for industrial and deluge water consumption.	NASA will evaluate future disposition options for this property.

Notes:
KSC = Kennedy Space Center
NASA = National Aeronautics and Space Administration
PWR = Pratt-Whitney Rocketdyne
SSC = Stennis Space Center
SSME = Space Shuttle main engine
SSP = Space Shuttle Program

4.6.1 Environmental Consequences Summary for SSC

The environmental resources that were evaluated and subsequently determined to have no potential for environmental impacts are provided in Exhibit 1-2. The environmental consequences for the resource areas present at SSC are summarized in Exhibit 4-12.

EXHIBIT 4-12
Summary of Environmental Consequences for SSC

Resource Area	Alternative	Overall Effects of Alternative	Impact
Air Quality	No Action and Proposed Action	The demolition or disposition of property would have the potential to temporarily increase emissions at SSC during the demolition or disposition operations.	Minimal to No Impact
Biological Resources	No Action and Proposed Action	**Vegetation.** Facilities on SSC are located in developed portions of the Fee Area, and little or no natural vegetation would be disturbed by the implementation of the Proposed Action and the No Action alternative.	Minimal Impact
		Wetlands. No facilities on SSC are located in wetlands.	No Impact
		Floodplains. No facilities on SSC are located in floodplains.	No Impact
		Wildlife. Facilities on SSC are located in developed portions of the Fee Area, and these developed areas do not provide quality habitat for the wildlife. Therefore, minimal impact of the Proposed Action and the No Action alternative on wildlife would be anticipated because the disposition of the property would occur in developed portions of the Fee Area.	Minimal Impact
		Protected Species. Facilities on SSC are located in developed portions of the Fee Area, and these developed areas do not provide quality habitat for wildlife. Therefore, minimal impact of the Proposed Action and the No Action alternative on protected species would be anticipated because the disposition of the property would occur in the developed portions of the Fee Area.	Minimal Impact
Cultural Resources	No Action and Proposed Action	**Archaeological Resources.** There would be minimal to no impact on archaeological resources under the Proposed Action and No Action Alternatives because no ground-disturbing activities are anticipated.	Minimal to No Impact

EXHIBIT 4-12
Summary of Environmental Consequences for SSC

Resource Area	Alternative	Overall Effects of Alternative	Impact
		Historic Resources. As the SSP approaches the end of its mission, a variety of buildings and facilities at several NASA installations will no longer be of use to SSP. Once SSP identifies and reports to a host installation that it no longer needs a building or facility, NASA will initiate the standard process for addressing excess infrastructure (as described in Section 2.1). Termination of SSP by NASA will not lead to a specific decision or action on the future of each infrastructure asset and the associated environmental impacts to that asset. NASA will conduct an appropriate level of federally mandated NEPA analysis before final decisions on the disposition of SSP infrastructure assets are made. If any such properties are listed in or eligible for listing in the NRHP, NASA will take no action that would affect any such property until the NHPA Section 106 process is complete.	Moderate Impact
Hazardous and Toxic Materials and Waste	Proposed Action	**Storage and Handling.** If the facilities were reutilized, mothballed, demolished, or released to the GSA, minimal impacts on the storage and handling of hazardous materials associated with real property would be expected because waste generation would be expected to remain at the same level.	Minimal to No Impact
		Waste Management. If the facilities were mothballed, demolished, or released to the GSA, it would be likely that the waste management procedures associated with the SSP would no longer be applicable, because no wastes would be generated.	Minimal to No Impact
		Contaminated Areas. If the facilities were mothballed, demolished, or released to the GSA, it could be possible that new contaminated areas might be identified during closure activities (such as the closure of ASTs or USTs). Newly identified contaminated areas would be addressed by the Center's restoration programs. However major impacts would be unlikely because the Center has undergone investigation efforts under RCRA/CERCLA.	Minimal Impact
		Asbestos and Lead-based Paint. If real property were to be demolished, it probably would contribute to the generation of asbestos waste or LBP. Asbestos and LBP surveys would be conducted before demolition. If ACMs were determined to be present, they would be removed appropriately before demolition. Such wastes would need to be disposed according to the hazardous waste classification determined.	Minimal to No Impact

EXHIBIT 4-12
Summary of Environmental Consequences for SSC

Resource Area	Alternative	Overall Effects of Alternative	Impact
	No Action	**Storage and Handling.** The property disposal process could become overwhelmed with the volume of property to disposition. The volume of property that would be processed could result in schedule and cost impacts if a structured disposal process were not implemented. In addition, the amount of hazardous waste that would require disposal could exceed landfill and less–than–90-day hazardous waste storage yard capacities.	Minimal Impact
		Waste Management. The property disposal process could become overwhelmed with the volume of property to disposition. The volume of property that would be processed could result in schedule and cost impacts if a structured disposal process were not implemented. In addition, the amount of hazardous waste that would require disposal could exceed landfill capacities.	Minimal Impact
		Contaminated Areas. If the facilities were mothballed, demolished, or released to the GSA, it would be possible that new contaminated areas could be identified during closure activities (such as the closure of ASTs or USTs). Newly identified contaminated areas would be addressed by the Center's restoration programs. However, major impacts would be unlikely because the Center has undergone investigation efforts under RCRA/CERCLA.	Minimal Impact
		Asbestos and Lead-based Paint. If real property were to be demolished, it probably would contribute to the generation of asbestos waste or LBP. Asbestos and LBP surveys would be conducted before demolition. If ACMs were determined to be present, they would be removed appropriately before demolition. Such wastes would need to be disposed according to the hazardous waste classification determined. The property disposal process could become overwhelmed with the volume of property to disposition. The volume of asbestos due to demolition operations could result in schedule and cost impacts if a structured disposal process were not implemented. In addition, the amount of ACM that would require disposal could exceed landfill capacities.	Minimal Impact

EXHIBIT 4-12
Summary of Environmental Consequences for SSC

Resource Area	Alternative	Overall Effects of Alternative	Impact
Health and Safety	No Action and Proposed Action	Health and safety risks associated with real property at SSC could include contamination or damage resulting from major spills or accidents. Buildings at SSC could contain asbestos and LBP. Employees conducting renovation or demolition work must meet the safety standards outlined in 29 CFR 1926.1101, OSHA's construction standard for asbestos, or 29 CFR 1962.62, OSHA's construction standard for lead, to prevent exposure. The appropriate level of PPE must be worn depending on the level of the abatement, in accordance with OSHA's construction standard to ensure worker safety.	Minimal Impact
Hydrology and Water Quality	No Action and Proposed Action	Disposition or removal of buildings or structures at SSC could result in minimal temporary soil disturbances, thus resulting in erosion during removal activities. The methods used for the removal of any designated facilities would vary in relation to the type of structure, its location, the materials encountered in demolition, and the contractor's experience. Best engineering practices, codes, specifications, and standards would be followed to prevent or limit potential impacts. These would include the implementation of erosion and turbidity controls. Storm water permits might need to be obtained and soil stabilization measures might need to be implemented.	Minimal Impact
Land Use	No Action and Proposed Action	If the existing facilities were destroyed or released to GSA, new facilities potentially could be constructed in their place. It is anticipated that they would be compatible with the existing land use categories.	Minimal Impact
Noise	No Action and Proposed Action	The demolition or disposition of property would have the potential to temporarily increase noise levels at SSC during the associated demolition or disposition operations. Any demolition or disposition activities would comply with the OSHA hearing protection standards for employees and other individuals in the vicinity.	Minimal Impact

EXHIBIT 4-12
Summary of Environmental Consequences for SSC

Resource Area	Alternative	Overall Effects of Alternative	Impact
Site Infrastructure	No Action and Proposed Action	The existing utilities are sufficient to support current activities at SSC. It is anticipated that most of the SSP property at SSC would be transferred to other NASA programs upon disposition of SSP real and personal property. In this case, it is assumed that the infrastructure would continue to operate at similar levels, because it is assumed that programs receiving SSP property would have similar infrastructure needs. If the property were not transferred but remained unused, a decreased load on the site infrastructure would result. Any impacts would result from decreased use.	Minimal Impact
Solid Waste	Proposed Action	If it were determined that the real property would be demolished, the overall impacts probably would include the generation of solid waste consisting of concrete, asphalt, glass, metals (conduit, piping, and wiring), lumber, asbestos, and LBP. Items and materials that could be reused would be salvaged to the extent possible for NASA's future use.	Minimal Impact
	No Action	The property disposal process could become overwhelmed with the volume of property to disposition. The volume of property that would be processed could result in schedule and cost impacts if a structured disposal process were not implemented. In addition, the amount of solid waste that would require disposal could exceed landfill capacities.	Minimal Impact
Transportation	No Action and Proposed Action	Real property demolition could generate more destruction-related truck trips. It would increase the traffic on the surrounding streets in the study area. A traffic control plan could be required to control the movement of truck traffic during the demolition.	Minimal Impact

Notes:
ACM = Asbestos-containing material
AST = Aboveground storage tank
CCSMP = Cape Canaveral Spaceport Master Plan
CERCLA = Comprehensive Environmental Response, Compensation, and Liability Act
CFR = Code of Federal Regulations
GSA = General Services Administration
LBP = Lead-based paint
NASA = National Aeronautics and Space Administration
NEPA = National Environmental Policy Act
OSHA = Occupational Safety and Health Administration
PPE = Personal protective equipment
PTE = Potential to emit
RCRA = Resource Conservation Recovery Act
SSC = Stennis Space Center
SSP = Space Shuttle Program
UST = Underground storage tank

4.7 Michoud Assembly Facility

Exhibit 4-13 outlines the major property at MAF.

EXHIBIT 4-13
Major Property at MAF

SSP Asset/Facility	Description	Disposition
41 buildings	MAF operates 41 buildings associated with machining, welding, and cleaning aluminum and aluminum-lithium panels and various parts, and 92 major tooling and unique equipment workstations to produce the ET's three major components.	Several facilities at MAF have been identified for use by the Constellation Program, including, but not limited to, the following: Manufacturing Building (103) – Ares Upper Stage structural welding, avionics, and common bulkhead assembly. Vertical Assembly Building (Building 110) – Ares Upper Stage and Orion Crew Module, Service Module, back shell, and heat shield fabrication. Acceptance and Preparation Building (Building 420) – Ares Upper Stage. Pneumatic Test Facility and Control Building (Buildings 451 and 452) – Pressure and dynamic test area. High Bay Addition (Building 114) – Ares I Upper Stage and Ares V Core Stage assembly and foam application.

Notes:
ET = External Tank
MAF = Michoud Assembly Facility

4.7.1 Environmental Consequences Summary for MAF

The environmental resources that were evaluated and subsequently determined to have no potential for environmental impacts are provided in Exhibit 1-2. The environmental consequences for the resource areas present at MAF are summarized in Exhibit 4-14.

EXHIBIT 4-14
Summary of Environmental Consequences for MAF

Resource Area	Alternative	Overall Effects of Alternative	Impact
Air Quality	No Action and Proposed Action	The demolition or disposition of property would have the potential to temporarily increase emissions at MAF during the demolition or disposition operations.	Minimal to No Impact

EXHIBIT 4-14
Summary of Environmental Consequences for MAF

Resource Area	Alternative	Overall Effects of Alternative	Impact
Biological Resources	No Action and Proposed Action	**Vegetation.** MAF has been heavily altered; the undeveloped areas are regularly maintained and little or no natural vegetation would be disturbed.	Minimal Impact
		Wetlands. No facilities on MAF are located in wetlands.	No Impact
		Floodplains. No existing development on MAF is located within the floodplains	No Impact
		Wildlife. Increased human activity and noise due to the disposition and demolition of real property temporarily could increase disturbance of wildlife.	Minimal Impact
		Protected Species. No protected species rely on the SSP NASA properties for habitat, and it is unlikely that protected species are present on MAF.	No Impact
Cultural Resources	No Action and Proposed Action	**Archaeological Resources.** There would be minimal to no impact on archaeological resources under the Proposed Action and No Action Alternatives because no ground-disturbing activities are anticipated.	Minimal to No Impact
		Historic Resources. As the SSP approaches the end of its mission, a variety of buildings and facilities at several NASA installations will no longer be of use to SSP. Once SSP identifies and reports to a host installation that it no longer needs a building or facility, NASA will initiate the standard process for addressing excess infrastructure (as described in Section 2.1). Termination of SSP by NASA will not lead to a specific decision or action on the future of each infrastructure asset and the associated environmental impacts to that asset. NASA will conduct an appropriate level of federally mandated NEPA analysis before final decisions on the disposition of SSP infrastructure assets are made. If any such properties are listed in or eligible for listing in the NRHP, NASA will take no action that would affect any such property until the NHPA Section 106 process is complete.	Moderate Impact
Hazardous and Toxic Materials and Waste	Proposed Action	**Storage and Handling.** If the facilities were reutilized, mothballed, demolished, or released to the GSA, minimal impacts on the storage and handling of hazardous materials associated with real property would be expected because waste generation would be expected to remain at the same level.	Minimal to No Impact

4. ENVIRONMENTAL CONSEQUENCES

EXHIBIT 4-14
Summary of Environmental Consequences for MAF

Resource Area	Alternative	Overall Effects of Alternative	Impact
		Waste Management. If the facilities were mothballed, demolished, or released to the GSA, it would be likely that the waste management procedures associated with the SSP would no longer be applicable, because no wastes would be generated.	Minimal to No Impact
		Contaminated Areas. If the facilities were mothballed, demolished, or released to the GSA, it would be possible that new contaminated areas might be identified during the closure activities (closure of ASTs or USTs). Newly identified contaminated areas would be addressed by the Center's restoration programs. However, significant impacts would be unlikely because the Center has undergone investigation efforts under RCRA/CERCLA.	Minimal Impact
		Asbestos and Lead-based Paint. If real property were to be demolished, it probably would contribute to the generation of asbestos waste or LBP. Asbestos and LBP surveys would be conducted before demolition. If ACMs were determined to be present, they would be removed appropriately before demolition. Such wastes would need to be disposed according to the hazardous waste classification determined.	Minimal to No Impact
	No Action	**Storage and Handling.** The property disposal process may become overwhelmed with the volume of property to disposition. The volume of property that would be processed could result in schedule and cost impacts if a structured disposal process were not implemented. In addition, the amount of hazardous waste that would require disposal could exceed landfill and less–than-90-day hazardous waste storage yard capacities.	Minimal Impact
		Waste Management. The property disposal process could become overwhelmed with the volume of property to disposition. The volume of property that would be processed could result in schedule and cost impacts if a structured disposal process were not implemented. In addition, the amount of hazardous waste that would require disposal could exceed landfill capacities.	Minimal Impact

EXHIBIT 4-14
Summary of Environmental Consequences for MAF

Resource Area	Alternative	Overall Effects of Alternative	Impact
		Contaminated Areas. If the facilities were mothballed, demolished, or released to the GSA, it would be possible that new contaminated areas could be identified during closure activities (such as the closure of ASTs or USTs). Newly identified contaminated areas would be addressed by the Center's restoration programs. However, major impacts would be unlikely because the Center has undergone investigation efforts under RCRA/CERCLA.	Minimal Impact
		Asbestos and Lead-based Paint. If real property were to be demolished, it probably would contribute to the generation of asbestos waste or LBP. Asbestos and LBP surveys would be conducted before demolition. If ACMs were determined to be present, they would be removed appropriately before demolition. Such wastes would need to be disposed according to the hazardous waste classification determined. The property disposal process could become overwhelmed with the volume of property to disposition. The volume of asbestos due to demolition operations could result in schedule and cost impacts if a structured disposal process were not implemented. In addition, the amount of ACM that would require disposal could exceed landfill capacities.	Minimal Impact
Health and Safety	No Action and Proposed Action	Health and safety risks associated with real property at MAF could include contamination or damage resulting from major spills or accidents. Buildings at MAF could contain asbestos and LBP. Employees conducting renovation or demolition work must meet the safety standards outlined in 29 CFR 1926.1101, OSHA's construction standard for asbestos, or 29 CFR 1962.62, OSHA's construction standard for lead, to prevent exposure. The appropriate level of PPE must be worn depending on the level of the abatement, in accordance with OSHA's construction standard to ensure worker safety.	Minimal Impact
Hydrology and Water Quality	No Action and Proposed Action	Disposition or removal of buildings or structures at MAF could result in minimal temporary soil disturbances, thus resulting in erosion during the removal activities. The methods used for removal of any designated facilities would vary in relation to the type of structure, its location, the materials encountered in demolition, and the contractor's experience. Best engineering practices, codes, specifications, and standards would be followed to prevent or limit potential impacts. These would include the implementation of	Minimal Impact

EXHIBIT 4-14
Summary of Environmental Consequences for MAF

Resource Area	Alternative	Overall Effects of Alternative	Impact
		erosion and turbidity controls. Storm water permits might need to be obtained and soil stabilization measures might need to be implemented.	
Land Use	No Action and Proposed Action	If the existing facilities were destroyed or released to the GSA, new facilities potentially could be constructed in their place. It is anticipated that they would be compatible with the existing land use categories.	Minimal Impact
Noise	No Action and Proposed Action	The demolition or disposition of real property would have the potential to temporarily increase noise levels at MAF during the associated demolition or disposition operations. Any demolition or disposition activities would comply with the OSHA hearing protection standards for employees and other individuals in the vicinity.	Minimal Impact
Site Infrastructure	No Action and Proposed Action	The existing utilities are sufficient to support the current activities at MAF. It is anticipated that most of the SSP property at MAF would be transferred to other NASA programs upon SSP T&R. In this case, it is assumed that infrastructure would continue to operate at similar levels because any programs receiving SSP property would have similar infrastructure needs. If the future programs at MAF required additional utility capacity, the new facilities would be required to undergo evaluation under NEPA, and the potential for any utility service incompatibilities would be identified at that time.	Minimal Impact
Socioeconomics	No Action and Proposed Action	The specific disposition methods selected for the SSP real and personal property probably would have minimal to no impact on the population, regional economy, and community services in the region surrounding MAF. MAF has been selected by NASA to manufacture large structures and composites for future vehicles. It is expected that most of the buildings at MAF that are used by the SSP would be reused for future space programs, with similar functions.	Minimal to No Impact
Solid Waste	Proposed Action	If it were determined that the real property would be demolished, the overall impacts probably would include the generation of solid waste consisting of concrete, asphalt, glass, metals (conduit, piping, and wiring), lumber, asbestos, and LBP. Items and materials that could be reused would be salvaged to the extent possible for NASA's future use. Non-RCRA solid wastes would be collected and sent to an offsite landfill.	Minimal Impact

EXHIBIT 4-14
Summary of Environmental Consequences for MAF

Resource Area	Alternative	Overall Effects of Alternative	Impact
	No Action	The property disposal process may become overwhelmed with the volume of property to disposition. The volume of property that would be processed could result in schedule and cost impacts if a structured disposal process were not implemented. In addition, the amount of solid waste that would require disposal could exceed landfill capacities.	Minimal Impact
Transportation	No Action and Proposed Action	Real property demolition could generate more destruction-related truck trips. It would increase traffic on the surrounding streets in the study area. A traffic control plan could be required to control the movement of truck traffic during the demolition.	Minimal Impact

Notes:
ACM = Asbestos-containing material
AST = Aboveground storage tank
CERCLA = Comprehensive Environmental Response, Compensation, and Liability Act
CFR = *Code of Federal Regulations*
GSA = General Services Administration
LBP = Lead-based paint
MAF = Michoud Assembly Facility
NASA = National Aeronautics and Space Administration
NEPA = National Environmental Policy Act
OSHA = Occupational Safety and Health Administration
PPE = Personal protective equipment
PTE = Potential to emit
RCRA = Resource Conservation Recovery Act
SSP = Space Shuttle Program
UST = Underground storage tank

4.8 Marshall Space Flight Center

Exhibit 4-15 outlines the major SSP property at MSFC.

EXHIBIT 4-15
Major SSP Property at MSFC

SSP Asset/Facility	Description	Disposition
Office Building (Building 4202)	Six project offices for major development projects and six directorates that embody the institutional capabilities of MSFC carry out NASA's missions. The directorates are Shuttle Propulsion, Space Transportation Programs/ Projects, Space Systems Programs/Projects, Engineering, Science and Technology, and Center Operations. These facilities	NASA will evaluate future disposition options for this property.

EXHIBIT 4-15
Major SSP Property at MSFC

SSP Asset/Facility	Description	Disposition
	and directorates do not contain significant personal property.	
Office Building (Building 4203)	These facilities do not contain significant personal property.	NASA will evaluate future disposition options for this property.
Communications Facility (Building 4207)	Provides multimedia services to MSFC.	NASA will evaluate future disposition options for this property.
Hardware Simulation Laboratory (Building 4436)	The facility was designed to test and verify the SSME avionics and software, control system, and mathematical models.	Building 4436 has been identified for use by the Constellation Program. Ares Upper Stage engine control system and software testing and avionics and systems integration will be conducted in the facility.
Test Control and Services Building (Building 4561)	Engine testing at the test stands are controlled at the Test Control and Services Building.	NASA will evaluate future disposition options for this property.
Office Building (Building 4566)	These facilities do not contain significant personal property.	NASA will evaluate future disposition options for this property.
Office Building (Building 4600)	These facilities do not contain significant personal property.	NASA will evaluate future disposition options for this property.
Shuttle Hardware Storage (Building 4625)	These facilities do not contain significant personal property.	NASA will evaluate future disposition options for this property.
Multi-purpose HB Facility and Neutral Buoyancy Simulator (Building 4705)	The facility was designed to provide a simulated zero-gravity environment in which engineers, designers, and astronauts could perform, for extended periods of time, the various phases of space development to gain a first-hand knowledge of design problems and operational characteristics. The tank is 75 feet in diameter and 40 feet deep and designed to hold 1.5 million gallons of water. There are four observation levels for underwater audio and video communications. The southwestern corner of Building 4705 that houses the facility has a completely equipped test control center for	Building 4705 has been identified for use by the Constellation Program for Ares Upper Stage fabrication.

EXHIBIT 4-15
Major SSP Property at MSFC

SSP Asset/Facility	Description	Disposition
	directing, controlling, and monitoring the simulation activities.	
National Center for Advanced Manufacturing (Building 4707)	The National Center for Advanced Manufacturing (NCAM) addresses the manufacturing requirements of space transportation systems. NASA partners with other government agencies, industry, and academia in support of NCAM to leverage assets and successfully meet the requirements of future systems to provide safe, low-cost, access to space.	Building 4707 has been identified for use by the Constellation Program for Ares Upper Stage support actions and evaluations.
Engineering and Development Laboratory (Building 4708)	Contains laboratory space used for SSP and ISS development.	Building 4708 has been identified for use by the Constellation Program for final assembly and preparation for Ares Upper Stage testing.
Developmental Process Laboratory (Building 4711)	The Process and Methods Development Laboratory occupies 12,000 feet of floor area in Building 4711. The facility is for the development and testing of new processes, techniques, materials, and mechanical manufacturing devices as they relate to fabrication and assembly.	NASA will evaluate future disposition options for this property.
SOFI Formulation Facility (Building 4739)	The SOFI Formation Facility serves as a laboratory for the development of improved foam for space vehicle insulation.	NASA will evaluate future disposition options for this property.

Notes:
HB = High bay
MSFC = Marshall Space Flight Center
NASA = National Aeronautics and Space Administration
SOFI = Spray-on foam insulation

4.8.1 Environmental Consequences Summary for MSFC

The environmental resources that were evaluated and subsequently determined to have no potential for environmental impacts are provided in Exhibit 1-2. The environmental consequences for the resource areas present at MSFC are summarized in Exhibit 4-16.

EXHIBIT 4-16
Summary of Environmental Consequences for MSFC

Resource Area	Alternative	Overall Effects of Alternative	Impact
Air Quality	No Action and Proposed Action	The demolition or disposition of property would have the potential to temporarily increase emissions at MSFC during the demolition or disposition operations.	Minimal to No Impact
Biological Resources	No Action and Proposed Action	**Vegetation.** Most of the NASA operational areas at MSFC are on developed landscapes and are devoid of natural vegetation. Natural vegetation may spread into the developed areas once the property has been disposed, thereby increasing the distribution of natural vegetation in the area. NASA operational areas that currently support natural vegetation would remain undisturbed with the disposition of NASA property.	Minimal Impact
		Wetlands. No facilities on MSFC are located in wetlands.	No Impact
		Floodplains. No SSP facilities on MSFC are located in floodplains.	No Impact
		Wildlife. Facilities on MSFC are located in developed areas that do not provide quality habitat for wildlife. Therefore, minimal impact of the Proposed Action and the No Action alternative on wildlife would be anticipated because the disposition of the property would occur in developed areas.	Minimal Impact
		Protected Species. Facilities on MSFC are located in developed areas that do not provide quality habitat for the wildlife. Therefore, minimal impact of the Proposed Action and the No Action alternative on protected species would be anticipated because the disposition of the property would occur in developed areas.	Minimal Impact
Cultural Resources	No Action and Proposed Action	**Archaeological Resources.** There would be minimal to no impact on archaeological resources under the Proposed Action and No Action Alternatives because no ground-disturbing activities are anticipated.	Minimal to No Impact
		Historic Resources. As the SSP approaches the end of its mission, a variety of buildings and facilities at several NASA installations will no longer be of use to SSP. Once SSP identifies and reports to a host installation that it no longer needs a building or facility, NASA will initiate the standard process for addressing excess infrastructure [as described in Section 2.1]. Termination of SSP by NASA will not lead to a specific decision or action on the future of each infrastructure asset and the associated environmental impacts to that asset. NASA will conduct an appropriate level of federally mandated NEPA analysis before final decisions on the disposition of SSP infrastructure assets are made. If any such properties are listed in or eligible for listing in the NRHP, NASA will take no action that would affect any such property until the NHPA Section 106 process is complete.	Moderate Impact

EXHIBIT 4-16
Summary of Environmental Consequences for MSFC

Resource Area	Alternative	Overall Effects of Alternative	Impact
Hazardous and Toxic Materials and Waste	Proposed Action	**Storage and Handling.** If the facilities were reutilized, mothballed, demolished, or released to the GSA, minimal impacts on the storage and handling of hazardous materials associated with real property would be expected because waste generation would be expected to remain at the same level.	Minimal to No Impact
		Waste Management. If the facilities were mothballed, demolished, or released to the GSA, it would be likely that the waste management procedures associated with the SSP would no longer be applicable, because no wastes would be generated.	Minimal to No Impact
		Contaminated Areas. If the facilities were mothballed, demolished, or released to the GSA, it is possible that new contaminated areas could be identified during closure activities (such as the closure of ASTs or USTs). Newly identified contaminated areas would be addressed by the Center's restoration programs. However, major impacts would be unlikely because the Center has undergone investigation efforts under RCRA/ CERCLA.	Minimal Impact
		Asbestos and Lead-based Paint. If real property were to be demolished, it probably would contribute to the generation of asbestos waste or LBP. Asbestos and LBP surveys would be conducted before demolition. If ACMs were determined to be present, they would be removed appropriately before demolition. Such wastes would need to be disposed according to the hazardous waste classification determined.	Minimal to No Impact
	No Action	**Storage and Handling.** The property disposal process could become overwhelmed with the volume of property to disposition. The volume of property that would be processed could result in schedule and cost impacts if a structured disposal process were not implemented. In addition, the amount of hazardous waste that would require disposal could exceed landfill and less-than-90-day hazardous waste storage yard capacities.	Minimal Impact
		Waste Management. The property disposal process could become overwhelmed with the volume of property to disposition. The volume of property that would be processed could result in schedule and cost impacts if a structured disposal process were not implemented. In addition, the amount of hazardous waste that would require disposal could exceed landfill capacities.	Minimal Impact
		Contaminated Areas. If the facilities were mothballed, demolished, or released to the GSA, it would be possible that new contaminated areas could be identified during closure activities (such as the closure of ASTs or USTs). Newly identified contaminated areas would be addressed by the Center's restoration programs. However, major impacts would be unlikely because the Center has undergone investigation efforts under RCRA/CERCLA.	Minimal Impact

EXHIBIT 4-16
Summary of Environmental Consequences for MSFC

Resource Area	Alternative	Overall Effects of Alternative	Impact
		Asbestos and Lead-based Paint. If real property were to be demolished, it probably would contribute to the generation of asbestos waste or LBP. Asbestos and LBP surveys would be conducted before demolition. If ACMs were determined to be present, they would be removed appropriately before demolition. Such wastes would need to be disposed according to the hazardous waste classification determined. The property disposal process could become overwhelmed with the volume of property to disposition. The volume of asbestos due to demolition operations could result in schedule and cost impacts if a structured disposal process were not implemented. In addition, the amount of ACM that would require disposal could exceed landfill capacities.	Minimal Impact
Health and Safety	No Action and Proposed Action	Health and safety risks associated with real property at MSFC could include contamination or damage resulting from major spills or accidents. Buildings at MSFC could contain asbestos and LBP. Employees conducting renovation or demolition work must meet the safety standards outlined in 29 CFR 1926.1101, OSHA's construction standard for asbestos, or 29 CFR 1962.62, OSHA's construction standard for lead, to prevent exposure. The appropriate level of PPE must be worn depending on the level of the abatement, in accordance with OSHA's construction standard to ensure worker safety.	Minimal Impact
Hydrology and Water Quality	No Action and Proposed Action	Disposition or removal of buildings or structures at MSFC could result in minimal temporary soil disturbances, thus resulting in erosion during removal activities. The methods used for removal of any designated facilities would vary in relation to the type of structure, its location, the materials encountered in demolition, and the contractor's experience. Best engineering practices, codes, specifications, and standards would be followed to prevent or limit potential impacts. These would include the implementation of erosion and turbidity controls. Storm water permits might need to be obtained and soil stabilization measures might need to be implemented.	Minimal Impact
Land Use	No Action and Proposed Action	If the existing facilities were destroyed or released to GSA, new facilities potentially could be constructed in their place. It is anticipated that they would be compatible with the existing land use categories.	Minimal Impact

4. ENVIRONMENTAL CONSEQUENCES

EXHIBIT 4-16
Summary of Environmental Consequences for MSFC

Resource Area	Alternative	Overall Effects of Alternative	Impact
Noise	No Action and Proposed Action	The demolition or disposition of property would have the potential to temporarily increase noise levels at MSFC during the associated demolition or disposition operations. Any demolition or disposition activities would comply with the OSHA hearing protection standards for employees and other individuals in the vicinity.	Minimal Impact
Site Infrastructure	No Action and Proposed Action	The existing utilities are sufficient to support current activities at MSFC. It is anticipated that most of the SSP property at MSFC would be transferred to other NASA programs upon disposition of SSP real and personal property. In this case, it is assumed that the infrastructure would continue to operate at similar levels, because it is assumed that programs receiving SSP property would have similar infrastructure needs. If the property were not transferred but remained unused, a decreased load on the site infrastructure would result. Any impacts would result from decreased use.	Minimal Impact
Socioeconomics	No Action and Proposed Action	The specific disposition methods selected for SSP real and personal property probably would has minimal to no impact on the population, regional economy, and community services in the region surrounding MSFC. MSFC has been given project management responsibility for the future programs, including vehicle systems engineering, vehicle systems integration and safety, and mission assurance activities. It is likely that at least some of the buildings at MSFC that currently are used by the SSP would be reused for future programs, with the same or similar functions.	Minimal to No Impact
Solid Waste	Proposed Action	If it were determined that the real property would be demolished, the overall impacts probably would include the generation of solid waste consisting of concrete, asphalt, glass, metals (conduit, piping, and wiring), lumber, asbestos, and LBP. Items and materials that could be reused would be salvaged to the extent possible for NASA's future use. Non-salvageable solid waste would be disposed in accordance with the applicable health and safety and environmental regulations at MSFC, the RSA inert landfill, or the City of Huntsville Refuse-to-Steam Plant. Therefore, minimal impacts to solid waste would be anticipated.	Minimal Impact
	No Action	The property disposal process may become overwhelmed with the volume of property to disposition. The volume of property that would be processed could result in schedule and cost impacts if a structured disposal process were not implemented. In addition, the amount of solid waste that would require disposal could exceed landfill capacities.	Minimal Impact

EXHIBIT 4-16
Summary of Environmental Consequences for MSFC

Resource Area	Alternative	Overall Effects of Alternative	Impact
Transportation	No Action and Proposed Action	Real property demolition could generate more destruction-related truck trips. It would increase the traffic on the surrounding streets in the study area. A traffic control plan could be required to control the movement of truck traffic during the demolition.	Minimal Impact

Notes:
ACM = Asbestos-containing material
AST = Aboveground storage tank
CCSMP = Cape Canaveral Spaceport Master Plan
CERCLA = Comprehensive Environmental Response, Compensation, and Liability Act
CFR = Code of Federal Regulations
GSA = General Services Administration
LBP = Lead-based paint
MAF = Michoud Assembly Facility
MSFC = Marshall Space Flight Center
NASA = National Aeronautics and Space Administration
NEPA = National Environmental Policy Act
OSHA = Occupational Safety and Health Administration
PPE = Personal protective equipment
PTE = Potential to emit
RCRA = Resource Conservation Recovery Act
SSP = Space Shuttle Program
UST = Underground storage tank

4.9 White Sands Test Facility

Exhibit 4-17 outlines the major SSP property at WSTF and the preliminary plans for their disposition.

EXHIBIT 4-17
Major SSP Property at WSTF

SSP Asset/Facility	Description	Disposition
300 Area	The 300 Area is used for propulsion testing of the forward and aft Reaction Control System of the Orbiter. In addition, this area is used to test the Orbiter's Improved Auxiliary Power Unit.	This property has not been identified for use by new programs.
400 Area	The 400 Area is used for propulsion testing of the Orbiter Maneuvering Subsystem and the PRCTs and VRCTs.	This property has not been identified for use by new programs.

4. ENVIRONMENTAL CONSEQUENCES

EXHIBIT 4-17
Major SSP Property at WSTF

SSP Asset/Facility	Description	Disposition
Analytical Chemistry and Metallurgy Laboratories	There are numerous laboratories and hazardous testing is conducted specifically for the SSP. Testing includes the following: • Materials and Components Testing with Hypergols • Materials Flammability in Oxygen-enriched Atmospheres • Standard Materials Testing per NASA Standard 6001 • High-pressure Oxygen Component Quality Testing • Materials and Components Testing in High-temperature, High-flow, Gaseous Oxygen and Hydrogen • Hypervelocity Impact Testing of Hazardous and Nonhazardous Materials and Assembled Items • Composite Overwrapped Pressure Vessel Safety Assessment Testing • Space Environment Simulation • Low-velocity Impact Testing	This property has not been identified for use by new programs.
Flight Component Depot for the SSP	These capabilities include the following processes and the associated personal property: • PRCS and VRCS Thruster Flushing, Valve R&R, Chamber R&R, and Other Repairs • PRCS and VRCS Thruster Valve Overhaul • OMS Engine, including Series Valve and Pneumatic Pack, Quad Check Valve, AC Motor Valve, Manual Valve, and Burst Disk/Relief Valve Overhaul • Rebuilt PRCS and VRCS Hot Firings performed at TS405 and 406 • Hydrogen and Oxygen Flow Control Valve ATP • LH2 Recirculation Pump Cryogenic ATP • Atmospheric Revitalization Pressure and Control Subsystem Panels Oxygen Wetting and Certification	This property has not been identified for use by new programs.
WSSH runways	WSSH is a back-up landing site for the Orbiter if the conditions at KSC or EAFB are not favorable for a landing. WSSH also is used to develop NAVAIDs to aid in Orbiter navigation, as well as to develop landing procedures. Three runways are maintained at WSSH on the dry gypsum lake bed.	This property has not been identified for use by new programs.

EXHIBIT 4-17
Major SSP Property at WSTF

SSP Asset/Facility	Description	Disposition
	NASA has equipment to support the Orbiter in the event of a landing. However, there are no major personal property assets located at WSSH.	
	WSSH is the primary training area for Space Shuttle pilots flying practice approaches and landings in the STA and T-38 chase aircraft. Three runways are maintained at WSSH on the dry gypsum lake bed. Two of the runways are 35,000 ft by 900 ft, which includes a 15,000-foot by 300-foot marked runway with 10,000-foot extensions on either end and 300 ft on either side. These long runways are positioned to simulate approaches at Edwards AFB and KSC and are the back-up runways in the event of an Orbiter landing. The third runway is shorter and is used for training for a TAL site.	

Notes:
AFB = Air Force Base
ATP = Acceptance Test Procedure
ft = Feet
KSC = Kennedy Space Center
LH2 = Liquid hydrogen
NASA = National Aeronautics and Space Administration
NAVAID = Navigational aid
OMS = Orbital Maneuvering System
PCRT = Primary Reaction Control Thrusters
PRCS = Primary Reaction Control System
R&R = Repair and Refurbishment
SSP = Space Shuttle Program
STA = Shuttle Training Aircraft
TAL = Transoceanic Abort Landing
VRCS = Vernier Reaction Control System
VRCT = Vernier Reaction Control Thrusters
WSSH = White Sands Space Harbor

4.9.1 Environmental Consequences Summary for WSTF

The environmental resources that were evaluated and subsequently determined to have no potential for environmental impacts are provided in Exhibit 1-2. The environmental consequences for the resource areas present at WSTF are summarized in Exhibit 4-18.

EXHIBIT 4-18
Summary of Environmental Consequences for WSTF

Resource Area	Alternative	Overall Effects of Alternative	Impact
Air Quality	No Action and Proposed Action	The demolition or disposition of property would have the potential to temporarily increase emissions at WSTF during the demolition or disposition operations.	Minimal to No Impact
Biological Resources	No Action and Proposed Action	**Vegetation.** Most of the facilities at WSTF are surrounded by common semi-desert vegetation or landscaped natural vegetation. Natural vegetation could spread into the developed areas once the property is dispositioned, thereby increasing the distribution of natural vegetation in the area. Therefore, an overall minimal impact of the Proposed Action and the No Action alternative on vegetation would be anticipated.	Minimal Impact
		Wildlife. Increased human activity and noise due to disposition and demolition of property temporarily could increase disturbance of wildlife. However, wildlife probably would return to the area after the disposition and demolition were complete. Therefore, the disposition of property on WSTF would have minimal impacts on wildlife.	Minimal Impact
		Protected Species. Increased human activity and noise due to disposition and demolition of property temporarily could increase disturbance of protected species. However, wildlife probably would return to the area after the disposition and demolition were complete. Therefore, the disposition of real property on WSTF would have minimal impacts on protected species.	Minimal Impact
Cultural Resources	No Action and Proposed Action	**Archaeological Resources.** There would be minimal to no impact on archaeological resources under the Proposed Action and No Action Alternatives because no ground-disturbing activities are anticipated.	Minimal to No Impact
		Historic Resources. As the SSP approaches the end of its mission, a variety of buildings and facilities at several NASA installations will no longer be of use to SSP. Once SSP identifies and reports to a host installation that it no longer needs a building or facility, NASA will initiate the standard process for addressing excess infrastructure [as described in Section 2.1]. Termination of SSP by NASA will not lead to a specific decision or action on the future of each infrastructure asset and the associated environmental impacts to that asset. NASA will conduct an appropriate level of federally mandated NEPA analysis before final decisions on the disposition of SSP infrastructure assets are made. If any such properties are listed in or eligible for listing in the NRHP, NASA will take no action that would affect any such property until the NHPA Section 106 process is complete.	Moderate Impact

EXHIBIT 4-18
Summary of Environmental Consequences for WSTF

Resource Area	Alternative	Overall Effects of Alternative	Impact
Hazardous and Toxic Materials and Waste	Proposed Action	**Storage and Handling.** If the facilities were reutilized, mothballed, demolished, or released to the GSA, minimal impacts on the storage and handling of hazardous materials associated with real property would be expected because waste generation would be expected to remain at the same level.	Minimal Impact
		Waste Management. If the facilities were mothballed, demolished, or released to the GSA, it would be likely that the waste management procedures associated with the SSP would no longer be applicable, because no wastes would be generated.	Minimal to No Impact
		Contaminated Areas. If the facilities were mothballed, demolished, or released to the GSA, it is possible that new contaminated areas could be identified during closure activities (such as the closure of ASTs or USTs). Newly identified contaminated areas would be addressed by the Center's restoration programs. However, major impacts would be unlikely because the Center has undergone investigation efforts under RCRA/CERCLA.	Minimal to No Impact
	No Action	**Storage and Handling.** The property disposal process could become overwhelmed with the volume of property to disposition. The volume of property that would be processed could result in schedule and cost impacts if a structured disposal process were not implemented. In addition, the amount of hazardous waste that would require disposal could exceed landfill and less-than-90-day hazardous waste storage yard capacities.	Minimal Impact
		Waste Management. The property disposal process could become overwhelmed with the volume of property to disposition. The volume of property that would be processed could result in schedule and cost impacts if a structured disposal process were not implemented. In addition, the amount of hazardous waste that would require disposal could exceed landfill capacities.	Minimal Impact
		Contaminated Areas. If the facilities were mothballed, demolished, or released to the GSA, it would be possible that new contaminated areas could be identified during closure activities (such as the closure of ASTs or USTs). Newly identified contaminated areas would be addressed by the Center's restoration programs. However, major impacts would be unlikely because the Center has undergone investigation efforts under RCRA/CERCLA.	Minimal Impact
Health and Safety	No Action and Proposed Action	Health and safety risks associated with real property at WSTF could include contamination or	Minimal Impact

EXHIBIT 4-18
Summary of Environmental Consequences for WSTF

Resource Area	Alternative	Overall Effects of Alternative	Impact
		damage resulting from major spills or accidents. Buildings at WSTF could contain asbestos and LBP. Employees conducting renovation or demolition work must meet the safety standards outlined in 29 CFR 1926.1101, OSHA's construction standard for asbestos, or 29 CFR 1962.62, OSHA's construction standard for lead, to prevent exposure. The appropriate level of PPE must be worn depending on the level of the abatement, in accordance with OSHA's construction standard to ensure worker safety.	
Hydrology and Water Quality	No Action and Proposed Action	Disposition or removal of buildings or structures at WSTF could result in minimal temporary soil disturbances, thus resulting in erosion during removal activities. The methods used for the removal of any designated facilities would vary in relation to the type of structure, its location, the materials encountered in demolition, and the contractor's experience. Best engineering practices, codes, specifications, and standards would be followed to prevent or limit potential impacts. These would include the implementation of erosion and turbidity controls. Storm water permits might need to be obtained and soil stabilization measures might need to be implemented.	Minimal Impact
Land Use	No Action and Proposed Action	If the existing facilities were destroyed or released to the GSA, new facilities potentially could be constructed in their place. It is anticipated that they would be compatible with the existing land use categories.	Minimal Impact
Noise	No Action and Proposed Action	The demolition or disposition of property would have the potential to temporarily increase noise levels at WSTF during the associated demolition or disposition operations. Any demolition or disposition activities would comply with the OSHA hearing protection standards for employees and other individuals in the vicinity.	Minimal Impact
Site Infrastructure	No Action and Proposed Action	The existing utilities are sufficient to support current activities at WSTF. It is anticipated that most of the SSP property at WSTF would be transferred to other NASA programs upon disposition of SSP real and personal property. In this case, it is assumed that the infrastructure would continue to operate at similar levels, because it is assumed that programs receiving SSP property would have similar infrastructure needs. If the property were not transferred but remained unused, a decreased load on the site infrastructure would result. Any impacts would result from decreased use.	Minimal Impact
Socioeconomics	No Action and Proposed Action	The specific disposition methods selected for SSP real and personal property probably would have	Minimal to No Impact

EXHIBIT 4-18
Summary of Environmental Consequences for WSTF

Resource Area	Alternative	Overall Effects of Alternative	Impact
		minimal to no impact on the population, regional economy, and community services in the region surrounding WSTF. It is anticipated that WSTF would be the Abort Test Booster test site for future space programs. WSTF also is being considered for future hazardous testing of system components such as vehicle reaction control systems.	
Solid Waste	Proposed Action	If it were determined that the real property would be demolished, the overall impacts probably would include the generation of solid waste consisting of concrete, asphalt, glass, metals (conduit, piping, and wiring), lumber, asbestos, and LBP. Items and materials that could be reused would be salvaged to the extent possible for NASA's future use.	Minimal Impact
	No Action	The property disposal process may become overwhelmed with the volume of property to disposition. The volume of property that would be processed could result in schedule and cost impacts if a structured disposal process were not implemented. In addition, the amount of solid waste that would require disposal could exceed landfill capacities.	Minimal Impact
Transportation	No Action and Proposed Action	Real property demolition could generate more destruction-related truck trips. It would increase the traffic on the surrounding streets in the study area. A traffic control plan could be required to control the movement of truck traffic during the demolition and disposition activities.	Minimal Impact

Notes:
ACM = Asbestos-containing material
AST = Aboveground storage tank
CCSMP = Cape Canaveral Spaceport Master Plan
CERCLA = Comprehensive Environmental Response, Compensation, and Liability Act
CFR = Code of Federal Regulations
GSA = General Services Administration
LBP = Lead-based paint
MSFC = Marshall Space Flight Center
NASA = National Aeronautics and Space Administration
NEPA = National Environmental Policy Act
OSHA = Occupational Safety and Health Administration
PPE = Personal protective equipment
PTE = Potential to emit
RCRA = Resource Conservation Recovery Act
SSP = Space Shuttle Program
UST = Underground storage tank
WSTF = White Sands Testing Facility

4.10 Dryden Flight Research Center

Exhibit 4-19 lists the SSP property at DFRC.

EXHIBIT 4-19
SSP Property at DFRC

SSP Asset/Facility	Description	Disposition
Buildings 4833 and 4680	The buildings total 32,755 ft^2 and function as the Shuttle hangar and shops when landings must occur at DFRC.	This property has not been identified for use by new programs.
Building 4860	Mate-Demate Device	This property has not been identified for use by new programs.

Notes:
DFRC = Dryden Flight Research Center
ft^2 = Square feet

4.10.1 Environmental Consequences Summary for DFRC

The environmental resources that were evaluated and subsequently determined to have no potential for environmental impacts are provided in Exhibit 1-2. The environmental consequences for the resource areas present at DFRC are summarized in Exhibit 4-20.

EXHIBIT 4-20
Summary of Environmental Consequences for DFRC

Resource Area	Alternative	Overall Effects of Alternative	Impact
Air Quality	No Action and Proposed Action	The demolition or disposition of property would have the potential to temporarily increase emissions at DFRC during the demolition or disposition operations.	Minimal to No Impact
Cultural Resources	No Action and Proposed Action	**Archaeological Resources.** There would be minimal to no impact on archaeological resources under the Proposed Action and No Action Alternatives because no ground-disturbing activities are anticipated.	Minimal to No Impact
		Historic Resources. As the SSP approaches the end of its mission, a variety of buildings and facilities at several NASA installations will no longer be of use to SSP. Once SSP identifies and reports to a host installation that it no longer needs a building or facility, NASA will initiate the standard process for addressing excess infrastructure [as described in Section 2.1]. Termination of SSP by NASA will not lead to a specific decision or action on the future of each infrastructure asset and the associated environmental impacts to that asset. NASA will conduct an appropriate level of federally mandated NEPA analysis before final decisions on the disposition of SSP infrastructure assets are	Moderate Impact

EXHIBIT 4-20
Summary of Environmental Consequences for DFRC

Resource Area	Alternative	Overall Effects of Alternative	Impact
		made. If any such properties are listed in or eligible for listing in the NRHP, NASA will take no action that would affect any such property until the NHPA Section 106 process is complete.	
Hazardous and Toxic Materials and Waste	Proposed Action	**Storage and Handling.** If the facilities were reutilized, mothballed, demolished, or released to the GSA, minimal impacts on the storage and handling of hazardous materials associated with real property would be expected because waste generation would be expected to remain at the same level.	Minimal to No Impact
		Waste Management. If the facilities were mothballed, demolished, or released to the GSA, it would be likely that the waste management procedures associated with the SSP would no longer be applicable, because no wastes would be generated.	Minimal to No Impact
		Contaminated Areas. If the facilities were mothballed, demolished, or released to the GSA, new contaminated areas potentially could be identified during closure activities (such as the closure of ASTs or USTs). Newly identified contaminated areas would be addressed by the Center's restoration programs. However, major impacts would be unlikely because the Center has undergone investigation efforts under RCRA/CERCLA.	Minimal Impact
		Asbestos and Lead-based Paint. If real property were to be demolished, it probably would contribute to the generation of asbestos waste or LBP. Asbestos and LBP surveys would be conducted before demolition. If ACMs were determined to be present, they would be removed appropriately before demolition. Such wastes would need to be disposed according to the hazardous waste classification determined.	Minimal to No Impact
	No Action	**Storage and Handling.** The property disposal process could become overwhelmed with the volume of property to disposition. The volume of property that would be processed could result in schedule and cost impacts if a structured disposal process were not implemented. In addition, the amount of hazardous waste that would require disposal could exceed landfill and less–than–90-day hazardous waste storage yard capacities.	Minimal Impact
		Waste Management. The property disposal process may become overwhelmed with the volume of property to disposition. The volume of property that would be processed could result in schedule and cost impacts if a structured disposal process were not implemented. In addition, the amount of hazardous waste that would require	Minimal Impact

4. ENVIRONMENTAL CONSEQUENCES

EXHIBIT 4-20
Summary of Environmental Consequences for DFRC

Resource Area	Alternative	Overall Effects of Alternative	Impact
		disposal could exceed landfill capacities.	
		Contaminated Areas. If the facilities were mothballed, demolished, or released to the GSA, it would be possible that new contaminated areas could be identified during closure activities (such as the closure of ASTs or USTs). Newly identified contaminated areas would be addressed by the Center's restoration programs. However, major impacts would be unlikely because the Center has undergone investigation efforts under RCRA/CERCLA.	Minimal Impact
		Asbestos and Lead-based Paint. If real property were to be demolished, it probably would contribute to the generation of asbestos waste or LBP. Asbestos and LBP surveys would be conducted before demolition. If ACMs were determined to be present, they would be removed appropriately before demolition. Such wastes would need to be disposed according to the hazardous waste classification determined. The property disposal process could become overwhelmed with the volume of property to disposition. The volume of asbestos due to demolition operations could result in schedule and cost impacts if a structured disposal process were not implemented. In addition, the amount of ACM that would require disposal could exceed landfill capacities.	Minimal Impact
Health and Safety	No Action and Proposed Action	Health and safety risks associated with real property at DFRC could include contamination or damage resulting from major spills or accidents. Buildings at DFRC could contain asbestos and LBP. Employees conducting renovation or demolition work must meet the safety standards outlined in 29 CFR 1926.1101, OSHA's construction standard for asbestos, or 29 CFR 1962.62, OSHA's construction standard for lead, to prevent exposure. The appropriate level of PPE must be worn depending on the level of the abatement, in accordance with OSHA's construction standard to ensure worker safety.	Minimal Impact

EXHIBIT 4-20
Summary of Environmental Consequences for DFRC

Resource Area	Alternative	Overall Effects of Alternative	Impact
Hydrology and Water Quality	No Action and Proposed Action	Disposition or removal of buildings or structures at DFRC could result in minimal temporary soil disturbances, thus resulting in erosion during removal activities. The methods used for removal of any designated facilities would vary in relation to the type of structure, its location, the materials encountered in demolition, and the contractor's experience. Best engineering practices, codes, specifications, and standards would be followed to prevent or limit potential impacts. These would include the implementation of erosion and turbidity controls. Storm water permits might need to be obtained and soil stabilization measures might need to be implemented.	Minimal Impact
Land Use	No Action and Proposed Action	DFRC would evaluate land use possibilities for the facilities at DFRC.	Minimal Impact
Noise	No Action and Proposed Action	The demolition or disposition of property would have the potential to temporarily increase noise levels at DFRC during the associated demolition or disposition operations. Any demolition or disposition activities would comply with the OSHA hearing protection standards for employees and other individuals in the vicinity.	Minimal Impact
Site Infrastructure	No Action and Proposed Action	The existing utilities are sufficient to support current activities at DFRC. It is anticipated that most of the SSP property at DFRC would be transferred to other NASA programs upon disposition of SSP real and personal property. In this case, it is assumed that the infrastructure would continue to operate at similar levels, because it is assumed that programs receiving SSP property would have similar infrastructure needs. If the property were not transferred but remained unused, a decreased load on the site infrastructure would result. Any impacts would result from decreased use.	Minimal Impact
Solid Waste	Proposed Action	Municipal-type solid wastes and nonhazardous wastes generated from property disposition would be disposed in accordance with the applicable health and safety and environmental regulations at the EAFB landfill. Non-RCRA wastes would be taken offsite by recyclers and reclaimers or sent to an appropriate offsite, permitted disposal facility, depending on the waste classification.	Minimal Impact
	No Action	The property disposal process could become overwhelmed with the volume of property to disposition. The volume of property that would be processed could result in schedule and cost impacts if a structured disposal process were not implemented. In addition, the amount of solid waste that would require disposal could exceed landfill capacities.	Minimal Impact

EXHIBIT 4-20
Summary of Environmental Consequences for DFRC

Resource Area	Alternative	Overall Effects of Alternative	Impact
Transportation	No Action and Proposed Action	Real property demolition could generate more destruction-related truck trips. It would increase the traffic on the surrounding streets in the study area. A traffic control plan could be required to control the movement of truck traffic during the demolition.	Minimal Impact

Notes:
ACM = Asbestos-containing material
AST = Aboveground storage tank
CCSMP = Cape Canaveral Spaceport Master Plan
CERCLA = Comprehensive Environmental Response, Compensation, and Liability Act
CFR = Code of Federal Regulations
DFRC = Dryden Flight Research Center
GSA = General Services Administration
LBP = Lead-based paint
MSFC = Marshall Space Flight Center
NASA = National Aeronautics and Space Administration
NEPA = National Environmental Policy Act
OSHA = Occupational Safety and Health Administration
PPE = Personal protective equipment
PTE = Potential to emit
RCRA = Resource Conservation Recovery Act
SSP = Space Shuttle Program
UST = Underground storage tank

4.11 Palmdale

The major SSP-related properties at Palmdale are listed in Exhibit 4-21.

EXHIBIT 4-21
SSP-related Property at Palmdale

SSP Asset/Facility	Description	Disposition
Orbiter Lifting Facility	Lifting fixture used to mate the Orbiter to the Shuttle Carrier Aircraft.	This property has not been identified for use by new programs.
Detail Manufacturing and Testing Facility	The individual parts, pieces, and systems of the Orbiter are assembled and tested in this facility. Contains two Orbiter bays.	This property has not been identified for use by new programs.

Note:
SSP = Space Shuttle Program

4.11.1 Environmental Consequences Summary for Palmdale

The environmental resources that were evaluated and subsequently determined to have no potential for environmental impacts are provided in Exhibit 1-2. The environmental consequences for the resource areas present at Palmdale are summarized in Exhibit 4-22.

EXHIBIT 4-22
Summary of Environmental Consequences for Palmdale

Resource Area	Alternative	Overall Effects of Alternative	Impact
Air Quality	No Action and Proposed Action	The demolition or disposition of property would have the potential to temporarily increase emissions at Palmdale during the demolition or disposition operations.	Minimal to No Impact
Cultural Resources	No Action and Proposed Action	**Archaeological Resources.** There would be minimal to no impact on archaeological resources under the Proposed Action and No Action Alternatives because no ground-disturbing activities are anticipated.	Minimal to No Impact
Cultural Resources	No Action and Proposed Action	**Historic Resources.** As the SSP approaches the end of its mission, a variety of buildings and facilities at several NASA installations will no longer be of use to SSP. Once SSP identifies and reports to a host installation that it no longer needs a building or facility, NASA will initiate the standard process for addressing excess infrastructure [as described in Section 2.1]. Termination of SSP by NASA will not lead to a specific decision or action on the future of each infrastructure asset and the associated environmental impacts to that asset. NASA will conduct an appropriate level of federally mandated NEPA analysis before final decisions on the disposition of SSP infrastructure assets are made. If any such properties are listed in or eligible for listing in the NRHP, NASA will take no action that would affect any such property until the NHPA Section 106 process is complete.	Moderate Impact
Hazardous and Toxic Materials and Waste	Proposed Action	**Storage and Handling.** If the facilities were reutilized, mothballed, demolished, or released to the GSA, minimal impacts on the storage and handling of hazardous materials associated with real property would be expected because waste generation would be expected to remain at the same level.	Minimal Impact
		Waste Management. If the facilities were mothballed, demolished, or released to the GSA, it would be likely that the waste management procedures associated with the SSP would no longer be applicable, because no wastes would be generated.	Minimal to No Impact
		Contaminated Areas. If the facilities were mothballed, demolished, or released to the GSA, new contaminated areas potentially could be identified during closure activities (such as the closure of ASTs or USTs). Newly identified contaminated areas would be addressed by the Center's restoration programs. However, major impacts would be unlikely because the Center has undergone investigation efforts under RCRA/CERCLA.	Minimal to No Impact

EXHIBIT 4-22
Summary of Environmental Consequences for Palmdale

Resource Area	Alternative	Overall Effects of Alternative	Impact
	No Action	**Storage and Handling.** The property disposal process could become overwhelmed with the volume of property to disposition. The volume of property that would be processed could result in schedule and cost impacts if a structured disposal process were not implemented. In addition, the amount of hazardous waste that would require disposal could exceed landfill and less–than–90-day hazardous waste storage yard capacities.	Minimal Impact
		Waste Management. The property disposal process could become overwhelmed with the volume of property to disposition. The volume of property that would be processed could result in schedule and cost impacts if a structured disposal process were not implemented. In addition, the amount of hazardous waste that would require disposal could exceed landfill capacities.	Minimal Impact
		Contaminated Areas. If the facilities were mothballed, demolished, or released to the GSA, it would be possible that new contaminated areas could be identified during closure activities (such as the closure of ASTs or USTs). Newly identified contaminated areas would be addressed by the Center's restoration programs. However, major impacts would be unlikely because the Center has undergone investigation efforts under RCRA/CERCLA.	Minimal Impact
Health and Safety	No Action and Proposed Action	Health and safety risks associated with real property at Palmdale could include contamination or damage resulting from major spills or accidents. Buildings at Palmdale could contain asbestos and LBP. Employees conducting renovation or demolition work must meet the safety standards outlined in 29 CFR 1926.1101, OSHA's construction standard for asbestos, or 29 CFR 1962.62, OSHA's construction standard for lead, to prevent exposure. The appropriate level of PPE must be worn depending on the level of the abatement, in accordance with OSHA's construction standard to ensure worker safety.	Minimal Impact
Noise	No Action and Proposed Action	The demolition or disposition of property would have the potential to temporarily increase noise levels at Palmdale during the associated demolition or disposition operations. Any demolition or disposition activities would comply with the OSHA hearing protection standards for employees and other individuals in the vicinity.	Minimal Impact

EXHIBIT 4-22
Summary of Environmental Consequences for Palmdale

Resource Area	Alternative	Overall Effects of Alternative	Impact
Site Infrastructure	No Action and Proposed Action	The existing utilities are sufficient to support current activities at Palmdale. It is anticipated that most of the SSP property at Palmdale would be transferred to other NASA programs upon disposition of SSP real and personal property. In this case, it is assumed that the infrastructure would continue to operate at similar levels, because it is assumed that programs receiving SSP property would have similar infrastructure needs. If the property were not transferred but remained unused, a decreased load on the site infrastructure would result. Any impacts would result from decreased use.	Minimal Impact
Solid Waste	Proposed Action	Items and materials that could be reused would be salvaged to the extent possible for NASA's future use. Non-RCRA solid waste would be disposed in accordance with the applicable health and safety and environmental regulations at an offsite landfill by a solid waste disposal contractor.	Minimal Impact
	No Action	The property disposal process could become overwhelmed with the volume of property to disposition. The volume of property that would be processed could result in schedule and cost impacts if a structured disposal process were not implemented. In addition, the amount of solid waste that would require disposal could exceed landfill capacities.	Minimal Impact
Transportation	No Action and Proposed Action	Real property demolition could generate more destruction-related truck trips. It would increase the traffic on the surrounding streets in the study area. A traffic control plan could be required to control the movement of truck traffic during the demolition.	Minimal Impact

Notes:
ACM = Asbestos-containing material
AST = Aboveground storage tank
CCSMP = Cape Canaveral Spaceport Master Plan
CERCLA = Comprehensive Environmental Response, Compensation, and Liability Act
CFR = Code of Federal Regulations
GSA = General Services Administration
LBP = Lead-based paint
MSFC = Marshall Space Flight Center
NASA = National Aeronautics and Space Administration
NEPA = National Environmental Policy Act
OSHA = Occupational Safety and Health Administration
PPE = Personal protective equipment
PTE = Potential to emit
RCRA = Resource Conservation Recovery Act
SSP = Space Shuttle Program
UST = Underground storage tank

4.12 Environmental Justice

EO 12898, *Federal Actions to Address Environmental Justice in Minority Populations and Low-Income Populations*, directs federal agencies to identify and address, as appropriate, disproportionately high and adverse health or environmental effects of their programs, policies, and activities on minority populations and low-income populations.

The CEQ has oversight responsibility for documentation prepared in compliance with NEPA (42 U.S.C. 4321 *et seq.*). In December 1997, the CEQ released its guidance on Environmental Justice (CEQ, 1997). The CEQ's guidance was adopted as the primary guide for the environmental justice analysis performed for this Programmatic EA for the disposition of SSP real and personal property.

This analysis provides the data necessary to assess the potential for disproportionately high and adverse human health or environmental effects on minority and/or low-income populations that may be associated with the disposition of SSP real and personal property.

4.12.1 Definitions

4.12.1.1 Minority Individuals and Minority Populations

During the Census of 2000, the U.S. Bureau of the Census collected population data in compliance with guidance adopted by the OMB (62 FR 58782). The OMB published its guidelines regarding the aggregation of multiple race data in March 2000 (OMB, 2000). Modifications to the definitions of minority individuals in the CEQ's guidance on Environmental Justice (CEQ, 1997) were made in this analysis to comply with the OMB's guidelines issued in March 2000. The following definitions of minority individuals and population are used in this environmental justice analysis:

Minority Individuals: Persons, as reported by the 2000 U.S. Census, who are members of any of the following population groups: Black or African American, Asian, Native Hawaiian or Other Pacific Islander, American Indian or Alaska Native, Multiracial (and at least one race which is a minority race under the 1997 CEQ guidance), or Hispanic or Latino (regardless of race).

Minority Population: The total number of minority individuals residing within a potentially affected area.

4.12.1.2 Low-income Individuals and Low-income Populations

Poverty thresholds are used to identify "low-income" individuals and populations (CEQ, 1997). The following definitions of low-income individuals and population are used in this analysis:

Low-income Individuals: Persons, as reported by the 2000 U.S. Census, whose self-reported income is below the poverty threshold.

Low-income Population: The total number of low-income individuals residing within a potentially affected area.

The population for whom poverty status is determined is based on all people except institutionalized people, people in military group quarters, people in college dormitories, and unrelated individuals under 15 years old.

4.12.1.3 Disproportionately High and Adverse Human Health Effects

Disproportionately high and adverse human health effects are those that are significant (per NEPA, 40 CFR 1508.27) or above generally accepted norms, and for which the risk of adverse effects to minority populations or low-income populations appreciably exceeds the risk to the general population.

4.12.1.4 Disproportionately High and Adverse Environmental Effects

Disproportionately high and adverse environmental effects are those that are significant (per NEPA, 40 CFR 1508.27), and that would adversely affect minority populations or low-income populations appreciably more than the general population.

4.12.2 Methodology

The purpose of this analysis is as follows: 1) to identify minority and low-income populations that potentially would be affected by the Proposed Action; and 2) to assess whether the implementation of the Proposed Action would result in disproportionately high and adverse effects on these populations. In the event that human health or environmental risks were found to be significant (as defined in 40 CFR 1508.27), then these risks to minority and low-income populations would be evaluated to determine if they are disproportionately high.

For this analysis, 2000 U.S. Census Bureau Block Groups that are located immediately adjacent to the facility were selected as the study area to identify the minority and low-income populations that may be affected adversely by the disposition of SSP real and personal property. This study area was selected because most of the environmental effects resulting from the disposition of SSP real and personal property are expected to occur within the boundaries of facility. The analysis also included a detailed review of the environmental effects that would result from the disposition of SSP real and personal property, relying principally on the information developed and documented in this Programmatic EA.

4.12.3 Population Characterization and Impact Analysis

For the minority and low-income population analyses, year 2000 U.S. Census data at the Block Group level for all Block Groups immediately adjacent to the boundaries of the potentially affected facilities, as well as Census data regarding the surrounding county or counties, were collected. The minority and low-income population characteristics of the Block Groups and the counties are illustrated in Exhibit 4-23. As listed in Exhibit 4-23, the majority of the adjacent Census Block

Groups have lower concentrations of minority and low-income individuals than those of the county or counties in which the Census Block Groups are located.

Exhibit 4-23 also indicates the range of potential environmental impacts for each of the facilities for the resource areas (air, biological, cultural, hazardous and toxic materials and waste, health and safety, land use, noise, site infrastructure, socioeconomics, solid waste, and traffic and transportation) for the Proposed Action and the No Action alternatives. The potential impacts were placed into one of the pre-determined classifications, which range from No Impact to Major-Environmental Impacts. Most of the potential impacts for all of the resource areas were identified as Minimal to No Impacts; therefore, they are not expected to be measurable, or would be too small to cause any changes to the environment.

On the basis of this analysis, no adverse impacts are expected as a result of the disposition of SSP real and personal property activities. Therefore, no disproportionately high or adverse impacts to minority and/or low-income populations are expected as a result of the disposition of SSP real and personal property at any of the facilities.

4.13 Cumulative Impacts

Cumulative impacts result from the incremental impact of actions when they are combined with other past, present, and reasonably foreseeable future actions, regardless of what agency or person undertakes such other actions (40 CFR 1508.7). Cumulative impacts can result from individually minor, but collectively significant, actions taking place over a period of time.

EXHIBIT 4-23
Minority and Low-Income Population Characteristics and Potential Environmental Impacts

Location	Census Block Groups Adjacent to Facilities		Surrounding County/Counties		Potential Impact of Proposed Action	Potential Impact of No Build Alternative
	Minority[1]	Population below Poverty[1]	Minority[1]	Population below Poverty[1]		
Kennedy Space Center	8.2	5.3	16.4 and 18.1[2]	9.5 and 11.6[2]	Minimal to no impact	Minimal to no impact
Johnson Space Center	28.3	5.2	57.9	15	Minimal to no impact	Minimal to no impact
Ellington Field	30.2	2.6	57.9	15	Minimal to no impact	Minimal to no impact
El Paso Forward Operation Location	90.4	24.4	83.0	23.8	Minimal to no impact	Minimal to no impact
Sonny Carter Training Facility	30.2	2.6	57.9	15	Minimal to no impact	Minimal to no impact
John C. Stennis Space Center	3.4	4.3	10.7 and 15.0[3]	14.4 and 9.7[3]	Minimal to no impact	Minimal to no impact
Michoud Assembly Facility	85.0	47.9	73.3	27.9	Minimal to no impact	Minimal to no impact
Marshall Space Flight Center	41.8	10.4	29.0	10.5	Minimal to no impact	Minimal to no impact
White Sands Testing Facility	46.7	9.9	67.5	25.4	Minimal to no impact	Minimal to no impact
Dryden Flight Research Center	32.2	2.8	50.6	20.8	Minimal to no impact	Minimal to no impact
Palmdale	79.8	21.1	69.1	17.9	Minimal to no impact	Minimal to no impact
Santa Susanna Field Laboratory	21.6	4.3	69.1 and 43.4[4]	17.9 and 9.2[4]	Minimal to no impact	Minimal to no impact

Notes:
[1] Numbers represent the percent of the population reported as minority and below the poverty threshold.
[2] Numbers are for Brevard County and Volusia County, respectively.
[3] Numbers are for Hancock County and St. Tammany Parish, respectively.
[4] Numbers are for Los Angeles County and Ventura County, respectively.

4. ENVIRONMENTAL CONSEQUENCES

The SSP T&R activities addressed in this EA would take place at various NASA Centers. SSP operations at NASA Centers represent a large portion of overall operations. The proposed project relevant for the consideration of cumulative impacts on a programmatic level is the Constellation Program, which is NASA's current plan for human space flight exploration. This program will entail the use of new space vehicles, as well as a new space capsule that will be developed similarly to the ones used during the Apollo program. The Constellation Program will allow for a variety of missions, from Space Station resupply to lunar landings.

It is anticipated that other NASA programs, such as the Constellation Program, would replace the SSP at many of the facilities and that this program would have similar operational requirements as the SSP. Because the use of many existing facilities probably would be transitioned to this new program, the cumulative effect on resource areas (site infrastructure, air quality, noise, geology and soils, hydrology and water quality, biological resources, cultural resources, hazardous materials and waste, health and safety, socioeconomic resources, transportation, and environmental justice) would be expected to be minimal. The evaluation of the NRHP-significant properties across the NASA Centers included in this Programmatic EA resulted in a cumulative impact of use conversions on historic properties that would be considered less than significant if the conversion affected only a small percentage of the total number of structures. It is anticipated that any new activity would be compatible with the existing land use categories or the future land use categories and that no impacts on land use would occur. As stated in the Final Cx PEIS (NASA, 2007t), NASA intends to retain a beneficial socioeconomic footprint in the regional economies surrounding the Centers and to preserve the critical and unique capabilities provided by each NASA Center. Meaningful estimates of the specific work allocations at each Center would be available once the prime contracts have been awarded for all of the Program's major projects and procurements.

If, in addition to SSP, other programs were to mothball or abandon their facilities, the emissions to air and water would decrease and the demand on environmental resources would decrease. Therefore, the cumulative environmental effect would be beneficial. However, if facilities used by many personnel were not replaced, adverse cumulative effects on local employment and related socioeconomic resources would result.

Demolishing facilities in addition to those currently operated by the SSP potentially would result in short-term impacts associated with the demolition. Air emissions would experience a short-term increase as a result of demolition activities and the generation of fugitive dust. Noise would increase temporarily as a result of demolition and the increase in traffic (from workers traveling to the site and the use of demolition equipment). In addition, traffic would increase during demolition. Regional economies would benefit temporarily as a result of contracts for demolition and hiring of the required workers. Demolition would increase the need for landfill space; however, most of the waste generated during demolition would be inert and either could be recycled or disposed in a landfill designated for construction and demolition waste.

The demolition of existing structures would result in similar cumulative impacts as those described above for conversion of uses of the structures. The evaluation of the total number of SSP-significant properties across the NASA Centers included in this EA showed that the cumulative impact of a single demolition would be considered less than significant, compared to the total number of NRHP-significant structures. The loss of multiple NRHP-significant structures at a single facility and across all of the facilities could have cumulative impacts to NRHP properties.

This page intentionally left blank.

5. List of Preparers

This programmatic EA for the decommissioning of the SSP was prepared by NASA. The individuals and organizations listed below contributed to the overall effort in the preparation of this document.

National Aeronautics and Space Administration

Donna Holland
Marshall Space Flight Center NEPA Coordinator

CH2M HILL (Contractor to NASA)

Jason Glasgow
SSP EA Project Manager

Mark Bennett
SSP EA Technical Lead

Karin Lilienbecker
NEPA Senior Consultant

Kristen Duda
Environmental Engineer- Noise, Health and Safety

Judith Dempsey
Environmental Scientist - Air, Site Infrastructure

Dave Golles
Environmental Engineer- Solid Waste, Hazardous Materials and Hazardous Waste

Beth Vaughan
Senior Environmental Engineer- Hazardous Materials and Hazardous Waste

Katy Oakes
Environmental Scientist – Biological Resources

Rob Price
Environmental Scientist – Water Resources

Mike Phillips
Project Planner - Transportation

Ginny Farris
Project Planner - Socioeconomics

Ed McCarthy
Environmental Scientist – Land Use

5. LIST OF PREPARERS

Jim Bard
Cultural Resources Specialist

Rob Thomson
Environmental Engineer – Water Resources

Loren Bloomberg
Senior Environmental Engineer - Transportation

Christopher Clayton
Senior Socioeconomics Specialist

Sara Orton
Environmental Planner – Cultural Resources

Rob Rodland
Environmental Planner – Environmental Justice

Lorraine Woodman and Elizabeth Calvit
Senior Cultural Resources Specialist

Vicky Potter
Editor

Denise Godwin
Document Processor

UNITeS (Contractor to NASA)

Scott Stevens
Graphics and GIS Support

6. References

Alabama Department of Environmental Management (ADEM). 2006. Draft 2006 303d List. http://www.adem.state.al.us/WaterDivision/WQuality/303d/Draf06303(d)List.pdf. Accessed February 5, 2007.

Advisory Council on Historic Preservation. April 1999. *Archive of Prominent Section 106 Cases: Alabama: Relocation of Equipment from the Neutral Buoyancy Space Simulator (Huntsville).* Accessed February 1, 2007. http://www.achp.gov/casearchive/cases4-99AL.html.

Advisory Council on Historic Preservation. February 1991. *Balancing Historic Preservation Needs with the Operation of Highly Technical of Scientific Facilities.* Washington, D.C.

Advocate. February 17, 2007. "LA Seeks More NASA Work."

Air Force Flight Test Center (AFFTC). October 2002. *Integrated Natural Resources Management Plan for Edwards Air Force Base, California.* Environmental Management Office (EMO). Accessed on February 14, 2007.

Alabama Department of Conservation and Natural Resources (ADCNR). August 21, 2006. County by County Listing of Alabama Species on the Federal List for Threatened and Endangered Species or Whose Status is a Concern. Madison County. Accessed January 25, 2007. (http://www.outdooralabama.com/watchable-wildlife/regulations/endangeredbycounty.cfm).

Alabama Department of Environmental Management. 2007. Regulations. Accessed May 5, 2007.

Alabama Natural Heritage Program (ANPH). June 2006. *Alabama Inventory List–The Rare, Threatened, & Endangered Plants & Animals of Alabama.* Accessed January 25, 2007. http://www.alnhp.org/track_2006.pdf.

Alexander and Avery. 2006. *The 2005 Phase I Archaeological Survey of the Marshall Space Flight Center, National Aeronautics and Space Administration, Madison County, AL.* August 2006.

ARCADIS. June 2004a. *Biological Survey for Sensitive Flora and Fauna, Michoud Assembly Facility.*

ARCADIS. June 2004b. *Wetland Delineation, Michoud Assembly Facility.*

6. REFERENCES

Archaeological Consultants, Inc. 2007a. *NASA-Wide Survey and Evaluation of Historic Facilities and Properties in the Context of the U.S. Space Shuttle Program- Canoga Park Facility, Canoga Park, California.*

Archaeological Consultants, Inc. 2007b. *NASA-Wide Survey and Evaluation of Historic Facilities and Properties in the Context of the U.S. Space Shuttle Program-John F. Kennedy Space Center, Brevard County, Florida.*

Archaeological Consultants, Inc. 2007c. *NASA-Wide Survey and Evaluation of Historic Facilities and Properties in the Context of the U.S. Space Shuttle Program – Lyndon B. Johnson Space Center, Houston, Texas.*

Archaeological Consultants, Inc. 2007d. *NASA-Wide Survey and Evaluation of Historic Facilities and Properties in the Context of the U.S. Space Shuttle Program – George Marshall Space Flight Center, Huntsville, Alabama.*

Archaeological Consultants, Inc. 2007e. *NASA-Wide Survey and Evaluation of Historic Facilities and Properties in the Context of the U.S. Space Shuttle Program – White Sands Test Facility, Las Cruces, New Mexico.*

Archaeological Consultants, Inc. 2007F. *Space Shuttle Program Historic Properties: NASA Dryden Flight Research Center.*

Archaeological Consultants, Inc. 2007g. *Survey of NASA-Owned Historic Facilities and Properties in the Context of the U.S. Space Shuttle Program-Air Force Plant 42, Site 1 North Palmdale, California.*

Archaeological Consultants, Inc. 2006. *Space Shuttle Program Historic Properties: NASA Glenn.*

BISON-M. 2007a. Database Query Dona Ana County Endangered/Threatened Species. Accessed February 8, 2007.

Bay Area Houston Economic Partnership (BAHEP). 2007. *Economic Impact NASA JSC.* Accessed August 17, 2007. http://www.bayareahouston.com/Home/NASA-JohnsonSpaceCente/EconomicImpact/.

Biota Information System of New Mexico (BISON-M). 2007b. Code Table List. Accessed February 8, 2007. http://www.bison-m.org/codetablelist.aspx?type=status.

Boeing. July 2006. *Air Force Plant 42 Site Environmental Review.*

Brevard County, Florida. 1988 Update. *1988 Brevard County Comprehensive Plan.* Prepared in accordance with Chapter 163, Part II, Florida Statutes, "The Local Government Comprehensive Planning and Land Development Regulation Act of 1985", and Florida Administrative Code Rule 9J-5, "Minimum Criteria for Review of Local Government Comprehensive Plans and Determination of Compliance." Replaces the 1981 version of the Brevard County Comprehensive Plan.

Bureau of Economic and Business Research (BEBR). April 2006. "4/1/2006 Official Population Estimate." University of Florida. Excel file, downloaded on March 8, 2007. http://www.edr.state.fl.us/conferences/population/demographic.htm.

Bureau of Business and Economic Research. University of New Mexico. 2004. "New Mexico Economic Development Department Projected Population New Mexico Counties July 1, 2000 to July 1, 2030." Released August 2002 and revised April 2004. Downloaded May 11, 2007. http://www.edd.state.nm.us/index.php?/data/C31/.

California Air Resources Board (CARB). 2007. "Area Designation Maps." Accessed June 19, 2007. http://www.arb.ca.gov/desig/adm/adm.htm.

California Department of Fish and Game (CDFG). 2004. *Habitat Conservation Planning Branch*. Accessed on February 14, 2007. http://www.dfg.ca.gov/hcpb/species/t_e_spp/tespp.shtml.

California Environmental Protection Agency (Cal-EPA). 2007a. California Permit Search Page. Accessed February 22, 2007. http://www.waterboards.ca.gov/losangeles/html/permits/permits.html.

California Environmental Protection Agency (Cal-EPA). 2007b. *Fact Sheet: National Pollutant Discharge Elimination System Permit for the Boeing Company (Santa Susana Field Laboratory)*. NPDES Permit Number: CA0001309. Public Notice Number: 05-068. January 19, 2006.

California Native Plant Society (CNPS). 2007. *Inventory of Rare and Endangered Plants* (Online Edition, v7-07a). Accessed on February 14, 2007. California Native Plant Society. Sacramento, California. http://www.cnps.org/inventory.

California Office of Historic Preservation. 2007. Office of Historic Preservation - Kern, California Historical Landmarks. Accessed February 21, 2007. [HAVE ACCESSED DATE — WHAT'S WEBSITE??]

Center for Business and Economic Research (CBER). August 2001. "Alabama County Population 2000 and Projections 2005-2025." University of Alabama. http://cber.cba.ua.edu/edata/est_prj.html.

Chamber of Commerce of Huntsville/Madison County Website. 2007. Accessed April 23, 2007. http://www.huntsvillealabamausa.com.

City of New Orleans Fire Department (NOFD). March 14, 2006. *Strategic Recovery and Reconstitution Planning Process for the City of New Orleans Fire Department Decimated by Hurricane Katrina*.

Clear Lake Area Chamber of Commerce Website. 2007. Accessed May 8, 2007. http://www.clearlakearea.com/living/communities.asp.

6. REFERENCES

Council on Environmental Quality. December 10, 1997. *Environmental Guidance under the National Environmental Policy Act.* Executive Office of the President. Washington, D.C.

Directorate of Environment and Safety, Environmental Services Division. January 1998. *White Sands Missile Range Range-Wide Environmental Impact Statement.* WSMR. New Mexico.

Dethloff, Henry C. 1993. *Suddenly Tomorrow Came . . . A History of the Johnson Space Center.* The NASA History Series.

Dick, Steven J. February 2005. *NASA Center Histories: Stennis Space Center.* Accessed February 1, 2007. http://history.nasa.gov/centerhistories/printFriendly/stennis.htm.

Doña Ana County Website. 2007. Accessed May 11, 2007. http://www.co.dona-ana.nm.us/development.

Edwards Air Force Base (EAFB). 2003. *Edwards Air Force Base General Plan.*

El Paso Community Development. 2007. "Community Profile." Downloaded May 11, 2007. http://www.elpasotexas.gov/econdev/default.asp.

Enterprise Florida, Inc. (eFlorida). 2006. County Profiles. http://www.eflorida.com/profiles.

Federal Emergency Management Agency (FEMA). 2007. Map Service Center. http://msc.fema.gov/webapp/wcs/stores/servlet/FemaWelcomeView?storeId=10001&catalogId=10001&langId=-1.

Florida Agency for Workforce Innovation. 2007. "Local Area Unemployment Statistics by County (not seasonally adjusted)". Excel file: "Unempl_LFSJAN1.xls." Downloaded March 12, 2007. http://www.labormarketinfo.com/library/ep.htm

Florida Department of Environmental Protection (FDEP). 2007a. "Air Quality." Accessed June 19, 2007. http://www.dep.state.fl.us/mainpage/programs/air.htm.

Florida Department of Environmental Protection (FDEP). 2007b. Air Permit Document Search. Accessed January 16, 2007. http://www.dep.state.fl.us/air/eproducts/apds/default.asp.

Florida Department of Environmental Protection (FDEP). 2007c. Central Air District Homepage. Accessed January 16, 2007. http://www.dep.state.fl.us/central/.

Florida Department of Environmental Protection (FDEP). 2007d. Industrial Wastewater Program Homepage. Accessed January 16, 2007. http://www.dep.state.fl.us/water/wastewater/iw/index.htm.

Florida Department of Environmental Protection (FDEP). 2006. Accessed June 2007. http://www.dep.state.fl.us/swapp/Aquifer.asp.

Florida Office of Economic and Demographic Research (EDR). January 2007. County Profiles. Downloaded March 8, 2007. (Last updated January 23, 2007.) http://edr.state.fl.us/county%20profiles.htm.

Florida Office of Economic and Demographic Research (EDR). October 26, 2006a. Economic Estimating Conference: Florida Economic Forecast.

Florida Office of Economic and Demographic Research (EDR). October 2002b. Florida Components of Growth, 1980 - 1990 and 1990 – 2000, Census 2000.

Greater Houston Partnership 2007a. Fact Sheets: "The Economy at a Glance-Houston"; "Population and Employment Forecast"; and "Largest Houston Employers." http://www.houston.org/media/dataSheets.asp.

Greater Houston Partnership. 2007b. Industry Guide. Accessed May 8, 2007. http://www.houston.org/industryGuide.

Greater New Orleans Community Data Center. April 12, 2007. *The Katrina Index – Tracking the Recovery of New Orleans and the Metro Area.* In collaboration with the Brookings Institution, Metropolitan Policy Program.

Hess, Mark and Ed Campion. 1994. *Space Shuttle Modification Work to Continue at Palmdale.* Washington, D.C. Accessed February 21, 2007. http://www.nasa.gov/home/hqnews/1994/94-044.txt.

King, Ronetta. February 11, 2007. "Launching the Future." *New Orleans Times-Picayune.*

Lockheed Martin Space Systems Company. May 1, 2006a. *Michoud Assembly Facility Environmental Permits.*

Lockheed Martin Space Systems Company. 2006b. *Michoud Assembly Facility Master Plan Study.* Michoud Operations. Prepared by Valerie Melancon (Facilities & Environmental Operations, Long Range Planning).

Lockheed Martin Space Systems Company. July 2004. *Spill Prevention, Control, and Countermeasures Plan.*

Louisiana Department of Environmental Quality (LDEQ). 2004. Louisiana Water Quality Standards. Accessed January 31, 2007. http://www.deq.louisiana.gov/portal/Portals/0/planning/regs/title33/33v09.pdf #page=53.

Louisiana Department of Natural Resources (LDNR). 2006. Coastal Management Division. Accessed January 24, 2007. http://dnr.louisiana.gov/crm/coastmgt/coastmgt.asp.

6. REFERENCES

Louisiana Department of Wildlife and Fisheries (LDWF). 2004. Threatened and Endangered Species Table. Accessed February 14, 2007. http://www.wlf.louisiana.gov/experience/threatened/threatenedandendangeredtable/.

Louisiana Natural Heritage Program (LNHP). May 2006. *Rare, Threatened, & Endangered Species & Natural Communities Tracked by the Louisiana Natural Heritage Program, Orleans Parish.* Accessed January 24, 2007.

Martin Associates. April 2007. *The Economic Impact of the Port of Houston–2006.* Port of Houston Authority Website. Accessed May 8, 2007. http://www.portofhouston.com/pdf/genifo/PHA.EIS2006.pdf.

Mississippi Department of Environmental Quality (MDEQ). 2003. *Water Quality Criteria for Intrastate, Interstate, and Coastal Waters.* Accessed February 5, 2006. http://www.deq.state.ms.us/newweb/MDEQRegulations.nsf/RN/WPC-2.

Mississippi Natural Heritage Program (MNHP). 2002. *Endangered Species of Mississippi.* Museum of Natural Science, Mississippi Department of Wildlife, Fisheries, and Parks, Jackson, Mississippi. 2 pp.

Montgomery Watson Herza (MWH). July. 2004. *RCRA Facility Investigation Program Report Surficial Media Operable Unit Santa Susan Field Laboratory, Ventura County, California.* Volume I.

National Aeronautics and Space Administration (NASA). 2007a. Space Shuttle Program Business Management Office. "Workforce Demographics, v. 1, Feb. 6, 2007" (PowerPoint presentation and supporting Excel file by the same name). Via e-mail from Dennis Davidson, dated February 9, 2007.

National Aeronautics and Space Administration (NASA). 2007b. John C. Stennis Space Center (SSC). *Environmental Assurance Program–Air Quality.* Accessed February 2, 2007. http://www.ssc.nasa.gov/environmental/resource_mngmnt/air_quality/airqual.html.

National Aeronautics and Space Administration (NASA). 2007c. John C. Stennis Space Center (SSC). *Environmental Assurance Program–Waste Management: Water Treatment.* Accessed February 5, 2007. http://www.ssc.nasa.gov/environmental/wastemngmnt/water_treatment/watertrtmt.html.

National Aeronautics and Space Administration (NASA). 2007d. "Current Missions." Accessed February 14, 2007. http://www.nasa.gov/mission_pages/shuttle/shuttlemissions/list_main.html.

National Aeronautics and Space Administration (NASA). 2007e. "Human Space Flight." Accessed June 20, 2007. http://history.nasa.gov/tindex.html#5.

National Aeronautics and Space Administration (NASA). 2007f. *2005 Annual Report: Kennedy Space Center.*

National Aeronautics and Space Administration (NASA). 2007g. *Apollo Expeditions to the Moon.* Chapter 3.2. Accessed January 23, 2007. http://history.nasa.gov/SP-350/ch-3-2.html.

National Aeronautics and Space Administration (NASA). 2007h. Environmental Management Division, Procedural Requirements. *Implementing the National Environmental Policy Act and Executive Order 12114.* Accessed June 20, 2007. http://nodis3.gsfc.nasa.gov/displayDir.cfm?t=NPR&c=8580&s=1.

National Aeronautics and Space Administration (NASA). 2007i. *Fact Sheets: Dryden Flight Research Center.* Accessed February 1, 2007.
http://www.nasa.gov/centers/dryden/news/FactSheets/FS-001-DFRC.html;
http://www.nasa.gov/centers/dryden/news/FactSheets/FS-015-DFRC.html; and
http://www.nasa.gov/centers/dryden/news/FactSheets/FS-091-DFRC.html.

National Aeronautics and Space Administration (NASA). 2007j. John C. Stennis Space Center (SSC). *Environmental Assurance Program–Water Quality: Surface Water.* Accessed February 5, 2007.
http://www.ssc.nasa.gov/environmental/resource_mngmnt/water_quality/surface_water/surfwater.html.

National Aeronautics and Space Administration (NASA). 2007k. John F. Kennedy Space Center (KSC). 2007. *Environmental Program Branch, Drinking Water Homepage.* Accessed January 16, 2007.
http://environmental.ksc.nasa.gov/permitting/waterDrink.htm.

National Aeronautics and Space Administration (NASA). 2007l. John F. Kennedy Space Center (KSC). *Environmental Program Branch, Domestic Wastewater Program Homepage.* Accessed January 16, 2007.
http://environmental.ksc.nasa.gov/permitting/waterDomestic.htm.

National Aeronautics and Space Administration (NASA). 2007m. John F. Kennedy Space Center (KSC). *Environmental Program Branch, National Pollutant Discharge Elimination System (NPDES) Permitting Homepage.* Accessed January 16, 2007.
http://environmental.ksc.nasa.gov/permitting/waterNpdes.htm.

National Aeronautics and Space Administration (NASA). 2007n. Marshall Space Flight Center Website. http://www.nasa.gov/centers/marshall.

National Aeronautics and Space Administration (NASA). 2007o. Space Shuttle Program Business Management Office. 2007. E-mail from Dennis Davidson, dated February 20, 2007.

6. REFERENCES

National Aeronautics and Space Administration (NASA). 2007p. *Space Shuttle Program, Shuttle Environmental Assurance, 2003 Annual Report.* Accessed June 15, 2007.

National Aeronautics and Space Administration (NASA). 2007q. *Space Shuttle System–External Tank. 2006.* Accessed June 20, 2007. http://www.nasa.gov/returntoflight/system/system_ET.html.

National Aeronautics and Space Administration (NASA). February 7, 2007r. *Highlights of NASA's FY 2008 Budget Request,* and *FY 2008 Budget Estimates.* Presented by NASA Administrator Michael D. Griffin on February 5, 2007. (http://www.nasa.gov/pdf/168652main_NASA_FY08_Budget_Request.pdf; and http://www.nasa.gov/pdf/168674main_mg_fy08_budget_rollout.pdf.

National Aeronautics and Space Administration (NASA). January 2007s. *Space Shuttle Program Site Environmental Summaries, Space Shuttle Program.*

National Aeronautics and Space Administration (NASA). August 2007t. *Final Constellation Programmatic Environmental Impact Statement.*

National Aeronautics and Space Administration (NASA). 2007u. John F. Kennedy Space Center Environmental Program Branch Website. Accessed January 4, 2007. http://environmental.ksc.nasa.gov/permitting/waterprog.htm.

National Aeronautics and Space Administration. 2007v. John F Kennedy Space Center Environmental Program Branch. Energy Program Overview. Accessed July 27, 2007.

National Aeronautics and Space Administration (NASA). 2007w. Shuttle Survey Historic Eligibility Report for the Stennis Space Center, Hancock Country, Mississippi.

National Aeronautics and Space Administration (NASA). February 2007x. *Cultural Resources Management Plan.* Marshall Space Flight Center. Huntsville, Alabama.

National Aeronautics and Space Administration (NASA). 2007y. Johnson Space Center Environmental Engineering Department. JSC Energy Purchases. Via e-mail from Dennis Klekar, dated February 9, 2007.

National Aeronautics and Space Administration (NASA). 2007z. *John F. Kennedy Space Center Environmental Branch Homepage.* http://environmental.ksc.nasa.gov.

National Aeronautics and Space Administration (NASA). 2007aa. *Kennedy Space Flight Center Environmental Program Branch Website.* Accessed 2007. //http://environmental.ksc.nasa.gov/permitting/reporting.htm.

National Aeronautics and Space Administration (NASA). November 2007bb. *Baseline Socioeconomic Resources, Space Shuttle Program, Fiscal Year 2006.* Prepared for the Space Shuttle Programmatic Environmental Assessment.

National Aeronautics and Space Administration (NASA). 2006a. *WSTF Environmental Program Overview.* White Sands Treatment Facility (WSTF). PowerPoint.

National Aeronautics and Space Administration (NASA). August 2006b. *Final Environmental Assessment for the Development of the Crew Exploration Vehicle.*

National Aeronautics and Space Administration (NASA). October 2006c. *Firing Room Gets New Look.* October. By Linda Herridge and Elaine Marconi, KSC Staff Writers. Accessed December 26, 2006. http://www.nasa.gov/mission_pages/constellation/main/fr1.html.

National Aeronautics and Space Administration (NASA). June 2006d. "SSC Plays Vital Role in NASA's Project Constellation." *Lagniappe.* Vol. 1, No. 6, p. 1.

National Aeronautics and Space Administration (NASA). January 31, 2006e. Procedural Requirements 4310.1, "Identification and Disposition of NASA Artifacts."

National Aeronautics and Space Administration (NASA). September 21, 2006f. "Shuttle Transition Environmental Support Team Current Status of WSTF Restoration Program" and "WSTF Environmental Overview." Presentations by Michael Zigmond.

National Aeronautics and Space Administration (NASA). March 2006g. *Economic Impact of NASA in Florida, Fiscal Year 2005.* Produced by the NASA Office of the CFO at Kennedy Space Center, Florida, with the support of W. Warren McHone, Ph.D., Transportation Economics Research Institute (TERI).

National Aeronautics and Space Administration. December 2005a. *John C. Stennis Space Center Environmental Resources Document, Revision Basic.* No. SCWI-8500-0026 ENV.

National Aeronautics and Space Administration (NASA). March 2005b. *Environmental Resources of Ellington Field, Houston, Texas.*

National Aeronautics and Space Administration (NASA). March 2005c. *Final Programmatic Environmental Impacts Statement for the Mars Exploration Program.* Science Mission Directorate.

National Aeronautics and Space Administration (NASA). April 20. 2005d. Dryden Centerwide Procedure, Code O, Aircraft Maintenance & Safety Manual (AM&SM) (DCP-O-001).

National Aeronautics and Space Administration (NASA). 2005e. *DFRC Range Safety.* Dryden Flight Research Center.

6. REFERENCES

National Aeronautics and Space Administration (NASA). December 2004a. *Environmental Resources of Lyndon B. Johnson Space Center, Houston, Texas.* Prepared by Team Dyncorp-Lynx Ltd. for Johnson Space Center.

National Aeronautics and Space Administration (NASA). December 2004b. *John C. Stennis Space Center Environmental Operations and Implementation Program Procedural Requirements.* SPR 8500.2 Revision A.

National Aeronautics and Space Administration (NASA). Revised March 2004c. *El Paso Forward Operation Location Environmental Resources Document.*

NASA Aeronautics and Space Administration (NASA). 2004d. *Integrated Cultural Resources Management Plan 2004-2009.* White Sands Missile Range.

National Aeronautics and Space Administration (NASA). 2004e. *Return to Flight.* Accessed June 2007. http://www.nasa.gov/returntoflight/system/.

National Aeronautics and Space Administration (NASA). January 2004f. Presidential Policy Directive, Vision for Space Exploration.

National Aeronautics and Space Administration (NASA). January 14, 2004g. "President Bush Announces New Vision for Space Exploration Program Fact Sheet: A Renewed Spirit of Discovery." Accessed August 1, 2007. http://www.whitehouse.gov/news/releases/2004/01/20040114-1.html.

National Aeronautics and Space Administration (NASA). August 2003a. *Environmental Resource Document, Division D.* John F. Kennedy Space Center (KSC-DF-3080).

National Aeronautics and Space Administration (NASA). July 2003b. *Sonny Carter Training Facility Environmental Resource Document.*

National Aeronautics and Space Administration (NASA). June 2003c. *Dryden Flight Research Center Environmental Resource Document.*

National Aeronautics and Space Administration (NASA). 2003d. *Edwards Air Force Base General Plan.*

National Aeronautics and Space Administration (NASA). January 2003e. *John C. Stennis Space Center, Historic Preservation Plan.*

National Aeronautics and Space Administration (NASA). January 2002a. *Marshall Space Flight Center Environmental Resource Document–Final.*

National Aeronautics and Space Administration (NASA). June 2002b. *Final Environmental Assessment for Launch of NASA Routine Payloads on Expendable Launch Vehicles from Cape Canaveral Air Force Station, Florida, and Vandenberg Air Force Base, California.*

National Aeronautics and Space Administration (NASA). July 2002c. *JSC Safety and Health Handbook.* (JPR 1700.1, Revision I).

National Aeronautics and Space Administration (NASA). November 2002d. *John C. Stennis Space Center, Pollution Prevention Plan.*

National Aeronautics and Space Administration (NASA). June 2002e. *Draft Integrated Cultural Resources Management Plan for Edwards Air Force Base, California.* Air Force Flight Test Center Plan 32-7065.

National Aeronautics and Space Administration (NASA). February 16, 2001a. *White Sands Test Facility Environmental Resources Document. Final.* RD-WSTF-0025. WSTF-ERD-4.

National Aeronautics and Space Administration (NASA). September 2001b. *Michoud Assembly Facility Environmental Resource Document.*

National Aeronautics and Space Administration (NASA). January 1998. *White Sands Missile Range Range-Wide Environmental Impact Statement.* Directorate of Environment and Safety, Environmental Services Division. New Mexico.

National Aeronautics and Space Administration (NASA). December 1997. *Environmental Justice Plan.* John F Kennedy Space Center. Prepared by KSC Biomedical Office; Environmental Program Office.

National Aeronautics and Space Administration (NASA). 1995. *Environmental Justice Strategy.* Washington, D.C.

National Aeronautics and Space Administration (NASA). 1992. *NASA Facts Online.* "NASA's Orbiter Fleet." Accessed June 20, 2007. http://www-pao.ksc.nasa.gov/kscpao/nasafact/orbiters.htm.

National Aeronautics and Space Administration (NASA). March 1989. *Final Environmental Impact Statement, Space Shuttle Advanced Solid Rocket Motor Program.* John C. Stennis Space Center; George C. Marshall Flight Center.

National Aeronautics and Space Administration (NASA). January 1978. *Final Institutional Environmental Impact Statement, Michoud Assembly Facility, New Orleans, Louisiana.*

National Aeronautics and Space Administration (NASA). Not Dated-a. *Human Space Flight Transition Plan.*

National Aeronautics and Space Administration (NASA). Not Dated-b. MSFC History Office. Accessed February 14, 2007. http://history.msfc.nasa.gov/landmark/historic_landmarks.html.

National Park Service (NPS). 2007a. *Aviation: From Sand Dunes to Sonic Booms.* Accessed February 12, 2007. http://www.cr.nps.gov/nr/travel/aviation/lab.htm.

6. REFERENCES

National Park Service (NPS). 2007b. *Aviation: From Sand Dunes to Sonic Booms: Rocket Propulsion Test Complex.* Accessed February 1, 2007. http://www.cr.nps.gov/nr/travel/aviation/roc.htm.

National Park Service (NPS). 2007c. *Aviation: From Sand Dunes to Sonic Booms, Apollo Mission Control Center.* Accessed February 1, 2007. http://www.cr.nps.gov/nr/travel/aviation/apo.htm.

National Park Service (NPS). 2007d. *Aviation: From Sand Dunes to Sonic Booms: Saturn V Dynamic Test Stand.* Accessed February 1, 2007. http://www.cr.nps.gov/nr/travel/aviation/sat.htm

National Park Service (NPS). 2007e. *National Historic Landmarks: Illustrating the Heritage of the United States.* Accessed February 14, 2007. http://www.cr.nps.gov/nhl/publications/bro1.htm.

National Park Service (NPS). 2007f. *National Register Bulletin: Guidelines for Evaluating and Documenting Aviation Properties.* Accessed March 8, 2007. http://www.cr.nps.gov/nr/publications/bulletins/aviation/nrb_aviation_IV.htm

National Park Service (NPS). 2007g. Nationwide Rivers Inventory Website. Accessed February 5, 2006. http://www.nps.gov/rtca/nri/.

National Park Service (NPS). 2006. *Aviation: From Sand Dunes to Sonic Booms. White Sands V-2 Launching Site.* Accessed on December 26, 2006. http://www.cr.nps.gov/nr/travel/aviation/whi.htm; and http://www.cr.nps.gov/nr/travel/aviation/jfk.htm.

National Park Service (NPS). 1989. *CRM Bulletin, Volume 12: No. 6 Cultural Resource Management, A Technical Bulletin for Parks, Federal Agencies, States, Local Governments, and the Private Sector.* Washington, D.C.

National Park Service. 1995. *National Register Bulletin: How to Apply National Register Criteria for Evaluation.* Washington, D.C.

National Park Service. 1998. *National Register Bulletin: Guidelines for Evaluating and Nominating Properties that Have Achieved Significance within the Last Fifty Years.* Washington, D.C.

National Space Transportation System. 1988. *The NSTS News Reference Manual.* Accessed February 1, 2007. http://science.ksc.nasa.gov/shuttle/technology/sts-newsref/sts-lcc.html#sts-ksc-towing.

New Mexico Environment Department. 2007. New Mexico Solid Waste Management Regulations 20.9.1 NMAC. Accessed July 16, 2007. http://www.nmcpr.state.nm.us/nmac/parts/title20/20.009.0001.htm.

New Mexico Environmental Department. 2001. "Air Quality Division." Accessed June 19, 2007. http://www.nmenv.state.nm.us/aqb/.

New Mexico Natural Heritage Program and Environment and Safety Directorate. November 2001. *White Sands Missile Range Integrated Natural Resource Management Plan.* Environmental Stewardship Division, White Sands Missile Range.

Office of Historic Preservation (OHP). 2007. Office of Historic Preservation Project Review. Accessed February 21, 2007. http://ohp.parks.ca.gov/default.asp?page_id=1071.

Office of Management and Budget (OMB). December 18, 2006. *OMB Bulletin No. 07-01: Update of Statistical Area Definitions and Guidance on Their Uses.* http://www.whitehouse.gov/omb/bulletins/fy2007/b07-01.pdf.

Page and Tunbull, Inc. 2007. *Space Shuttle Program: NASA Ames Research Center Mott Field, California.*

Personal Communication. 2007a. "WSTF Employees by Zip Code.xls." From Harry Johnson (WSTF-RA) to Amy Keith (MSFC-AS10), via e-mail dated March 1, 2007.

Personal Communication. 2007b. "Employees by Zip as of 7-27-06." Excel Spreadsheet, via e-mail. From Barbara Naylor, KSC, to Donna Holland, via e-mail. August 1, 2007.

Personal Communication. 2007c. "Headcount_By Zip_021207.xls." From Daniel Sword, NASA MAF, to Donna Holland. Excel spreadsheet, via e-mail. February 12, 2007.

Personal Communication. 2007d. "MSFC_State_County_Count_Contractors_OnSite.xls" and "MSFC_State_County_Count_CivilService_OnSite.xls." From David Jeffreys (MSFC-HS20) to Donna Holland. Excel Spreadsheets with Summary Data from MSFC Security Database, via e-mail dated February 13, 2007.

Personal Communication. 2007e. "SBS_ZipCodes 02-15-2007.xls." From Carolyn Kennedy (SSC-RA02) to Donna Holland. Excel Spreadsheet, via e-mail dated February 9, 2007.

Personal Communication. 2007F. "WSTF Data Needed for Socioeconomics.doc." From Harry Johnson (WSTF-RA) to Amy Keith (MSFC-AS10), via e-mail dated February 28, 2007.

Personal Communication. 2007g. From Annette Dittmer, NASA KSC, Human Resources Office, to Sue Leibert, via e-mail. February 12, 2007.

Personal Communication. 2007h. JSC Residence Data. From Perri Fox (JSC-JA) to Donna Holland (MSFC-AS10), via e-mail. February 15, 2007.

Personal Communication. 2007i. JSC Space Shuttle NASA Workforce Data. From Eameal Holstien (JSC-LB), via e-mail to Sue Leibert (JSC-AH/MD). January 1, 2007.

6. REFERENCES

Personal Communication. 2007j. National Aeronautics and Space Administration, Space Shuttle Business Office. Teleconference dated January 30, 2007.

Personal Communication. May 23, 2006. Personal communication with Louisiana Department of Environmental Quality. Lockheed Martin Space Systems Company.

Preserve America. 2007. Accessed May 17, 2007. http://www.preserveamerica.gov/.

Science Applications International Corporation. 2007. *Space Shuttle Program Historic Eligibility Report for NASA Langley Research Center.*

Sites Southwest & Bohannon Huston, Inc. October 2005. *Otero County Comprehensive Plan.* http://www.co.otero.nm.us/Otercocomplan-final-10-05%20small.pdf.

State of New Mexico Environment Department (NMED) Air Quality Bureau. January 14, 1997. "Area 400 Test Facility Air Quality Permit No. 629-Area 400-M-1."

State of New Mexico Environment Department (NMED) Air Quality Bureau. November 19, 1993. "Area 300 Altitude Simulation Air Quality Permit No. 29-M-3."

Tennessee Advisory Commission on Intergovernmental Relations (TACIR). December 2003. *Population Projections for the State of Tennessee 2005-2025.* In cooperation with the University of Tennessee Center for Business and Economic Research. Accessed 2007. http://cber.bus.utk.edu/download.htm.

Texas Commission on Environmental Quality (TCEQ). 2007a. Air Pollutant Watch List. Accessed June 19, 2007. http://www.tceq.state.tx.us/implementation/tox/AirPollutantMain/APWL.html.

Texas Commission on Environmental Quality (TCEQ). 2007b. Central Permit Registry Website for Johnson Space Center. Accessed February 6, 2007. http://www4.tceq.state.tx.us/crpub/index.cfm?fuseaction=cust.showSingleCust&requesttimeout=120&princ_id=339137452002006.

Texas Commission on Environmental Quality (TCEQ). 2005. State of Texas 2004 303d list.

Texas State Data Center. October 2006. *Projections of the Population of Texas and Counties in Texas by Age, Sex and Race/Ethnicity for 2000-2040.* In cooperation with the Office of the State Demographer, Institute for Demographic and Socioeconomic Research, University of Texas at San Antonio. Produced for the Population Estimates and Projections Program. Accessed 2007. http://txsdc.utsa.edu/tpepp/2006projections/.

The Weather Channel (TWC). 2007a. "Monthly Averages, Ellington Field, Texas." Accessed February 7, 2007.

http://www.weather.com/outlook/travel/businesstraveler/wxclimatology/monthly/graph/EFD:9?from=hrly_bottomnav_business.

The Weather Channel (TWC). 2007b. Monthly Averages for Kennedy Space Center, Florida. Accessed January 16, 2007.
http://www.weather.com/weather/wxclimatology/monthly/graph/USFL0339?from=36hr_bottomnav_undeclared

The Weather Channel (TWC). 2007c. "Monthly Averages, Palmdale, California." Accessed 2007.
http://www.weather.com/outlook/travel/businesstraveler/wxclimatology/monthly/graph/USCA0829?from=36hr_bottomnav_business.

The Weather Channel (TWC). 2007d. "Monthly Averages for Canoga Park., California." Accessed February 22, 2007.
http://www.weather.com/outlook/travel/businesstraveler/wxclimatology/monthly/graph/91304?from=36hr_bottomnav_business.

Thiokol Corporation. 1992. "Chemical Rocket Propulsion and the Environment." Presented to: 28th AIAA/SAE/ASME Joint Propulsion Conference and Exhibit. Opryland Hotel, Nashville, Tennessee. July 6 through July 8, 1992.

Tidwell, Lisa. Not Dated. "Practice Makes Perfect." *Space Center Round Up.*

TRC. May 2007. *Evaluation of Resources Associated with the Space Shuttle Program, Michoud Assembly Facility, New Orleans, Louisiana.* Prepared for Lockheed Martin Corporation, New Orleans, Louisiana.

TRC Garrow Associates, Inc. May 2001. *Architectural Survey of the NASA Michoud Assembly Facility, New Orleans, Louisiana.* Addendum Report: Survey and Evaluation of the Orion, the Pearl River, and the Poseidon. Prepared for the National Aeronautics and Space Administration. Authors: Cleveland, M. Todd and Mark D. Chancellor.

U.S. Air Force. 1995. Eastern/Western Range (EWR) 127-1. "Range Safety Requirements."

U.S. Army Corps of Engineers (USACE). January 1987. *Final Environmental Impact Statement of the Proposed Ground Based Free Electron Laser Technology Integration Experiment, White Sands Missile Range, New Mexico.* Prepared for the U.S. Army Strategic Defense Command. Prepared by the USACE, Huntsville Division, Huntsville, Alabama.

U.S. Bureau of Economic Analysis (BEA). 2007a. "State Personal Income: Third Quarter of 2006" (Table 1) in the January 2007 *Survey of Current Business.*

U.S. Bureau of Economic Analysis (BEA). May 2007b. *Local Area Personal Income for 2005.*

6. REFERENCES

U.S. Bureau of Economic Analysis (BEA). 2006a. "Personal Income for Metropolitan Areas for 2005" (Table 1) in the September 2006 *Survey of Current Business*.

U.S. Bureau of Economic Analysis (BEA). 2006b. Regional Economic Information System, Table CA30. Last updated April 2006. Downloaded March 8, 2007. http://www.bea.gov/regional/reis/CA30fn.cfm.

U.S. Bureau of Economic Analysis (BEA). 2006c. Regional Economic Information System, Table CA25 (NAICS). Last updated April 2006. Downloaded April 23, 2007. http://www.bea.gov/regional/reis.

U.S. Bureau of Economic Analysis (BEA). 2005. Frequently Asked Questions: "What are the effects of Hurricanes Katrina, Rita, and Wilma on Monthly Personal Income?" Created December 5, 2005. Last Updated January 19, 2006. http://faq.bea.gov/cgi-bin/bea.cfg/php/enduser/std_alp.php.

U.S. Bureau of Economic Analysis (BEA). 2004. *BEA Economic Profiles for Kennedy Space Center*. CA30 Regional Economic Profiles and CA04 Personal Income and Employment Summary.

U.S. Bureau of Labor Statistics. 2006. Local Area Unemployment Statistics. Downloaded April 24, 2006. http://data.bls.gov/PDQ/outside.jsp?survey=la.

U.S. Census Bureau. April 5, 2007a. "County total population, population change and estimated components of population change: April 1, 2000 to July 1, 2006." Population Division. http://www.census.gov/popest/datasets.html.

U.S. Census Bureau. 2007b. State & County QuickFacts. Last revised January 12, 2007; downloaded March 8, 2007 and April 23, 2007. http://quickfacts.census.gov/qfd/states/12/12009.html.

U.S. Census Bureau. July 2006a. American Community Survey (ACS): ACS Special Product for the Gulf Coast Area. http://www.census.gov/acs/www/Products/Profiles/gulf_coast/index.htm.

U.S. Census Bureau. May 25, 2006b. "January 1, 2006, Special Population Estimates for Impacted Counties in the Gulf Coast Area." http://www.census.gov/Press-Release/www/emergencies/impacted_gulf_estimates.html

U.S. Census Bureau. 2006c. "Table B-2. Metropolitan Areas — Components of Population Change." *IN: State and Metropolitan Area Data Book: 2006*.

U.S. Census Bureau. 2005. American Community Survey (ACS). Gulf Coast Area Data Profiles, New Orleans-Metairie-Kenner, LA, Metropolitan Statistical Area. ACS Special Product for the Gulf Coast Area. http://www.census.gov/acs/www/Products/Profiles/gulf_coast/tables/tab1_katrinaK0100US2203v.htm.

U.S. Census Bureau. 2002. Economic Census. "All Sectors: Geographic Area Series: Economy-Wide Key Statistics: 2002." Downloaded June 14, 2007. http://www.census.gov/econ/census02/index.html.

U.S. Census Bureau. 2000. Census American Factfinder. Data tables: downloaded March through May 2007. http://factfinder.census.gov.

U.S. Department of Defense (DoD). June. 1994. *Coming in from the Cold: Military Heritage and the Cold War. Report on the Department of Defense Legacy Cold War Project.* Prepared for the Legacy Program. Washington, D.C.

U.S. Department of Energy (DOE). 2007 http://apps.em.doe.gov/etec/ETEC-Background/Characteristics.html.

U.S. Department of the Interior, Fish & Wildlife Service. November 2006. *Merritt Island National Wildlife Refuge Draft Comprehensive Conservation Plan and Environmental Assessment.*

U.S. Environmental Protection Agency. 2007a. "Air Permits." Accessed March 8, 2007. http://www.epa.gov/air/oaqps/permjmp.html.

U.S. Environmental Protection Agency. 2007b. "Clean Air Act, Title VI." Accessed January 10, 2007. http://www.epa.gov/air/caa/title6.html.

U.S. Environmental Protection Agency. 2007c. "General Conformity Rule, *De Minimis* Levels." Accessed March 8, 2007. http://www.epa.gov/air/genconform/deminimis.htm.

U.S. Environmental Protection Agency. 2007d. *General Conformity Rule, Frequent Questions.* Accessed March 7, 2007. http://www.epa.gov/air/genconform/faq.htm#0.

U.S. Environmental Protection Agency. 2007e. *Green Book, Criteria Pollutant Reports.* Accessed January 22, 2007. http://www.epa.gov/oar/oaqps/greenbk/multipol.html.

U.S. Environmental Protection Agency. 2007f. *New Source Review, Basic Information.* Accessed March 7, 2007. http://www.epa.gov/nsr/info.html.

U.S. Environmental Protection Agency. 2007g. National Estuary Program Website: Accessed February, 6, 2007. http://www.epa.gov/owow/estuaries/ccmp/index.htm.

U.S. Environmental Protection Agency. 2007h. "Envirofacts Database Search." Accessed January 31, 2007. Lockheed Martin Corporation. http://oaspub.epa.gov/enviro/pcs_det_reports.detail_report?npdesid=LA0052256.

U.S. Environmental Protection Agency. 1993. "40 CFR Part 82, Protection of Stratospheric Ozone." Accessed January 22, 2007.

6. REFERENCES

U.S. Environmental Protection Agency. November 20, 1990. "40 CFR, Chapter 61, Subpart M. National Emission Standard for Asbestos."
http://www.epa.gov/oar/caa/caa601.txt.

U.S. Fish and Wildlife Service (USFWS). 2006. *Listed and Sensitive Species in Dona Ana County.* Accessed February 8, 2007.
http://www.fws.gov/southwest/es/NewMexico/SBC_view.cfm?spcnty=Dona%20Ana.

U.S. Fish and Wildlife Service (USFWS). April 2003. Lists of Endangered, Threatened, Proposed and Candidate Species for the Southeast Region, Louisiana. Accessed January 24, 2007.
http://www.fws.gov/southeast/es/county%20lists.htm.

U.S. Geological Survey (USGS). 2007. *USGS Ground Water Atlas of the United States: Arizona, Colorado, New Mexico, Utah HA 730-C – USGS.* Accessed February 6, 2007.
http://capp.water.usgs.gov/gwa/ch_c/C-text4.htm.

Appendix A: Acronyms and Abbreviations and Common Metric/British System Equivalents

AA	Associate administrator
ACHP	Advisory Council on Historic Preservation
ACM	Asbestos-containing material
ADCNR	Alabama Department of Conservation and Natural Resources
ADEM	Alabama Department of Environmental Management
AETF	Advanced Engine Test Facility
AFB	Air Force Base
AFFTC	Air Force Flight Test Center
AFP	Air Force Plant
AIAA	American Institute of Aeronautics and Astronautics
AM&SM	Aircraft Maintenance and Safety Manual
AMCOM	Army Aviation Missile Command
ANHP	Alabama Natural Heritage Program
ANSI	American National Standards Institute
AOC	Area of concern
APU	Auxiliary power unit
APZ	Accident Potential Zones
ARF	Assembly and Refurbishment Facility
AST	Aboveground storage tank
ASTM	American Society for Testing and Materials
ATK	ATK Launch Systems
AVEK	Antelope Valley-East Kern Water Agency
BAHEP	Bay Area Houston Economic Partnership
bhp	Brake horsepower
BLM	Bureau of Land Management
BLS	U.S. Bureau of Labor Statistics
BMP	Best management practice
BNSF	Burlington Northern Santa Fe
BOD	Biochemical oxygen demand
BSB	Building setback
BTEX	Benzene, toluene, ethylbenzene, and xylenes
CAA	Clean Air Act
Cal-EPA	California Environmental Protection Agency
CAP	Corrective Action Program
CARB	California Air Resources Board
CBER	Center for Business and Economic Research
CCAFS	Cape Canaveral Air Force Station

CCMP	Comprehensive Conservation and Management Plan
CCS	Cape Canaveral Spaceport
CCSMP	Cape Canaveral Spaceport Master Plan
CEQ	Council on Environmental Quality
CERCLA	Comprehensive Environmental Response, Compensation, and Liability Act
CFC	Chlorofluorocarbon
CFR	*Code of Federal Regulations*
cm	Centimeter
CMS	Corrective Measures Study
CNS	Canaveral National Seashore
CO	Carbon monoxide
COD	Chemical oxygen demand
CWA	Clean Water Act
Cx PEIS	Constellation Programmatic Environmental Impact Statement
CZM	Coastal Zone Management
DA	Department of Army
dB	Decibel
dBA	decibel A-rated
DFRC	Dryden Flight Research Center
DNAPL	Dense non-aqueous phase liquid
DO	Dissolved oxygen
DoD	U.S. Department of Defense
DOE	U.S. Department of Energy
DOT	Department of Transportation
DPD	Dryden Policy Directive
DTSC	Department of Toxic Substances Control
EA	Environmental Assessment
EAFB	Edwards Air Force Base
ECHO	Environmental and compliance History Online (EPA)
EDR	Economic and Demographic Research
EDS	Earth Departure Stage
eFlorida	Enterprise Florida, Inc.
EF	Ellington Field
ELS	Emergency Landing Site
ELV	Expendable Launch Vehicle
EMD	Environmental Management Division
EMO	Emergency Management Office
EMS	Environmental management system
EO	Executive Order
EPA	U.S. Environmental Protection Agency
EPB	Environmental Program Branch
EPCRA	Emergency Planning and Community Right-to-Know Act

EPFOL	El Paso Forward Operation Location
EPIA	El Paso International Airport
ERD	Environmental Resources Document
ERP	Environmental Restoration Program
ESE	Environmentally sensitive electronics
ESO	Environmental Services Office
ET	External Tank
ETU	Evaporation Tank Unit
°F	Degrees Fahrenheit
F.A.C.	Florida Administrative Code
FAA	Federal Aviation Administration
FDEP	Florida Department of Environmental Protection
FEMA	Federal Emergency Management Agency
FIRM	Floodplain Insurance Rate Map
FOS	Facility Operating Services
FPO	Federal Preservation Officer
FR	*Federal Register*
ft	Feet
ft^2	Square feet
FTE	Full-time equivalent
FTU	Fuel treatment unit
FY	Fiscal year
GH_2	Gaseous hydrogen
GIWW	Gulf Intracoastal Waterway
GNOCDC	Greater New Orleans Community Data Center
GO/CO	Government owned/contractor operated
GO/GO	Government owned/government-operated
gpd	Gallons per day
gpm	Gallons per minute
GSA	General Services Administration
GSE	Ground support equipment
GSFC	Goddard Space Flight Center
ha	Hectare
HAP	Hazardous Air Pollutant
HAZMAT	Hazardous material
HAZWOPER	Hazardous Waste Operations and Emergency Response Standards
HB	High bay
HCFC	Hydrochlorofluorocarbon
HD	Historic District
HEBF	High-energy Blast Facility
HMCA	Hypergol Maintenance and Checkout Area
HOSC	Huntsville Operations Support Center

hp	Horsepower
HPIW	High-pressure industrial water
HPO	Historic Preservation Officer
HPWG	Historic Preservation Working Group
HQ	Headquarters
HSL	Hardware Simulation Laboratory
HSWA	Hazardous and Solid Waste Amendments of 1984
HVAC	Heating, ventilating, and air conditioning
I	Interstate
ILS	Instrument Landing System
IRLS	Indian River Lagoon System
IRP	Installation Remediation Program
ISS	International Space Station
IWTF	Industrial water treatment facility
JBOSC	Joint Base Operations Support Contract
JER	Joranada Experimental Range
JP	Jet propellant
JPC	Jet Propellants Contractor
JPTS	Jet Propulsion Thermally Stable
JSC	Lyndon B. Johnson Space Center
KARS	Kennedy Athletic, Recreational and Social Organization
kg	Kilogram
km	Kilometer
km/h	Kilometer per hour
km^2	Square kilometer
km^3/d	Cubic kilometers per day
$KMnO_4$	Potassium permanganate
KNPR	Kennedy NASA Procedural Requirements
KSC	Kennedy Space Center
kV	Kilovolt
kVA	Kilovolt ampere
kW	Kilowatt
L	Liter
LBP	Lead-based paint
LC	Launch Complex
LCC	Launch Control Center
LDEQ	Louisiana Department of Environmental Quality
LDNR	Louisiana Department of Natural Resources
LDWF	Louisiana Department of Wildlife and Fisheries
LH2	Liquid hydrogen
LHe	Liquid helium
LN2	Liquid nitrogen
LNHP	Louisiana Natural Heritage Program

LOX	Liquid oxygen
LPDES	Louisiana Pollutant Discharge Elimination System
LQG	Large-quantity generator
LTM	Long-term monitoring
m	Meter
m^2	Square meter
MACT	Maximum achievable control technology
MAF	Michoud Assembly Facility
MMBtu/hr	Million British thermal units per hour
MDAB	Mojave Desert Air Basin
MDAH	Mississippi Department of Archives and History
MDEQ	Mississippi Department of Environmental Quality
MDWFP	Mississippi Department of Wildlife, Fisheries and Parks
mgd	Million gallons per day
mgy	Million gallons per year
µg/L	Micrograms per liter
MINWR	Merritt Island National Wildlife Refuge
MLP	Mobile launch platform
MNHP	Mississippi National Heritage Program
MOA	Memorandum of Agreement
MOM	Maintenance and Base Operations Management
MOU	Memorandum of Understanding
mph	Miles per hour
MPR	Marshall Procedural Requirement
MRGO	Mississippi River Gulf Outlet
MSA	Metropolitan Statistical Area
MSAAP	Mississippi Army Ammunition Plant
MSFC	Marshall Space Flight Center
msl	Mean sea level
MVA	Mega volt amps
MW	Megawatt
NAAQS	National Ambient Air Quality Standards
NAICS	North American Industry Classification System
NASA	National Aeronautics and Space Administration
NAVSCIATTS	Naval Small Craft Instruction and Technical Training School
NBL	Neutral Buoyancy Laboratory
NBS	Neutral Buoyancy Simulator
NEP	National Estuary Program
NEPA	National Environmental Policy Act
NESHAP	National Emission Standards for Hazardous Air Pollutant
NFA	No further action
NHL	National Historic Landmark

NHPA	National Historic Preservation Act
NMAC	New Mexico Administrative Code
NMED	New Mexico Environmental Department
NOFD	New Orleans Fire Department
NOI	Notice of Intent
NOx	Nitrogen oxide
NPD	NASA Policy Directive
NPDES	National Pollutant Discharge Elimination System
NPL	National Priorities List
NPR	NASA Procedural Requirements
NPS	National Park Service
NRHP	National Register of Historic Places
NRI	Nationwide Rivers Inventory
NSPS	National Stationary Performance Standards
NSR	New Source Review
NSSC	NASA Shared Services Center
NWI	National Wetland Inventory
O&M	Operations and maintenance
ODS	Ozone depleting substance
OMB	U.S. Office of Management and Budget
OPF	Orbiter Processing Facility
OSB	Operations Support Building
OSHA	Occupational Health and Safety Administration
OU	Operable unit
OV	Orbiter Vehicle
P.L.	Public Law
P2	Pollution prevention
PAFB	Patrick Air Force Base
PAH	Polycyclic aromatic hydrocarbon
PBR	Permit-by-Rule
PCB	Polychlorinated biphenyl
PES	Preliminary Environmental Survey
PM	Particulate matter
POTW	Publicly owned treatment work
ppb	Parts per billion
PPE	Personal protective equipment
ppm	Parts per million
PSD	Prevention of Significant Deterioration
psi	Pounds per square inch
psig	Pounds per square inch gauge
QD arc	Quantity-separation distance
R&D	Research and Development
RR	Rural route

RACM	Regulated Asbestos Containing Materials
RASA	RSA Support Agency
RAW	Removal Action Work
RCM	Refrigerant Compliance Manager
RCRA	Resource Conservation and Recovery Act
RFA	RCRA Facility Assessment
RFI	RCRA Facility Investigation
ROI	Region of influence
ROW	Right-of-way
RP	Rocket propellant
RSA FESD	Redstone Arsenal Fire and Emergency Services Department
RSA	Redstone Arsenal
RSRM	Reusable Solid Rocket Motor
S&MA	Safety and Mission Assurance
S&WB	Sewerage and Water Board
SAIL	Avionic Systems Laboratory
SANWR	San Andres National Wildlife Refuge
SCS	Soil Conservation Service
SCTF	Sonny Carter Training Facility
SDWA	Safe Drinking Water Act
SESL	Space Environmental Simulation Laboratory
SHPO	State Historical Preservation Officer
SID	State Indirect Discharge
SJRWMD	St. John's River Watershed Management Division
SLA	Superlight Ablator
SLF	Shuttle Landing Facility
SMR	Self-monitoring report
SOFI	Spray-on foam insulation
SOMD	Space Operations Mission Directorate
SOP	Standard operating procedure
SPL	Sound pressure level
SPR	SSC Procedural Requirement
SQG	Small-quantity generator
SR	State Route
SRB	Solid Rocket Booster
SRM	Solid Rocket Motor
SSC	Stennis Space Center
SSFL	Santa Susana Field Laboratory
SSME	Space Shuttle Main Engine
SSP	Space Shuttle Program
STA	Shuttle Training Aircraft
STP	Sewage treatment plant
STS	Space Transportation System

SWMU	Solid waste management unit
SWP3	Storm Water Pollution Prevention Plan
T&R	Transition and retirement
TAC	Texas Administrative Code
TAL	Transoceanic Abort Landing
TCE	Trichloroethylene
TCEQ	Texas Council on Environmental Quality
TCS	Thermal Control System
TDS	Total dissolved solids
TEAL	Triethylaluminum
TEB	Triethylborane
TF	Test facility
TOC	Total organic carbon
TPDES	Texas Pollutant Discharge Elimination system
TPS	Thermal protection system
TRI	Toxic Release Inventory
TS	Test stand
TSCA	Toxic Substances Control Act
TSDF	Treatment Storage and Disposal Facility
TSS	Total suspended solids
TVA	Tennessee Valley Authority
TWA	Time-weighted average
TWC	The Weather Channel
TxANG	Texas Air National Guard
U.S.	United States
U.S.C.	United States Code
UAV	Unmanned Aerial Vehicle
UIC	Underground Injection Control
USA	United States Alliance
USACE	U.S. Army Corps of Engineers
USAF	U.S. Air Force
USDA	U.S. Department of Agriculture
USDI	U.S. Department of the Interior
USFWS	U.S. Fish and Wildlife Services
USGS	U.S. Geological Survey
UST	Underground storage tank
VAB	Vehicle Assembly Building
VAFB	Vandenberg Air Force Base
VPP	Voluntary Protection Program
WNWR	Wheeler National Wildlife Refuge
WRPA	Waste Reduction Policy Act
WSMR	White Sands Missile Range
WSSH	White Sands Space Harbor

WSTF	White Sands Test Facility
WTP	Water treatment plant
WWTP	Wastewater treatment plant

This page intentionally left blank.

Common Metric/British System Equivalents

Linear

1 centimeter (cm) = 0.3937 inch	1 inch = 2.54 cm
1 centimeter = 0.0328 foot (ft)	1 foot = 30.48 cm
1 meter (m) = 3.2808 feet	1 ft = 0.3048 m
1 meter = 0.0006 mile (mi)	1 mi = 1609.3440 m
1 kilometer (km) = 0.6214 mile	1 mi = 1.6093 km
1 kilometer = 0.53996 nautical mile (nmi)	1 nmi = 1.8520 km
	1 mi = 0.87 nmi
	1 nmi = 1.15 mi

Area

1 square centimeter (cm^2) = 0.1550 square inch (in^2)	1 in^2 = 6.4516 cm^2
1 square meter (m^2) = 10.7639 square feet (ft^2)	1 ft^2 = 0.09290 m^2
1 square kilometer (km^2) = 0.3861 square mile (mi^2)	1 mi^2 = 2.5900 km^2
1 hectare (ha) = 2.4710 acres (ac)	1 ac = 0.4047 ha
1 hectare (ha) = 10,000 square meters (m^2)	1 ft^2 = 0.000022957 ac

Volume

1 cubic centimeter (cm^3) = 0.0610 cubic inch (in^3)	1 in^3 = 16.3871 cm^3
1 cubic meter (m^3) = 35.3147 cubic feet (ft^3)	1 ft^3 = 0.0283 m^3
1 cubic meter (m^3) = 1.308 cubic yards (yd^3)	1 yd^3 = 0.76455 m^3
1 liter (l) = 1.0567 quarts (qt)	1 qt = 0.9463264 l
1 liter = 0.2642 gallon (gal.)	1 gal. = 3.7845 l
1 kiloliter (kl) = 264.2 gal.	1 gal. = 0.0038 kl

Weight

1 gram (g) = 0.0353 ounce (oz)	1 oz = 28.3495 g
1 kilogram (kg) = 2.2046 pounds (lb)	1 lb = 0.4536 kg
1 metric ton (mt) = 1.1023 tons	1 ton = 0.9072 metric ton

Energy

1 joule = 0.0009 British thermal unit (BTU)	1 BTU = 1054.18 joule
1 joule = 0.2392 gram-calorie (g-cal)	1 g-cal = 4.1819 joule

Pressure

1 newton/square meter (N/m^2) = 0.0208 pound/square foot (psf)   1 psf = 48 N/m^2

Force

1 newton (N) = 0.2248 pound-force (lbf)   1 lbf = 4.4478 N

Radiation

1 becquerel (Bq) = 2.703 x 10^{-11} curies (Ci)   1 Ci = 3.70 x 10^{10} Bq

1 sievert (Sv) = 100 rem   1 rem = 0.01 Sv

Appendix B: Distribution List

NEWSPAPERS

Florida Today Newspaper
c/o Kathy Cicala
P.O. Box 419000
Melbourne, FL 32941-9000

The Citizen Bay Area
c/o Angie Holman
523 N. Sam Houston Parkway
Houston, TX 77060

Houston Chronicle
c/o Ana I. Meares
801 Texas Avenue, 4th Floor
Houston, TX 77002

El Paso Times
c/o Belia Duenes
300 N. Campbell St.
El Paso, Texas 79901

Antelope Valley Press
c/o Alison Adams
P.O. Box 4050
Palmdale, CA 93550-9343

Ridgecrest Daily Independent
c/o Elaine Jones
224 E. Ridgecrest Blvd.
Ridgecrest, CA 93555

Times Picayune
c/o Natalie Retreage
3800 Howard Avenue
New Orleans, LA 70125-1429

The Sea Coast Echo
c/o Annette Lee
124 Court St.
P.O. Box 2009
Bay St. Louis, MS 39521

The Sun Herald
c/o Lisa
P.O. Box 4567
Biloxi, MS 39535

Las Cruces Sun-News
c/o Heather Barry
256 West Las Cruces Ave.
Las Cruces, NM 88001

El Dia
c/o Carlos Romero
6120 Tarnef Drive
Houston, TX 77074

Los Angeles Daily News
c/o Jacqueline White
21221 Oxnard Street
Woodland Hills, CA 91367

The Huntsville Times
2317 South Memorial Parkway
Huntsville, AL 35801

APPENDIX B: DISTRIBUTION LIST

Aerotech News
c/o Gail Ellis
456 E. Ave. K-4, Suite 8
Lancaster, CA 93535

OTHER PARTIES
DFRC

Office of Planning and Research
California State Clearinghouse
P.O. Box 3044
Sacramento CA 95812-3044

U.S. Department of the Interior
Fish and Wildlife Service
Ventura Field Office
2493 Portola Road, Suite B
Ventura, CA 93003-7726

Environmental Protection Agency
Region 9
EIS Review Section
75 Hawthorne Street
San Francisco, CA 94105

Federal Aviation Administration
Western Pacific Region
PO Box 92007
Los Angeles, CA 90009

China Lake Naval Air Warfare
Center Public Affairs, China Lake
Code 750000D
1 Administration Circle
China Lake, CA 93555-6100

City of Lancaster
Planning Department
44933 N. Fern Ave.
Lancaster, CA 93534

City of Palmdale
Planning Department
38250 N. Sierra Highway
Palmdale, CA 93550-4798

Los Angeles County
Planning Department
Room 150 Hall of Records,
13th Floor
320 W. Temple Street
Los Angeles, CA 90012

San Bernardino County
Land Use Services Department
Planning Division
385 N. Arrowhead Ave., 1st Floor
San Bernardino, CA 92415-0182

Kern County
Department of Planning and
Development Services
2700 M Street, Suite 100
Bakersfield, CA 93301-2323

Kern County APCD
2700 M Street, Suite 302
Bakersfield, CA 93301-2370

Bureau of Land Management
Barstow Area Office
2601 Barstow Road
Barstow CA 92311-3221

Bureau of Land Management
Ridgecrest Area Office
300 S. Richmond Road
Ridgecrest, CA 93555-4436

California Department of
Fish and Game
1416 Ninth Street
Sacramento, CA 95814

CALTRANS
Department of Transportation
District 9
500 South Main Street
Bishop, CA 93514

OTHER PARTIES
JSC, EF, SCTF, EPFOL

Mr. Dale R. Hoff
Federal Emergency Management
Agency Region VI
800 North Loop 288
Denton, Texas 76201-3698

Mr. Michael Jansky
Regional Environmental Review
Coordinator
U.S. Environmental Protection
Agency
1445 Ross Avenue, Suite 1200
Dallas, Texas 75202-2733

Ms. Christine Maylath
National Park Service, IMDE-PE
12795 W. Alameda Parkway
Denver, Colorado 80225

Mr. Michael D. Talbott, P.E.
Harris County Flood Control
District
9900 Northwest Freeway
Houston, Texas 77092

Mr. Sheldon M. Kindall
Regional Director
Texas Archeological Society
414 Pebblebrook
Seabrook, Texas 77586

Mr. Al Davis
Harris County Historical
Commission
929 Waxmyrtle
Houston, Texas 77079

Mr. Alan C. Clark
MPO Director
Houston-Galveston Area Council
3555 Timmons Lane, Suite 120
Houston, Texas 77227-2777

Mr. Sam Brown
U.S. Department of Agriculture
Natural Resource Conservation Service
101 South Main
Temple, Texas 76501-7682

Ms. Edith Erfling
U.S. Fish and Wildlife Service
Division of Ecological Services
17629 El Camino Real, Suite 211
Houston, Texas 77058

Ms. Cathy Mayes
Texas Commission on Environmental Quality (TCEQ)
Office of Policy and Regulatory Development
P.O. Box 13087 - MC-205
Austin, Texas 78711-3087

Mr. Roy G. Frye
Texas Parks and Wildlife Department
Wildlife Habitat Assessment Program
4200 Smith School Road
Austin, Texas 78744

Mr. Rick Beverlin
Houston-Galveston Area Council
3555 Timmons Lane, Suite 120
Houston, Texas 77227-2777

Dr. James E. Bruseth
Deputy State Historical Preservation Officer
Texas Historic Commission
P.O. Box 12276
Austin, Texas 78711-2276

Mr. Tom Nuckols
Texas General Land Office
1700 North Congress Avenue
Austin, Texas 78701-1495

OTHER PARTIES MAF

Elaine Williams Givens
17 Reynolds Street
Natchez, MS 39120-2340

Honorable Walt Leger, III
LA House of Reps District 91
600 Carondlet St., 9th Floor
New Orleans, LA 70130

Andrew Rodgers
Eng. Mgr. Citywide Testing & Inspection
3305 Tchoupitoulas St.
New Orleans, LA 70115

APPENDIX B: DISTRIBUTION LIST

Andrea Calvin
P.O. Box 128
Belle Chasse, LA 70037

Evelyn M. Kingston
Our Lady of Prompt Succor School
2320 Paris Road
Chalmette, LA 70043-5026

Honorable Jim Tucker
LA House Reps District 86
732 Behrman Hwy #C2
Terrytown, LA 70056

Honorable Barbara M. Norton
LA State Senate District 3
3245 Hollywood Avenue
Shreveport, LA 71109

H. P. Vaughan
P.O. Box 740
Westwego, LA 70096-0740

Barry Kaufman
Construct Gen Labor UN 689
400 Soniat St.
New Orleans, LA 70115

Stanford Caillouet
577 W Hoover St.
Destrehan, LA 70047-9999

J. Sellers
P.O. Box 3
Luling, LA 70070

Honorable Walker Hines
LA House of Reps District 95
5500 Prytania St, #626
New Orleans, LA 70115

Honorable Arthur A. Morrell
LA House of Reps District 97
6305 Elysian Fields Ave.,
Suite 405
New Orleans, LA 70122-4284

Honorable Cameron Henry
LA House of Reps District 82
201 Evans Rd., Suite 101
Harahan, LA 70123

Honorable Nicholas J. Lousso
LA House of Reps District 94
4431 Canal St., Suite B
New Orleans, LA 70119

Tom Budelman
BOH Bros Construction Co.
730 South Tonti
New Orleans, LA 70119

Honorable Timothy G. Burns
LA House of Reps District 89
1 Sanctuary Blvd., Suite 306
Mandeville, LA 70471

Beth Boquet
P.O. Box 107
Houma, LA 70361

Raymond Gendron
48 Country Club Drive
LaPlace, LA 70068

APPENDIX B: DISTRIBUTION LIST

Honorable Patrick Connick
LA House of Reps District 84
1335 Barataria Blvd., Suite B
Marrero, LA 70072

Rolland A. Mura
9421 Liberty Ct.
River Ridge, LA 70123-2542

Michael E. Neal
600 Carondelet St.
New Orleans, LA 70130-3587

Simon, Peragine, Smith, and
Redfearn, LLP
1100 Poydras St., 30th Floor
New Orleans, LA 70163

Honorable Warren Triche, Jr.
907 Jackson Street
Thibodaux, LA 70301

Fred T. Mayer
501 Cheyenne Drive
Houma, LA 70360-6065

Earl J. Eues
P.O. Box 2768
Houma, LA 70361

Honorable Reggie P. Dupre, Jr.
P.O. Box 3893
Houma, LA 70361

State Representative Wilfred Pie
P.O. Box 91705
Lafayette, LA 70509-1705

Mary Brasseaux
219 Taylor Avenue
Crowley, LA 70526-2647

Lynn Palazzo
WARN
P.O. Box 1312
Bogalusa, LA 70429

John R. Rochelle
1438 Bayou Blue Road
Houma, LA 70364

Marilyn Duet
1438 Bayou Blue Road
Houma, LA 70364

Honorable Damon J. Baldone
162 New Orleans BV
Houma, LA 70364

Warren Gonzales
266 Hwy 1012
Napoleonville, LA 70390

Charlotte Gray
512 Main Street
Patterson, LA 70392

Jewel J. Johnson
59072 Borgne Ave.
Bogalusa, LA 70427

Rose Graham
18376 Bennett Road
Bogalusa, LA 70427

Bertha Hanks
P.O. Box 37
Crowley, LA 70527-0037

Brandon Arabie
9829 Clopha Road
Abbeville, LA 70510

Michael Phelps
10694 Sims Road
Denham Springs, LA 70706

WET, Inc.
P.O. Box 81
Duson, LA 70529-0081

Blaine A. LaCombe
418 Ferry Road
Egan, LA 70531-3608

Honorable Jack D. Smith
500 Main Street, Room 304
Franklin, LA 70538

Honorable Craig F. Romero
300 Iberia Street #B150
New Iberia, LA 70560

Honorable Mike Michot
P.O. Box 80372
Lafayette, LA 70598

Julius Pierce
3014 Lafanette Road
Lake Charles, LA 70605

Joe Hutchins
Cheryl Hutchins
4831 Troon Drive
Lake Charles, LA 70605

Jerome Summers
2723 Phils Lane
Lake Charles, LA 70611

Honorable John R. Smith
LA State Senate District 30
P.O. Box 94183
Baton Rouge, LA 70804

Honorable Herman Hill
529 Tramel Road
Dry Creek, LA 70637

Louisiana Rural Water Association
P.O. Box 180
Kinder, LA 70648

Donald Braxton
Patsy Braxton
850 Willow Springs Road
Sulphur, LA 70663

Maribeth Dietz
1686 White Acres Drive
Sulphur, LA 70663

Patricia Hemphill
405 White Oak Drive
Sulphur, LA 70663-6264

Justin LeJeune
309 Audubon Avenue
Sulphur, LA 70663-9200

Peggie Sullivan
6707 Oak Lake Drive
Sulphur, LA 70665-0665

George Foster
P.O. Box 4
Baker, LA 70714

APPENDIX B: DISTRIBUTION LIST

Darryl Sanderson
Dow Chemical Company
P.O. Box 150
Plaquemine, LA 70764

Robert Poche
43239 Weber City Road
Gonzales, LA 70737-7835

Scott Prejean
11924 Indigo Drive
St. Francisville, LA 70775

Murray McMillan
P.O. Box 11
St. Gabriel, LA 70776

Melissa Sellers
Press Secretary Office of the Governor
P.O. Box 94004
Baton Rouge, LA 70804-9004

Alvin Perkins
22670 Hwy 964
Zachary, LA 70791

Dane Revette
P.O. Box 94185
Baton Rouge, LA 70804-9185

Kathy C. Bretz
7175 Pride Pt. Hudson Road
Zachary, LA 70791

James Womack
5888 Airline Highway
Baton Rouge, LA 70805

Esteban Herrera
Sandra Edwards
P.O. Box 3513
Baton Rouge, LA 70821

Victor Kirk
5177 Greenwell Springs Road
Baton Rouge, LA 70806

Laborers International Union of North America
1233 Government Street
Baton Rouge, LA 70802

Marylee Orr
P.O. Box 66323
Baton Rouge, LA 70806

Honorable John A. Alario, Jr.
LA State Senate District 8
P.O. Box 94183
Baton Rouge, LA 70804

John Arbuthnot
13351 Scenic Highway
Baton Rouge, LA 70807

Honorable Nita Rusich Hutter
LA House of Reps District 104
P.O. Box 275
Chalmette, LA 70044

Roland Selig, Jr.
542 Hillgate Place
Baton Rouge, LA 70808

George Guidry
450 Laurel St. #1420
Baton Rouge, LA 70801

APPENDIX B: DISTRIBUTION LIST

Danny Smith
P.O. Box 84380
Baton Rouge, LA 70884-4380

Edward Jackson
12102 Hwy 73
Geismar, LA 70734

Clyde Heard
275 Middle Road
Dubberly, LA 71024

Harihara Mehendale
700 Univ. AV Sugar Hall, Room 306B
Monroe, LA 71209-0470

James Johnson
P.O. Box 1015
Minden, LA 71058-1015

Harold R. Riggin
P.O. Box 275
Fairbanks, LA 71240

Brian Benson
P.O. Box 239
Sibley, LA 71073-0239

Senator Robert J. Barham
P.O. Box 249
Oak Ridge, LA 71264-0249

Honorable Max T. Malone
610 Marshall
Shreveport, LA 71101

Honorable Richard Gallot, Jr.
LA House of Reps District 11
P.O. Box 1117
Ruston, LA 71273

Kai David Midboe
8270 Bontura Ct.
Baton Rouge, LA 70808

Gordon Moore
Ron Gray
2001 E. 70th St. #503
Shreveport, LA 71105

M. Caire
221 McMillian Road
West Monroe, LA 71291

Honorable Wayne Waddell
LA House of Reps District 5
P.O. Box 6772
Minden, LA 71136-6772

Clyde M. Todd, Jr.
P.O. Box 5067
Alexandria, LA 71307

Honorable Jane H. Smith
LA House of Reps District 8
P.O. Box 72624
Bossier City, LA 71172

Honorable Donald E. Hines
P.O. Box 262
Bunkie, LA 71322

Joy Bradford
P.O. Box 880
Jena, LA 71342-0880

Billy Lapriarie
155 Noble Wiley Road
Jonesville, LA 71343

Honorable Noble E. Ellington
LA House of Reps District 20
4272 Front St.
Winnsboro, LA 71418

APPENDIX B: DISTRIBUTION LIST

FW4 EN Lafayette
USF&WS Lafayette Field Office
Ecological Services
646 Cajundome Blvd., Suite 400
Lafayette, LA 70506

Mr. Mark Thompson
NOAA Fisheries – Southeast
Fisheries Science Center
Panama City Laboratory
3500 Delwood Beach Road
Panama City, Florida 32408

W. R. Stringfield
P.O. Box 128
Belle Chasse, LA 70037

Wayne Martin
P.O. Box 240
St. Gabriel, LA 70776

Derbigny D. Murrell
30080 Hwy 405
Bayou Goula, LA 70788

Ralph King
P.O. Box 120
White Castle, LA 70788

Honorable Richard T. Burford
LA State Senate District 7
671 Highway 171, Suite E
Stonewall, LA 71078

Honorable Austin J. Badon, Jr.
LA House of Reps District 100
3212 Prytania
New Orleans, LA 70115

Mr. Jim Rives
Louisiana Department of
Natural Resources
Office of Coastal Restoration
and Management
P.O. Box 44487
Baton Rouge, LA 70804-4487

Mr. Erick Hawk
NOAA Fisheries – Southeast
Fisheries Science Center
Panama City Laboratory
3500 Delwood Beach Road
Panama City, Florida 32408

Paul Andrews
10543 Oakley Trace Drive
Baton Rouge, LA 70809

Honorable Melvin L. "Kip" Holden
P.O. Box 2843
Baton Rouge, LA 70821-2843

R. Charles Ellis
737 Woodstone Drive
Baton Rouge, LA 70810

Vaughn Benoit
P.O. Box 4448
Baton Rouge, LA 70821-4448

William B. Daniel, IV
17170 Perkins Road
Baton Rouge, LA 70810-3817

Glen Hasse
P.O. Box 74040
Baton Rouge, LA 70874-4040

Honorable Juan A. LaFonta
LA House of Reps District 96
6305 Elysian Fields Ave.
New Orleans, LA 70122

Honorable Cedric Richmond
LA House of Reps District 101
5630 Crowder Blvd., Suite 205
New Orleans, LA 70126

Soumaya Ghosn
636 Bancroft Way
Baton Rouge, LA 70808

Paul Jenkins
6200 Harris Technology Blvd.
Charlotte, NC 28269-3732

R. Martin Guidry
Doris Grego
560 Hwy 44
LaPlace, LA 70068-6908

Honorable Scott Simon
LA House of Reps District 74
P.O. Box 1297
Covington, LA 70420

Ann Spell
943 Ellis Street
Franklinton, LA 70438

Honorable Reed S. Henderson
LA House of Reps District 103
8201 W. Judge Perez
Chalmette, LA 70043

Michael Kohn
10555 Airline Highway
Baton Rouge, LA 70816

Honorable Erich E. Ponti
LA House of Reps District 69
7341 Jefferson Hwy, Suite J
Baton Rouge, LA 70806

Larry Daigle
Earthnet Labs
7117 Belle Candice
Baton Rouge, LA 70817-4864

Dynea USA
344 Tannehill Road
Dodson, LA 71422-3263

Honorable Billy R. Chandler
P.O. Box 100
Dry Prong, LA 71423

Glenrose Pitt
5513 Highway 6
Natchitoches, LA 71457

Honorable Regina Ashford Barrow
LA State Senate District 29
4305 Airline Highway
Baton Rouge, LA 70805

Karen L. Oberlies
U.S. Army Corps of Engineers
New Orleans District
P.O. Box 60267
New Orleans, LA 70160-0267

APPENDIX B: DISTRIBUTION LIST

Honorable MP Schneider, III
P.O. Box 669
Slidell, LA 70459

Tim Leger
P.O. Box 51729
Lafayette, LA 70505

C.H. Fenstermaker & Associates
P.O. Box 52106
Lafayette, LA 70505-2106

Page Kraemer Environmental
1426 Eraste Landry Road
Lafayette, LA 70506

Charlie Voinche
333 E. Kaliste Saloom Road
Lafayette, LA 70508

Chris Accardo
U.S. Army Corps of Engineers
New Orleans District
Operations Division,
Regulatory Functions Branch
P.O. Box 60267
New Orleans, LA 70160

Mr. Jonathan Fricker
State Historic Preservation Officer
Department of Culture, Recreation, and Tourism
P.O. Box 44247
Baton Rouge, LA 70804

Ms. Cheryl Nolan
Louisiana Department of Environmental Quality
Office of Environmental Services
P.O. Box 4313
Baton Rouge, LA 70821-4313

Ms. Linda Levy
Louisiana Department of Environmental Quality
Office of Environmental Services
P.O. Box 4313
Baton Rouge, LA 70821-4313

Mr. David W. Fruge'
Louisiana Department of Natural Resources
Coastal Management Division
617 North 3rd Street, Suite 1048
Baton Rouge, LA 70804

Karen Roy
333 E. Kaliste Saloom Road
Lafayette, LA 70508

Honorable Jonathan W. Perky
LA House of Reps District 47
407 Charity St. 102
Abbeville, LA 70510-5111

Randa Jones
Gary L. Jones
25197 Zeigler Cemetery Road
Livingston, LA 70754

Armand S. Abay
P.O. Box 37
Convent, LA 70723

Honorable James Fannin
LA State Senate District 13
320 6th Street
Jonesboro, LA 71251
318-259-6620

Darrel Walton
11077 Hummingbird Drive
Denham Springs, LA 70726

Mike Dowty
P.O. Box 1529
Denham Springs, LA 70727-1529

U.S. Department of Commerce
NOAA, National Marine
Fisheries Service
Southeast Regional Office
263 13th Avenue South
St. Petersburg, FL 33701

Safety Coordinator
P.O. Box 190
Erwinville, LA 70729

Anne J. Crochet
P.O. Box 2471
Baton Rouge, LA 70821

Jimmy Bello
P.O. Box 290
New Roads, LA 70760

Danny Roddy
17564 Vaughn Lane
Livingston, LA 70754

Lynn Watts
P.O. Box 1032
Livingston, LA 70754

OTHER PARTIES
MSFC

Senator Richard Shelby
Huntsville International Airport
1000 Glenn Hearn Boulevard
20127
Huntsville, AL 35824

Onis "Trey" Glenn III, Director
Alabama Department of
Environmental Management
1400 Coliseum Blvd.
Montgomery, AL 36110-2059

Honorable Loretta Spencer
Mayor of Huntsville
308 Fountain Circle
Huntsville, AL 35801

Honorable Mike Gillispie,
Chairman
Madison County Commission
100 Northside Square
Courthouse 700
Huntsville, AL 35801

Representative Howard Sanderford
908 Tannahill Drive, SE
Huntsville, AL 35802-1971

Representative Laura Hall
100 St. Clair
Huntsville, AL 35810

Senator Jeff Sessions
AmSouth Center Suite 802
200 Clinton Avenue, NW
Huntsville, AL 35801-4932

Honorable Sandy Kirkindall
Mayor of Madison
100 Hughes Rd
Madison, AL 35758

J.I. Palmer, Jr., Regional
Administrator
Environmental Protection
Agency
Region 4
61 Forsyth St., SW
Atlanta, GA 30303

Representative Sue Schmitz
4649 Jeff Road
Toney, AL 35773-0012

Representative Randy
Hinshaw
P.O. Box 182
Meridianville, AL 35759

Senator Tom Butler
136 Hartington Dr.
Madison, AL 35758

Senator Lowell Barron
P.O. Box 65
Fyffe, AL 35971

The Honorable Bob Riley
State Capitol
600 Dexter Avenue
Montgomery, Alabama 36130

NASA/MSFC
Mail Code: CS30
ATTN: Ms. Monika Vest
MSFC, AL 35812

NASA/MSFC
Mail Code: CS20
ATTN: Ms. Kim Newton
MSFC, AL 35812

Senator Hinton Mitchem
412-A Gunter Avenue
Guntersville, AL 35976

NASA/MSFC
Mail Code: CS20
ATTN: Mr. Mike Wright
MSFC, AL 35812

Refuge Manager
USFWS Wheeler Wildlife Refuge
Rt. 4 Box 250
Decatur, AL 35603

NASA/MSFC
Mail Code CS20
ATTN: Mr. Dom Amatore
MSFC, AL 35812

Dr. Lee Warner
SHPO
Alabama Historical Commission
468 South Perry Street
Montgomery, AL 36130-0901

NASA/MSFC
Mail Code: CS01
MSFC, AL 35812

NASA/MSFC
Mail Code: CS30
ATTN: Ms. Rosa Kilpatrick
MSFC, AL 35812

Department of the Interior
Office of Environmental Affairs
MS 2340
18th and C Streets, NW
Washington, DC 20240

Representative Butch Taylor
224 Taylor Avenue
New Hope, AL 35760

Rep. Robert Aderholt
Federal Building, Ste 107
600 Broad Street
Gadsden, AL 35901

Congressman Robert Cramer
5th Congressional District of Alabama
200 Pratt Avenue, N.E., Suite A
Huntsville, AL 35801

APPENDIX B: DISTRIBUTION LIST

Senator Parker Griffith
101 Lowe STE 3-A
Huntsville, AL 35801

Representative Mac McCutchen
100 St. Clair
Huntsville, AL 35801

Senator Arthur Orr
P.O. Box 305
Decatur, AL 35602

Mr. Terry Hazle
Directorate of Environmental Management
U.S. Army Aviation and Missile Command
(AMSAM-RA-DEM)
Building 4488
Redstone Arsenal, AL 35898

U.S. Department of the Interior
Planning and External Affairs
National Park Service Southeast Regional Office
75 Spring Street, SW
Atlanta, GA 30303

Honorable Marvalene Freeman
Mayor of Triana
640 Sixth St
Triana, AL 35756

Elizabeth Ann Brown, Deputy SHPO
Alabama Historical Commission
468 South Perry St
Montgomery, AL 36130-0900

Alabama State Clearinghouse
Department of Economic and Community Affairs
P.O. Box 2929
3645 Norman Bridge Rd.
Montgomery, AL 36105-0939

Representative Mike Ball
P.O. Box 6302
Huntsville, AL 35213

APPENDIX B: DISTRIBUTION LIST

OTHER PARTIES
Palmdale

NASA Palmdale Office
P.O. Box 901240
Palmdale, CA 93590

Palmdale:
scott.l.mcclay@nasa.gov
wayne.r.vanlandingham@boeing.com

OTHER PARTIES
SSC

U.S. Army Corps of Engineers
4155 Clay Street
Vicksburg, MS 39183

U.S. Fish and Wildlife Service
Mississippi Office
6578 Dogwood View Parkway
Jackson, MS 39213

Mississippi Department of
Environmental Quality
P.O. Box 10385
Jackson, MS 39289

Mississippi Department of
Archives and History
P.O. Box 571
Jackson, MS 39205

Mississippi Department of
Marine Resources
1141 Bayview Avenue
Biloxi, MS 39530

Kevin Cobble
Refuge Manager
U.S. Fish & Wildlife Service,
San Andres National Wildlife
Refuge
P.O. Box 756
Las Cruces, NM 88004

Garrison Commander
Office of the Garrison
Commander, White Sands
Missile Range
100 Headquarters Ave.
WSMR, NM 8800

OTHER PARTIES
WSTF

Mr. Ron Curry, Secretary
New Mexico Environment
Department
1190 St. Francis Drive
Santa Fe, NM 87502

The Honorable Bill Richardson
Governor of New Mexico
State of New Mexico, Office of the
Governor
490 Old Santa Fe Trail
State Capitol, Room 400
Santa Fe, NM 87501

Mr. Patrick Lyons
Commissioner of Public Lands
New Mexico State Land Office
310 Old Santa Fe Trail
Santa Fe, NM 87501

Ms. Katherine Slick
State Historic Preservation Officer,
Director
Department of Cultural Affairs,
Historic Preservation Division
Bataan Memorial Building
407 Galisteo Street, Suite 236
Santa Fe, NM 87501

Frank J. Benz
Manager
NASA White Sands Test Facility
RA, Managers Office
P.O. Box 20
Las Cruces, NM 88004

Radel Bunker-Farrah
Environmental Program
Manager
NASA White Sands Test Facility
RA, Managers Office
P.O. Box 20
Las Cruces, NM 88004

Timothy J. Davis
NEPA Manager
NASA White Sands Test Facility
RA, Managers Office
P.O. Box 20
Las Cruces, NM 88004

Kris Havstad
Supervisory Range Scientist
USDA, ARS, Jornada
Experimental Range
P.O. Box 30003, MSC 3JER,
NMSU
Las Cruces, NM 88003-8003

Luis Rios, Supervisor
New Mexico Game and Fish,
Las Cruces Office
2715 Northrise Drive
Las Cruces, NM 88011

APPENDIX B: DISTRIBUTION LIST

Mr. Brian Haines
County Manager
Dona Ana County
845 N. Motel Road
Las Cruces, NM 88007

Ed Roberson
District Manager
Las Cruces District Office,
Bureau of Land Management
1800 Marquess Street
Las Cruces, NM 88005-3370

Mayor William Ken Miyagishima
Mayor, City of Las Cruces
200 N. Church Street
Las Cruces, NM 88004

LIBRARIES

Cocoa Beach Public Library
550 North Brevard Ave
Cocoa Beach, FL 32931

Melbourne Public Library
540 E. Fee Ave
Melbourne, FL 32901

Merritt Island Public Library
1195 North Courtenay Parkway
Merritt Island, FL 32953

Port St. John Public Library
6500 Carole Ave.
Port St. John, FL 32927

Titusville Public Library
2121 S. Hopkins Ave.
Titusville, FL 32780

Los Angeles County Library
Quartz Hill Branch
42018 N. 50th Street W.
Quartz Hill, CA 93536

Trona Library
82805 Mountain View St.
Trona, CA 93562

Kern County Library
Boron Branch
26967 20 Mule Team Road
Boron, CA 93516

Kern County Library
California City Branch
9507 California City Boulevard
California City, CA 93505

Los Angeles County Library
Lancaster Branch
601 W. Lancaster Boulevard
Lancaster, CA 93534

Kern County Library
Mojave Branch
16916-1/2 Highway 14
Mojave, CA 93501

APPENDIX B: DISTRIBUTION LIST

Clear Lake City
County Freeman Branch Library
16616 Diana Lane
Houston, Texas 77062

El Paso Public Library
Main (Downtown) Library
501 N. Oregon
El Paso, Texas 79901-1103

Palmdale City Library
700 East Palmdale Blvd.
Palmdale, CA 93550

AFFTC Technical Library
812 TSS/ENTL
Edwards AFB, CA 93524

Edwards Base Library
95 SPTG/SVMG
5 West Yeager Blvd.
Building 2665
Edwards AFB, CA 93524-1295

Kiln Public Library
17065 Highway 603
Kiln, MS 39556

Margaret Reed Crosby
Memorial Library
900 Goodyear Blvd.
Picayune, MS 39466

St. Tammany Parish Library
555 Robert Blvd.
Slidell, LA 70458

Kern County Library
Tehachapi Branch
450 West F Street
Tehachapi, CA 93561

Kern County Library
Wanda Kirk Branch
3611 Rosamond Boulevard
Rosamond, CA 93560

AFFTC Technical Library
412 TW/TSDL
Edwards AFB, CA 93524

Maury Oceanographic Library
Building 1003
Stennis Space Center, MS 39529

Bay St. Louis
Hancock County Library
312 Highway 90
Bay St. Louis, MS 39520

Kern County
Beale Memorial Library
701 Truxtun Avenue
Bakersfield, CA 93301

Huntsville Madison County
Main Library
915 Monroe Street
Huntsville, AL 35801

Inyo County Free Library
168 N. Edwards Street
Post Office Drawer K
Independence, CA 93526

NASA Headquarters Library
c/o Stephen McConnell
NASA FOIA Officer
300 E Street SW
Washington, DC 20546-0005

Branigan Memorial Library
Attn: Reference Desk
200 East Picacho Ave.
Las Cruces, NM 88001

New Orleans Library
Main Branch
219 Loyola Avenue
New Orleans, LA 70112

Scientific Technical Library
NASA JSC
2101 NASA Parkway
Mail Code IS23
Attention: Martha Giles
Houston, TX 77058

Central Brevard Public Library
& Reference Center
308 Forrest Ave
Cocoa, FL 32922

Edwards Base Library
95 SPTG/SVMG
Edwards AFB, CA 93524-1295

Kern River Valley Library
7054 Lake Isabella Blvd.
Lake Isabella, CA 93240-9205

Joanna Arnold, MSLIS
Technical Librarian,
Lockheed Martin
NASA MSFC Michoud
Assembly Facility
Bldg 102 | 1st Floor | Col. EH58
13800 Old Gentilly Road
New Orleans, LA 70129

OTHER PARTIES
KSC

Mr. James L. Quinn
Department of Environmental
Protection
3900 Commonwealth Blvd.
Mail Station 47
Tallahassee, FL 32399

U.S. Fish & Wildlife Service
North Florida Field Office
6620 Southport Dr. South
Suite 310
Jacksonville, FL 32216

Ms. Carol Clark
U.S. National Park Service
Superintendent
Canaveral National Seashore
212 S. Washington Ave.
Titusville, FL 32796

Ms. Stacey M. Zee
Environmental Specialist
FAA, Commercial Space Transportation
800 Independence Ave. SW #331
Washington, DC 20591

Patrick Blucker
Director Plans and Programs
45 SW/XP
1201 Edward H. White II Street
Patrick AFB, FL 32925

Alexander Stokes
Environmental Flight Chief
45 CES/CEE
1225 Jupiter Street
Patrick AFB, FL 32925

Robert Van Vonderen
Chief, Cape Engineer Flight
45 CES/CEL
CCAFS
185 Skid Strip Road, Room 120
Patrick AFB, FL 32925

Mr. Ron Hight
U.S. Fish & Wildlife Service
Refuge Manager
Merritt Island National Wildlife Refuge
P.O. Box 6504
Titusville, FL 32782

Regional Director
Southeast Region
National Park Service
100 Alabama Street, SW
1924 Building
Atlanta, GA 30303

NOAA Fisheries
1315 East West Highway
SSMC3
Silver Spring, MD 20910

Mr. Steve Kokkinakis
NOAA Program Planning and Integration
SSMC3
Room 15723 (PPI)
Silver Spring, MD 20910

Governor Charles Crist, Jr.
Governor of Florida
State of Florida, Office of Governor
The State Capitol
400 South Monroe Street
Tallahassee, FL 32399

Rep. Stan Mayfield
District 80
State Representative
1053 20th Place
Vero Beach, FL 32960

APPENDIX B: DISTRIBUTION LIST

Florida State Clearinghouse
Florida DEP
3900 Commonwealth Blvd.
Mail Station 47
Tallahassee, FL 32399

Senator Bill Posey
District 24
State Senate
1802 S. Fiske Blvd.
Suite 108
Rockledge, FL 32955

Senator M. Mandy Dawson
District 29
State Senate
33 N.E. 2nd Street
Suite 209
Ft. Lauderdale, FL 33301

Rep. Ralph Poppell
District 29
State Representative
400 South St.
Suite 1C
Titusville, FL 32780

Rep. Thad Altman
District 30
State Representative
7025 North Wickham Road, Suite 108
Melbourne, FL 32940

Rep. Bob Allen
District 32
State Representative
321 Magnolia Avenue
Merritt Island, FL 32952

Ms. Peggy Busacca
County Manager
Brevard County
County Manager's Office
2725 Judge Fran Jamieson Way
Building C
Viera, FL 32940

Mr. Mel Scott
Assistant County Manager
Development and
Environmental Services
2725 Judge Fran Jamieson Way
Building C
Viera, FL 32940

Mr. Robert S. Lay
Director
Broward County Emergency
Operations Center
1746 Cedar Street
Rockledge, FL 32955

Mr. William Farmer
Director
Public Safety – Fire and Rescue
1040 South Florida Avenue
Rockledge, FL 32955

Ms. Robin Sobrino
Planning and Zoning
Brevard Co. Government Center
2725 Judge Fran Jamieson Way
Building A, Room 202
Viera, FL 32940

Truman G. Scarborough
Commissioner, District 1
Brevard County
400 South Street
Titusville, FL 32780

Mr. Ernest Brown
Natural Resources Mgt. Office
Brevard County Government Center
2725 Judge Fran Jamieson Way
Building A – Suite 219
Viera, FL 32940

Helen Voltz
Commissioner, District 3
Brevard County
1311 East New Haven Avenue
Melbourne, FL 32901

Commander
7th Coast Guard District
Brikell Plaza, Federal Building
909 SE First Ave.
Miami, FL 33131

County Manager
Osceola County
1 Courthouse Square, Suite 4700
Kissimmee, FL 34741

County Manager
Seminole County
1101 E. First Street
Sandford, FL 32771

County Manager
Volusia County
Thomas C. Kelly Administration Center
123 W. Indiana Ave.
DeLand, FL 32720

County Manager
Lake County
315 West Main Street
Tavares, FL 32778

Mayor Mike Blake
City of Cocoa
Office of the Mayor
603 Brevard Ave.
Cocoa, FL 32922

Mary Bolin
Commissioner, District 4
Brevard County
2725 Judge Fran Jamieson Way
Building C
Viera, FL 32940

Jackie Colon
Commissioner, District 4
Brevard County
1515 Sarno Road, Building B
Melbourne, FL 32935

Rep. Mitch Needelman
District 31
State Representative
1565 Sarno Road, Suite A
Melbourne, FL 32935

Office of the Mayor
City of Merritt Island
2575 N. Courtenay Parkway
Merritt Island, FL 32953

Mayor Leon Beeler
City of Cocoa Beach
Office of the Mayor
2 South Orlando Avenue
Cocoa Beach, FL 32932

Mayor Harry Goode
City of Melbourne
Office of the Mayor
900 E. Strawbridge Ave.
Melbourne, FL 32901

Aphidalin Fancon
Environmental Planner
City of Titusville
555 Washington Avenue
Titusville, FL 32781

Mayor Buddy Dyer
City of Orlando
Office of the Mayor
400 S. Orange Ave.
Orlando, FL 32802

Mr. J. Stanley Payne
Chief Executive Officer
Canaveral Port Authority
200 George J. King Boulevard
Cape Canaveral, FL 32920

Mayor James L. Vandergrifft
City of New Smyrna Beach
Office of the Mayor
210 Sams Ave.
New Smyrna Beach, FL 32168

County Administrator
Orange County
Administration Building,
5th Floor
201 S. Rosalind Ave.
Orlando, FL 32801

Ms. Maureen Rupe
President
Partnership for a Sensible
Future, Inc.
7185 Bright Avenue
Cocoa, FL 32927

Ms. Linda Weatherman
Economic Development
Commission of Florida's Space
Coast
597 Haverty Court, Suite 100
Rockledge, FL 32955

Congressman Tom Feeney
323 CHOB
Washington, DC 20515

NASA Ames Research Center
c/o Kelly Garcia
FOIA Manager
NASA Ames Research Center
Moffett Field, CA 94035

APPENDIX B: DISTRIBUTION LIST

Congressman Tom Feeney
12424 Research Parkway
Suite 135
Orlando, FL 32826

Rep. David Weldon
2347 RHOB
Washington, DC 20515

Rep. David Weldon
2725 Judge Fran Jamieson Way
Building C
Melbourne, FL 32940

Senator Mel Martinez
SH-356
Washington, DC 20510-0903

Senator Mel Martinez
315 E. Robinson St.
Landmark Center 1
Suite 475
Orlando, FL 32801

Rocky Randels
Mayor
City of Cape Canaveral
Office of the Mayor
105 Polk Avenue
Cape Canaveral, FL 32920

NASA Dryden Flight Research Center
c/o Kim Lewis
DFRC FOIA Manager
NASA Dryden Flight Research Center
Edwards, CA 93523

NASA Goddard Space Flight Center
c/o Joan Belt
GSFC FOIA Manager
NASA Goddard Space Flight Center
Greenbelt, MD 20771

NASA Johnson Space Center
c/o Stella Luna
JSC FOIA Officer
NASA Johnson Space Center
Houston, TX 77058

NASA Kennedy Space Center
c/o Penny Myers
KSC FOIA Officer
NASA Kennedy Space Center
Kennedy Space Center, FL 32899

NASA Langley Research Center
c/o Cheryl Cleghorn
LaRC FOIA Officer
NASA Langley Research Center
Hampton, VA 23681

Glen Curtis
Director
ATK-Launch Systems Group
9160 North Highway 83
Corrine, UT 84307

APPENDIX B: DISTRIBUTION LIST

Mayor Ronald G. Swank
City of Titusville
Office of the Mayor
555 S. Washington Ave.
Titusville, FL 32781

NASA Marshall Space Flight Center
c/o Judi Hollingsworth
MSFC FOIA Officer
NASA Marshall Space Flight Center
Huntsville, AL 35812

Mayor Larry L. Schultz
City of Rockledge
Office of the Mayor
1600 Huntington Lane
Rockledge, FL 32955

Mayor Shirley Bradshaw
City of West Melbourne
Office of the Mayor
2285 Minton Road
West Melbourne, FL 32904

Dave Gosen, Director
ATK-Launch Systems Group
9160 North Highway 83
Corrine, UT 84302

Johnny Nguyen
Shuttle Processing Transition
NASA/KSC
Mail Code PH-82
Kennedy Space Center, FL 32899

Mr. Kran Kilpatrick
Natural Resources Manager
Ames Research Center
Mail Code: 218-1
Moffett Field, CA 94035

Ms. Ann Clarke
Ames Research Center
Mail Code QE
Moffett Field, CA 94035

Mr. Dan Morgan
Environmental Manager
Dryden Flight Research Center
Mail Code SH
Edwards, CA 93523

Ms. Trudy Kortes
NEPA Program Manager
Glenn Research Center
Mail Code QSEO
21000 Brookpark Road
Cleveland, OH 44135

Ms. Christie Myers
Environmental Engineer
Air Missions Program Manager
Glenn Research Center
Mail Code QSEO
21000 Brookpark Road
Cleveland, OH 44135

Ron Caswell
Principal Senior Engineer
Barrios Technologies
Mail Code OC-KSC
Kennedy Space Center, FL 32899

APPENDIX B: DISTRIBUTION LIST

Senator Bill Nelson
SH-716
Washington, DC 20510-0905

Senator Bill Nelson
225 E. Robinson St.
Suite 410
Orlando, FL 32801

Fred Krupp
President
Environmental Defense
National Headquarters
257 Park Avenue South
New York, NY 10010

Ms. Perri Fox
Historic Preservation Officer
Johnson Space Center
Mail Code JP/Planning & Integration Office
2101 NASA Parkway
Houston, TX 77058

Ms. Barbara Naylor
Environmental Program Specialist
Kennedy Space Center
Mail Code TA-C3
Kennedy Space Center, FL 32899

Mr. Barbara Naylor
Environmental Manager
Kennedy Space Center
Mail Code TA-C3
Kennedy Space Center, FL 32899

Mr. Francis Celino
Environmental Manager
Michoud Assembly Facility
Mail Stop SA39
13800 Old Gentilly Rd.
New Orleans, LA 70129

NASA John C. Stennis Space Center
c/o Joy Smith
SSC FOIA Officer
NASA Stennis Space Center
Stennis Space Center, MS 39529

NASA Glenn Research Center
c/o Angela Pierce
GRC FOIA Officer
NASA Information Center,
Glenn Research Center
Cleveland, OH 44135

NASA Jet Propulsion Laboratory
c/o Dennis Mahon
JPL FOIA Officer
Jet Propulsion Laboratory
Pasadena, CA 91109

Byron Whiteman
455W/XPR
16460 Hanger Road
CCAFS
Patrick AFB, FL 32925

Lt. Col. Joseph A. Szewc
USAF (retired)
742 Bayside Drive, #205
Cape Canaveral, FL 32920

APPENDIX B: DISTRIBUTION LIST

Mr. Roger Ferguson
Environmental Manager
Langley Research Center
Mail Code 318
Hampton, VA 23861-0001

Mr. Steve Glover
Marshall Space Flight Center
Mail Code MP71
Huntsville, AL 35812

Ms. Donna Holland
CLV Environmental Manager
Marshall Space Flight Center
Mail Code JS
Huntsville, AL 35812

Ron Schaub
Remote Sensing Analyst
Dynamac Corp
DYN-6
Kennedy Space Center, FL 32899

Ms. Vicky Ryan
Group Supervisor
Launch Approval Engineering Group
Jet Propulsion Laboratory
Mail Code 180-801B
4800 Oak Grove Dr.
Pasadena, CA 91109

E. Ray Gann
Hazmat Coordinator
Brevard County Emergency Management
1746 Cedar Street
Rockledge, FL 32955

Col. Michael Bedard
Chief Weather Office
45 WS/CC
1201 Edward H. White II Street
Patrick AFB, FL 32925

Col. Dave Nuckles
Chief Wing Safety
45 SW/SE
1201 Edward H. White II Street
 Patrick AFB, FL 32925

Charles Kilgore
Ground Operations
Kennedy Space Center
Mail Code LX-D
Kennedy Space Center, FL 32899

Ms. Lizabeth Montgomery
NEPA Program Manager
Goddard Space Flight Center
Mail Stop: 250
8800 Greenbelt Rd.
Greenbelt, MD 20771

Mr. Dave Hickens
Environmental Management Officer
Johnson Space Center
Mail Code JE/Environmental
2101 NASA Parkway
Houston, TX 77058

Gregory Sakala
19 Corriente St.
Merritt Island, FL 32952

APPENDIX B: DISTRIBUTION LIST

Ruth Gardner
Manager
Cx Ground Systems Project Office
Mail Code LX-D
Kennedy Space Center, FL 32899

Sue Gaines
CX Range POC
NASA/KSC
Mail Stop LX-I
Kennedy Space Center, FL 32899

Ms. Carolyn Kennedy
Environmental Specialist
Stennis Space Center
Mail Code RA02
Stennis Space Center, MS 39529

Ms. Tina Norwood
Historic Preservation
NASA Headquarters 5A33
300 E Street, SW
Washington, DC 20546-0001

Mr. Bob Tancig
State Coordinator
Florida Coalition for Peace and
Justice, County Road 18
10665 SW 89th Avenue
Hampton, FL 32044

Burt Summerfield
Kennedy Space Center
Mail Code TA
Kennedy Space Center, FL 32899

M. Rebecca Bolt
Wildlife Ecologist
Dynanac Corp
Mail Code Dyn-5
Kennedy Space Center, FL 32899

Bruce Vu, Ph.D.
Aerospace Engineer
Kennedy Space Center
Mail Stop NE-M1
Kennedy Space Center, FL 32899

Rosaly Santos
NASA/KSC
Mail Code TA-C3
Kennedy Space Center, FL 32899

Mr. Timothy J. Davis
NEPA Manager
White Sands Test Facility
12600 NASA Road
Las Cruces, NM 88012

Mr. Steve Kohler
President
Space Florida
MS: SPFL M6-306
Room 9030
Kennedy Space Center, FL 32899

Ravi Margasahayam
Kennedy Space Center
Mail Stop SA-B1
Kennedy Space Center, FL 32899

Alan Dumont
KSC Range Safety Manager
NASA/KSC
Mail Code SA-G
Kennedy Space Center, FL 32899

Kurt Geber
Health Physicist
Dynamac Corp
DYN-4
Kennedy Space Center, FL 32899

APPENDIX B: DISTRIBUTION LIST

Johnny Nguyen
Shuttle Processing Transition
NASA/KSC
Mail Code PH-82
Kennedy Space Center, FL 32899

Scott Skinner
Environmental Mgt. Consultant
Reynolds Smith & Hill, Inc.
10748 Deerwood Park Blvd.
Jacksonville, FL 32256

Federation of American Scientists
1717 K Street, NW
Suite 209
Washington, DC 20036

Linda Herridge
Public Affairs Writer
InDyne
Mail Code IDI-010
Kennedy Space Center, FL 32899

James Taffer
CHS, Inc.
Mail Code CHS-022
Kennedy Space Center, FL 32899

Carl Murphy
AFL/CIO
Bldg. Trades Dept.
P.O. Box 22257
Lake Buena Vista, FL

James E. Hildebrand
Training Director
Plumbers & Pipefitters Local
Union 295
743 N. Beach St.
Daytona Beach, FL 32174

Ron Caswell
Principal Senior Engineer
Barrios Technologies
Mail Code OC-KSC
Kennedy Space Center, FL 32899

American Institute of
Aeronautics and Astronautics
1801 Alexander Bell Drive,
Suite 500
Reston, VA 20191

Bruce Buckingham
NASA/KSC
CX 39 Press Site
Kennedy Space Center, FL 32899

Gary Latchworth
Engineering Manager
NASA/KSC
Mail Code LX-C
Kennedy Space Center, FL 32899

Don Kraemer
Diamondhead Property Owners
Association
President, Board of Directors
5300 Diamondhead Circle
Diamondhead, MS 39525

Oliver Winn
Business Agent
Plumbers & Pipefitters Local
Union 295
743 N. Beach St.
Daytona Beach, FL 32174

Aerospace Industries
Association
1000 Wilson Blvd., Ste. 1700
Arlington, VA 22209

APPENDIX B: DISTRIBUTION LIST

Mr. Erich Pica
Director Economic Programs
1717 Massachusetts Ave., NW
Suite 600
Washington, DC 20036

Mr. John Pike
Global Security
300 N. Washington Street
Suite B-100
Alexandria, VA 22314

Mr. John Flicker
National Audubon Society
700 Broadway
New York, NY 10003

Mr. David Brunner
National Fish and Wildlife
1120 Connecticut Ave., NW
Suite 900
Washington, DC 20036

Mr. Jerry Pardilla
National Wildlife Federation
2501 Rio Grande Blvd., NW
Albuquerque, NM 87104

Ms. Maureen Rupe
Partnership for Sustainable Future
7185 Bright Avenue
Cocoa, FL 32927

Mr. Richard Moore
South West Network for
Environmental and Economic Justice
P.O. Box 7399
Albuquerque, NW 87194

Mr. Bruce K. Gagnon
P.O. Box 652
Brunswich, ME 40011

Mr. Jim Ricco
Greenpeach International
702 H Street, NW
Suite 300
Washington, DC 20001

Ms. Jacqueline Johnson
Executive Director
1301 Connecticut Ave., NW
Suite 200
Washington, DC 20036

Mr. Robert Rivera
National Hispanic
Environmental Council
106 N. Fayette St.
Alexandria, VA 22314

Mr. Eric Goldstein
National Resources Defense
Council
40 West 20th Street
New York, NY 10011

Sierra Club
National Headquarters
85 2nd Street
Second Floor
San Francisco, CA 94105

Dr. Robert Zubrin
The Mars Society
P.O. Box 273
Indian Hills, CO 91106

Mr. George Whitesides
The National Space Society
1620 I Street, NW, Suite 615
Washington, DC 20006

The Planetary Society
65 North Catalina Avenue
Pasadena, CA 91106

The American Association for the Advancement
1200 New York Avenue, NW
Washington, DC 20005

Mr. Alden Meyer
Union of Concerned Scientists
1707 H. Street, NW, Suite 600
Washington, DC 20006

Ms. Joy Singfield, Director
National Society of Black Engineers
205 Daingerfield Road
Alexandria, VA 22314

Mr. Thomas Cassidy
The Nature Conervancy
4245 North Fairfax Drive
Arlington, VA 22203

The Space Foundation
310 S. 14th Street
Colorado Springs, CO 80904

Mr. Tom Bancroft
The Wilderness Society
1615 M Street, NW
Washington, DC 20036

Mr. Bob Werb
Space Frontier Foundation
16 First Avenue
Nyack, NY 10960

APPENDIX B: DISTRIBUTION LIST

This page intentionally left blank.

Appendix B-1 Responses to Draft EA Public Review Comments

The Draft Environmental Assessment (EA) was available for public comment beginning February 11, 2008, through March 14, 2008. The availability of the document was advertised in the newspapers listed in Appendix B, as well as on the NASA website. A Notice of Availability for the Draft EA was published in the *Federal Register* on February 25, 2008 (73 FR 10067). The EA was provided in both hard copy and electronic format to the information repositories listed in the newspaper articles. An electronic version of the Draft EA was mailed to the remaining recipients on the distribution list provided in Appendix B and hard copies were provided to the parties upon request. In addition, an electronic version of the EA was available at http://www.hq.nasa.gov/osf/relatedlinks.htm for the public to access. The comments received by the National Aeronautics and Space Administration (NASA) on the contents of the Draft EA are provided in this appendix.

Comments from the State of Alabama Historical Commission

STATE OF ALABAMA
ALABAMA HISTORICAL COMMISSION
468 SOUTH PERRY STREET
MONTGOMERY, ALABAMA 36130-0900

March 28, 2008

TEL: 334-242-3184
FAX: 334-240-3477

Donna L. Holland
Environmental Engineering & Occupational Health Office
NASA MSFC
Marshall Space Flight Center, Alabama 35812

Re: AHC 08-0453
Space Shuttle Program
Programmatic Environmental Assessment
Transition & Program Property Disposition
Marshall Space Flight Center
Madison County, Alabama

Dear Ms. Holland:

Thank you for forwarding the Draft Assessment to our office. Our review of this document has indicated that there are several issues which need to be addressed.

1. ES 4.2, Page ES 10, Personal Property: It is our opinion that some of this property may be eligible for the National Register of Historic Places (NRHP) due to its context with particular program activities. An example of this would be the space suites, supporting hardware, and other equipment associated with the Neutral Buoyancy Facility (NBS) as George C. Marshall Space Flight Center (MSFC). This equipment is part of the context of the research that was carried out for multiple space program initiatives. As the NBS is a National Historic Landmark, this equipment would be equally significant as it was part of that research. There may be many more personal property issues which need to be addressed as well and NASA, the Advisory Council on Historic Preservation (ACHP), and the State Historic Preservation Office (SHPO) should review these and determine their merit.

2. Furthermore, relating to the personal property transferred to Johnson Space Center (JSC) from MSFC and the NBS, we relayed at the time of transfer that these items were part of the significance of the NBS. By letter dated February 9, 1999, we stated this transfer would be an adverse effect. A letter from NASA dated March 16, 1999, to John Fowler, the Executive Director of the ACHP indicated that NASA's JSC would take care of the equipment and return it to MSFC's NBS when it was no longer needed. It came to our attention in the spring of 2001 that the equipment transferred from MSFC's NBS to JSC was not compatible and the equipment was in storage. By letter to NASA dated May 14, 2001, we requested

THE STATE HISTORIC PRESERVATION OFFICE
www.preserveala.org

NASA Programmatic EA
Shuttle Program Transition & Disposition
AHC 08-0453
Page 2

3. that the equipment be returned as agreed to by JSC and we requested a timetable for the return of this equipment. To date, we have not received a response. We renew our request that this material be returned. We have attached a copy of the aforementioned letters for your review.

4. ES 7.3.7, page 15, Disposition Methods: We would have to disagree with the assessment that demolition of NRHP listed or eligible buildings or National Historic Landmarks would be a "moderate" effect. In our opinion, based on the significance of the structures at MSFC, the effect would be catastrophic, and clearly not something with which we could concur.

We understand the issues which face NASA in trying to develop new program initiatives with limited budgets and the need to make the operations as efficient as possible. We also realize that addressing the potential significance of personal property is an arduous task. However, the significance of many of the structures, facilities, and personal properties cannot be understated. We are willing to work with NASA to address these issues and move forward. However, while the outsourcing of properties may be acceptable with the proper covenants, the demolition of these unique resources is unacceptable.

Should you have any questions, the point of contact for this matter is Greg Rhinehart at (334) 230-2662. Please have the AHC tracking number referenced above available and include it with any correspondence.

Truly yours,

Elizabeth Ann Brown
Deputy State Historic Preservation Officer

EAB/GCR/gcr
cc: Rep. Bud Cramer
 Ms. Jody Cook, NPS
 Mr. Don Klima, ACHP

STATE OF ALABAMA
ALABAMA HISTORICAL COMMISSION
468 South Perry Street
MONTGOMERY, ALABAMA 36130-0900

F. LAWERENCE OAKS
EXECUTIVE DIRECTOR

February 9, 1999

TELEPHONE NUMBER
334-242-3184
FAX: 334-240-3477

Pete Allen
Director, Facilities Service Office
George C. Marshall Space Flight Center
AB01
Marshall Space Flight Center, Alabama 35812

Re: AHC 99-0275
Neutral Buoyancy Facility
Transfer of Equipment to Johnson Space Center
Marshall Space Flight Center
Madison County, Alabama

Dear Mr. Allen:

Upon close review of the documentation supplied by your office and Johnson Space Center as well as information supplied by interested parties, the Alabama Historical Commission has determined the following. Marshall Space Flight Center, it's scientists, staff, and administrators has a long and storied history in the field of space exploration. From the testing of rockets to launch America's first satellite to the current space station mission, Marshall has been at the cutting edge of all these endeavors. While there are many people and places which warrant merit, the Redstone Rocket which launched Alan Shepherd into space, the Neutral Buoyancy Facility which trained our astronauts to walk and work in space, and the Saturn V Rocket which launched our astronauts to successful landings on the moon are of extraordinary significance. Perhaps no other form of human endeavor has had the profound and positive effect as did our nation's landing a man on the moon and returning him safely home.

Alabama's heritage from prehistoric Native Americans through the historic period is second to none. But there are few events, if any, to which all Alabamians can point to with pride as our contribution to the exploration of space. While this may be an achievement of national and international achievement, it is uniquely ours. Unfortunately, Alabama has lost archaeological sites, standing structures, and even components of space exploration in the name of progress or relocation. While we can not save everything, it is our office's responsibility to protect and preserve significant sites and features with all the resources available to us.

The State Historic Preservation Office
http://preserveala.org

We understand NASA's mission and we applaud your efforts and achievements. We also understand the need to recycle materials in these days of budget restrictions. However, the material requested by Johnson Space Center from Marshall's Neutral Buoyancy Facility is part of the justification for this facility being a National Historic Landmark. We cannot simply look at the building, the tank, the control center, or the implements used as individual items. Each of these is a part of the extraordinary significance of the facility. To remove or relocate any of these items would not only be an adverse effect to this National Landmark but it would also be a crime against the people of Alabama, Marshall Space Flight Center, the engineers, and the astronauts who trained so rigorously at this facility that we might all be proud to be Americans.

We can not let this shining moment of our heritage slip through our fingers. For this reason, our office has determined that the removal of the items requested by Johnson Space Center from the Neutral Buoyancy Facility constitutes an adverse effect of the highest order to this National Landmark. Therefore our office cannot concur with this action.

Sincerely,

Elizabeth Ann Brown
Deputy State Historic Preservation Officer

EAB/GCR

cc: Senator R. Shelby
     Representative B. Cramer

National Aeronautics and
Space Administration

Headquarters
Washington, DC 20546-0001

Reply to Attn of: JE

MAR 8 1999

Mr. John M. Fowler
Executive Director
Advisory Council on Historic Preservation
1100 Pennsylvania Avenue, NW
Washington, DC 20004

Dear Mr. Fowler:

Thank you for your letter dated February 23, 1999, concerning NASA's plan to temporarily relocate certain hardware from the Neutral Buoyancy Space Systems Tank (NBSST) at Marshall Space Flight Center (MSFC) to Johnson Space Center (JSC). Based on the recommendations and comments of your letter and NASA's program needs, we have decided to proceed with the temporary relocation of this hardware. However, please be assured that NASA has no intention of creating a long-term situation that might adversely affect the status of the NBSST as a National Historic Landmark.

The Advisory Council on Historic Preservation (ACHP) recommended that NASA investigate the logistics of conducting training in the future at the NBSST at MSFC rather than temporarily relocating the equipment to another facility. A few years ago NASA conducted an Agency-wide review of all of its capital assets to determine the mix of facilities that would meet the Agency's foreseeable needs while identifying redundant assets. NASA determined that due to size and capacity constraints at NBSST, all of its requirements for neutral buoyancy tank type training should be consolidated at JSC. Therefore, NASA itself has no present need for the capabilities of the NBSST. However, there is the potential that other parties in the future may have an interest in using the NBSST. NASA is encouraging MSFC to pursue such non-NASA work.

In your letter, there was a further recommendation that NASA document its commitment to return the hardware to MSFC and produce a reasonable timetable for its return and reintegration into the NBSST. At the end of the training requirements at JSC, NASA commits itself to promptly provide the funds for and return the hardware to and reintegrate it into the NBSST at MSFC. NASA's JSC is presently working on developing a more precise schedule for the training requirements that will use the NBSST hardware. On the basis of presently available information, we anticipate that the need will span five to seven years.

2

Because of pressing program needs, NASA has decided to move forward as soon as possible with the transfer of the hardware from MSFC to JSC. Within two weeks, NASA shall provide a more detailed schedule for the hardware's use at JSC to the ACHP and the Alabama State Historic Preservation Officer (SHPO). If in the future NASA determines that there is a need to extend the use of the hardware at JSC, we will promptly notify the ACHP and Alabama SHPO of our intentions and reopen consultation. Moreover, as indicated in previous information supplied, NASA will take prudent measures to ensure that the relocated hardware is used and maintained in a manner that will minimize the possibility of degrading its usefulness upon return to MSFC.

Thank you very much for your prompt attention to this NASA proposal. If you have any questions, please contact Kenneth Kumor, NASA's Federal Preservation Officer, at 202-358-1112.

Sincerely,

Olga M. Dominguez, Director
Environmental Management
Division

Richard J. Wisniewski
Deputy Associate Administrator
for Space Flight

STATE OF ALABAMA
ALABAMA HISTORICAL COMMISSION
468 SOUTH PERRY STREET
MONTGOMERY, ALABAMA 36130-0900

LEE H. WARNER
EXECUTIVE DIRECTOR

TEL: 334-242-3184
FAX: 334-240-3477

May 14, 2001

Pete Allen
Director, Facilities Service Office
George C. Marshall Space Flight Center
AB01
Marshall Space Flight Center, Alabama 35812

Re: AHC 99-0275
     MSFC Neutral Buoyancy Facility
     Multiple Actions
     Madison County, Alabama

Dear Mr. Allen:

It has come to our attention that there are issues which need to be addressed relating to activities that are being proposed for the MSFC Neutral Buoyancy Simulator, a structure which is listed as a National Historic Landmark and relating to equipment transferred from this facility to Johnson Space Center (JSC). The first item for discussion is to re-use some of the office space in the building which houses the NBS facility. As this should not affect the tank facility or the control room, we do not believe this will be a problem. However, we shall need to see the proposals, in writing, along with photographs and drawings depicting the areas of concern before we can formally approve these activities.

A more significant issue relates to the equipment transferred form the NBS to JSC. At the time of the transfer proposals, our office, along with others, relayed that much of this equipment was an integral part of the National Landmark. NASA expressed their agreement with our assessment but that the material was needed at the JSC facility. In the agreement worked out with NASA and the ACHP, it was stated that the equipment would be returned to the MSFC NBS if it was found not to be useful or when it was no longer necessary. This was made clear by letter dated March 8, 1999, from Olga M. Dominguez and Richard J. Wisniewksi of NASA's headquarters in Washington, D. C.

It is now our understanding that the MSFC equipment is not only no longer in use, but that it was never put to use due to incompatibility with JSC equipment. Therefore, we request clarification on the status of this equipment. If it is no longer in use, please provide a schedule for the agreed upon return to the MSFC NBS. If it was never put to use, please provide a detailed explanation as to why it was not used, why it was not returned to MSFC as agreed to, and how the equipment has been maintained to ensure its integrity.

THE STATE HISTORIC PRESERVATION OFFICE
www.preserveala.org

The relics of the United States exploration of space are slipping away and each item that remains becomes even more significant and preservation of those items which are a part of Alabama's history are our responsibility and our mission. This is why our concern is so great over the MSFC NBS and the equipment which was integral to its operation. We look forward to receiving your response at your earliest convenience. Should you have any questions or comments, please contact Greg Rhinehart at our office.

Yours truly,

Elizabeth Ann Brown
Deputy State Historic Preservation Officer

EAB/GCR

cc: B. Cramer/US Congressman
R. Shelby/US Senator
J. Fowler/ACHP

Response to the State of Alabama Historical Commission

NASA is working with the Alabama (AL) State Historic Preservation Office (SHPO) to address concerns regarding artifacts to ensure that culturally significant personal property is identified and reviewed for determination of eligibility. In addition, NASA has a Memorandum of Agreement (MOA) with Smithsonian regarding the disposition of significant property. NASA acknowledges the agreement made to return and reinstall personal property transferred from Marshall Space Flight Center (MSFC) to Johnson Space Center (JSC) in February 1999. MSFC has reopened discussions with JSC and Headquarters (HQ) and is working to resolve this issue. NASA MSFC is committed to keeping the AL SHPO involved and informed of the progress toward resolution of this comment.

Although NASA acknowledges that demolition is one option of property disposition, there are no plans for the demolition of property at MSFC listed or eligible for listing in the National Register of Historic Places (NRHP). As custodians of such historic property, NASA makes every effort to reutilize and preserve historic property per Section 110 of the National Historic Preservation Act (NHPA). Should demolition of a historic property appear to become necessary, NASA is committed to working with the AL SHPO, providing an opportunity for AL SHPO to comment and advise regarding the proposed action. Implementation of the appropriate mitigation measures developed in cooperation with the AL SHPO (and memorialized in an NHPA Section 106 MOA) may well reduce the level of adverse environmental impact associated with demolition. In any event, before any final action is taken toward the proposed demolition of a property listed or eligible for listing in the NRHP, NASA would complete both the associated National Environmental Policy Act (NEPA) and NHPA processes.

Comments from the State of Texas Historical Commission

TEXAS HISTORICAL COMMISSION
The State Agency for Historic Preservation

RICK PERRY, GOVERNOR
JOHN L. NAU, III, CHAIRMAN
F. LAWERENCE OAKS, EXECUTIVE DIRECTOR

25 March 2008

Donna L. Holland,
NEPA Coordinator
Marshall Space Flight Center,
National Aeronautics and Space Administration
AS10 / Environmental Engineering and Occupational Health Office
Building 4249 / 100C
Marshall Space Flight Center, AL 35812

Re: Draft Programmatic Environmental Assessment, Space Shuttle Program Transition and Property Disposition, National Aeronautics and Space Administration (NASA)
[including Johnson Space Center and Ellington Field, Houston, Harris County, Texas, and El Paso Forward Operation Location, El Paso, El Paso County, Texas]

Dear Ms. Holland:

Thank you for your submission of the Draft Programmatic Environmental Assessment (EA) for the Space Shuttle Program Transition and Property Disposition. This letter serves as comment from F. Lawerence Oaks, Executive Director of the Texas Historical Commission and the State Historic Preservation Officer.

Texas Historical Commission (THC) staff has completed its review of the submitted Draft EA. Last month, THC staff completed its review of a related document, the "Survey and Evaluation of Historic Facilities and Properties in the Context of the U.S. Space Shuttle Program at NASA Johnson Space Center, Houston, Harris County, Texas." At that time, THC **concurred** with NASA's determination of individual **eligibility** for listing in the National Register of Historic Places of 8 buildings and 3 structures at the Johnson Space Center (JSC) for their association with the Space Shuttle program as described in the report:

- 5, Jake Garn Mission Simulator and Training Facility, 1965, Building
- 7, Crew Systems Laboratory, 1964, Building
- 9, Systems Integration Facility, 1966, Building
- 16, Avionics Systems Laboratory (SAIL), 1964, Building
- 30, Mission Control Center, 1965, Building (Designated NHL)
- 44, Communications and Tracking Development Lab, 1966, Building
- 222, Atmospheric Reentry Materials and Structures Evaluation Facility, 1966, Building
- 920N, Sonny Carter Training Facility / Neutral Buoyancy Laboratory (NBL), 1993/1996, Building
- OV-103, *Discovery*, 1983, Structure
- OV-104, *Atlantis*, 1985, Structure
- OV-105, *Endeavour*, 1990, Structure

P.O. BOX 12276 · AUSTIN, TX 78711-2276 · 512/463-6100 · FAX 512/475-4872 · TDD 1-800/735-2989
www.thc.state.tx.us

Texas Historical Commission
NASA Space Shuttle Program Transition and Property Disposition
Draft Programmatic Environmental Assessment Comments
2

THC staff also **concurred** with the preliminary finding that there may be 2 or more potentially eligible thematic historic districts within the JSC. However, staff believes that **more information is needed** in order to effectively determine the scope of these potential districts and **further demonstrate the ineligibility** of 19 other resources related to the Space Shuttle Program as described in that report. Detailed information demonstrating the lack of significant historical association and photographs conveying the lack of integrity of these 19 resources are still needed in order to complete our review.

In a letter to Perri E. Fox, NASA Historic Preservation Officer, dated 20 February 2008, THC staff expressed its concern about the very narrow scope of the survey report, which we acknowledge was devoted solely to historic resources with clear relationships to the Space Shuttle Program. At that time, we suggested that consideration should be given to the possible eligibility of the entire JSC complex as a single historic district associated with space exploration. Gregory Smith, National Register Coordinator; A. Elizabeth Butman, Project Reviewer; and I would welcome the opportunity to visit the JSC facilities and meet with NASA historic preservation staff to discuss our concerns.

Until further clarification is provided regarding the ineligibility of the 19 additional resources associated with the Space Shuttle Program at JSC, THC staff is registering its concern about the potential loss of important historic resources due to the proposed undertaking as described in this Draft EA.

If you have any questions concerning our review or if we can be of further assistance, please contact Rachel Leibowitz at 512/463-6046. We look forward to further consultation with your office and hope to maintain a partnership that will foster effective historic preservation. Thank you for your cooperation in this federal review process, and for your efforts to preserve the irreplaceable heritage of Texas.

Sincerely,

F. Lawrence Oaks,
State Historic Preservation Officer

cc: Tina B. Norwood, Federal Preservation Officer, NASA
Perri E. Fox, Historic Preservation Officer, Lyndon B. Johnson Space Center
Patrick Van Pelt, Chairman, Harris County Historical Commission

Response to the State of Texas Historical Commission

A letter developed and submitted by JSC to the Texas SHPO in response to a similar inquiry dated February 20, 2008, is included below for the comment response.

National Aeronautics and
Space Administration

Lyndon B. Johnson Space Center
2101 NASA Road 1
Houston, Texas 77058-3696

CERTIFIED MAIL

Reply to Attn of: JP-08-017

APR 0 4 2008

Mr. F. Lawerence Oaks
State Historic Preservation Officer
Executive Director
Texas Historical Commission
P.O. Box 12276
Austin, TX 78711-2276

Attention: Greg Smith
           National Register Coordinator

Subject: Survey and Evaluation of Historic Facilities and Properties in the context of
          the US Space Shuttle Program at NASA Johnson Space Center (JSC)

Dear Mr. Oaks,

Thank you for your letter dated 20 February 2008. After careful review of your letter and telephone discussions with Ms. Elizabeth Butman, Project Reviewer at the Texas Historical Commission, NASA JSC would like to clarify that we have completed the survey as the facilities relate to the Space Shuttle Program. The survey was completed in accordance with the inventory requirements of Section 110 of the National Historic Preservation Act (NHPA) of 1966 [Public Law 89-665], as amended. Thus our letter to you was requesting your concurrence with the decision NASA JSC made on the historical significance of the properties surveyed.

Your letter concurs with the finding of eight (8) buildings and three (3) structures determined to be eligible for listing in the National Register of Historic Places (NRHP). These findings will now be used for any Section 106 consultations to be initiated if future undertakings involve these eligible assets.

NASA JSC has received clarification from Archaeological Consultants, Inc., (ACI) who conducted the eligibility survey and prepared the report. They have pointed out that National Park Service regulations state, "Parts of buildings . . . are not eligible independent of

JP-08-017   2

the rest of the existing building. The whole building must be considered, and its significant features must be identified." While JSC understands it is not typical to focus on a portion of a building, I would like to point out that JSC has two examples; the Mission Control Center, a National Historic Landmark (NHL) located within Building 30, as well as Chamber A and B are stated as the NHL boundaries within Building 32. JSC believes that Building 920N derives its exceptional significance primarily from the Neutral Buoyancy Laboratory (NBL) within this building. However, the NBL has not played a direct role in the SSP as it is considered a training facility for the International Space Station (ISS). Therefore, JSC does not consider Building 920N to be eligible for the NRHP under this SSP study. JSC will instead reconsider the eligibility of Building 920N in 2016 when the ISS is scheduled to retire.

The reference to historic districts at NASA JSC or considering the entire JSC complex as a historic district was not included in the scope of this survey. The report was focused only on the facilities or structures directly related to the SSP. NASA JSC acknowledges that the question of potential historic districts needs to be addressed. While potential historic districts were considered as part of the SSP survey, they were not identified and evaluated, given this study's examination of only a single NASA program. We therefore plan to conduct another survey when NASA JSC will attain fifty years in 2014, at which time the historic context will not be focused on only one program.

As discussed, I believe it would be helpful if you could see some of the assets included in the report. I hope you will accept my invitation to tour NASA JSC. Please contact us at your earliest convenience to schedule a tour.

Sincerely,

Abdul Hanif
Historic Preservation Officer

cc:
Perri Fox, Chief, Planning and Integration Office
Jennifer Ross-Nazzal, Ph. D., JSC Historian
OJE/T. Norwood
OJE/K. Kumor
AL/R. Bresnik

Comment from the City of Madison Alabama

OFFICE OF THE MAYOR

ARTHUR S. "SANDY" KIRKINDALL
MAYOR

CITY OF MADISON

100 HUGHES ROAD
MADISON, ALABAMA 35758

(256) 772-5602/5603
FAX (256) 772-3828

March 3, 2008

Ms. Donna L. Holland
Environmental Engineering and
　Occupational Health Office
NASA
Marshall Space Flight Center, Alabama 35812

Dear Ms. Holland,

Thank you for the draft copy of the Space Shuttle Program Programmatic Environmental Assessment dated February 2008.

The City of Madison has no comment on the draft PEA.

Thank you for the opportunity to comment. Please feel free to contact my office if I can be of further assistance.

Sincerely,

Arthur S. Kirkindall
Mayor

WWW.MADISONAL.GOV

Response to the City of Madison Alabama

We appreciate your review.

APPENDIX B-1 RESPONSES TO DRAFT EA PUBLIC REVIEW COMMENTS

Comments from the Department of Army New Orleans District COE

DEPARTMENT OF THE ARMY
NEW ORLEANS DISTRICT, CORPS OF ENGINEERS
P. O. BOX 60267
NEW ORLEANS, LOUISIANA 70160-0267

REPLY TO
ATTENTION OF

Operations Division
Operations Manager
Completed Works

Ms. Donna L. Holland
Environmental Engineering and Occupational Health Office
National Aeronautics and Space Administration
Marshall Space Flight Center, Alabama 35812

Dear Ms. Holland:

This is in response to your Solicitation of Views request dated March 10, 2008, concerning the Space Shuttle Program Transition and Property Disposition Programmatic Environmental Assessment at National Aeronautics and Space Administration facility in Louisiana.

We have performed a cursory review of your request for potential Department of the Army regulatory requirements and impacts on any Department of the Army projects.

We do not anticipate any adverse impacts to any Corps of Engineers projects.

Based on the limited information provided with this request, it appears that a Department of the Army permit under Section 404 of the Clean Water Act will not be required for the projects as proposed. Projects involving work (raising or demolishing) within the footprint of an existing structure will generally not require a permit unless this work involves impacts to wetlands or jurisdictional waters. Please note that any mechanized land clearing or deposition of fill material or debris in a wetland or other waters of the United States would require a Department of the Army permit under Section 404 of the Clean Water Act. All work in navigable or tidal waters will also require a Department of the Army permit pursuant to Section 10 of the Rivers and Harbors Act. Additional information will be needed before a final determination can be made.

-2-

You are advised that you must obtain a permit from the Orleans Levee District for any work within 300 feet of a federal flood control structure such as a levee. You must apply by letter to the Orleans Levee District including full-size construction plans, cross sections, and details of the proposed work. Concurrently with your application to the Orleans Levee District, you must also forward a copy of your letter and plans to Operations Division, Operations Manager for Completed Works of the Corps of Engineers and to the Louisiana Department of Transportation and Development (LA DOTD) in New Orleans for their review and comments concerning the proposed work. The Orleans Levee District will not issue a permit for the work to proceed until they have obtained letters of no objection from both of these reviewing agencies. For further information regarding permit requests affecting federal flood control levees and structures, please contact Ms. Amy Powell, Operations Manager for Completed Works at (504) 862-2241.

Off-site locations of activities such as borrow, disposals, haul-and detour-roads and work mobilization site developments may be subject to Department of the Army regulatory requirements and may have an impact on a Department of the Army project.

Should it be determined that a Department of the Army permit is required, you should apply for the said permit well in advance of the work to be performed so that an adequate jurisdictional determination can be performed. The application should include sufficiently detailed maps including the longitude and latitude coordinates, drawings, photographs, and descriptive text for accurate evaluation of the proposal. The permit application should be addressed to our Eastern Evaluation Section of Regulatory Branch, organization code CEMVN-OD-SE.

Please contact Dr. John Bruza, of our Regulatory Branch by telephone at (504) 862-1288, or by e-mail at John.D.Bruza@usace.army.mil for questions concerning wetlands determinations or need for on-site evaluations. Questions concerning regulatory permit requirements may be addressed to Mr. Michael Farabee by telephone at (504) 862-2292 or by e-mail at Michael.V.Farabee@usace.army.mil.

This determination of permit requirements is valid for a period of five years from the date of this letter unless new information warrants a revision prior to the expiration date. In addition, any changes or modifications to the proposed project may require a revised determination.

Sincerely,

Karen L. Oberlies
Solicitation of Views Manager

Response to the Department of Army New Orleans District COE

Thank you for your input. Your comments are appreciated and noted.

Comments from the Department of Army Redstone Arsenal

From: Fisher, Christine E Ms CTR USA IMCOM [mailto:christine.fisher2@us.army.mil]
Sent: Friday, March 28, 2008 2:03 PM
To: HQ-NASA-SSPEA
Subject: Redstone Arsenal EMD Comments for EA (UNCLASSIFIED)

Classification: **UNCLASSIFIED**
Caveats: NONE

Ms. Holland,

The Environmental Management Division of the Redstone Arsenal Garrison would like to submit comments regarding the Draft PEA for NASA Facilities (attached to e-mail). Feel free to contact me with any questions, comments, or concerns.

Sincerely,

Christine Fisher

NEPA Specialist, Biologist

Cultural/Natural Resources - Environmental Management Division

US Army Garrison - Redstone; Office A332

4488 Martin Road

Redstone Arsenal, AL 35898

I. General Comments for Programmatic EA

Purpose of Proposed Action

Section 1.3.3: "The disposition of common parts has no potential for significant impacts to the environment." In order to provide basis for this determination, the term "common parts" should be defined.

Comparison of Alternatives

Section 2.4: Use of the term "not substantial" to describe impacts instead of relevance to "significant" as required by CEQ implies that only those effects of large proportion are significant. Therefore, use of the term "Major" for "Environmental impacts that, individually or cumulatively, could be substantial" implies that none of the other assigned levels of impact (No Impact, Minimal, Minor, Moderate) could possibly lead to cumulative impacts. This is not logical and does not meet 40 CFR 1508.27 (b) 7.

Socioeconomic Effects of Federal Agency Actions

Section 4.1.2.1: As the breakpoint for significance was not defined, the numbers/percentages presented in this section may not represent "less than significant impacts" despite being low percentages. Additionally, if the loss of ouput and employment dollars that are spent locally is not evaluated in addition to losses to private services and goods that will no longer be obtained locally (e.g legal and physician services), the local impact is not derived accurately.

Overview of Cult. Res. & Socioeconomics

Section 4.12.2: What is the definition of "NEPA significant"?

II. Specific Comments for Marshall Space Flight Sections of the Programmatic EA

Environmental Sites

Section 3.8.4 and Env Consequences for MSFC: Any areas utilized for intrusive or non-intrusive activities by the Marshall Space Flight Center (MSFC) on Redstone Arsenal (RSA), Alabama must also comply with the local U.S. Army RSA Regulation 200-7; RSA Environmental Sites Access Control Program, which includes environmental sites within the MSFC boundary. Coordination through the MSFC (AS-10) Environmental Office is recommended.

Point of contact is Mr. Troy W. Pitts, Garrison Environmental Division, Directorate of Public Works, IMSE-RED-PWE, 4488 Martin Road, Redstone Arsenal, Alabama, 35898, troy.pitts@redstone.army.mil, 256-842-2836.

Cultural Resources

Section 3.8.3: There is no mention of archaeological resources in this section of the EA for Marshall Space Flight Center. The property was surveyed for archaeological resources in 2005 and the survey was published in Alexander and Alvey 2006. ("The 2005 Phase I Archaeological Survey of the Marshall Space Flight Center, National Aeronautics and Space Administration, Madison County, AL; submitted to MSFC August 2006; Contract Number DAMD17-01-2-0015-0024)

Entire document: Have all of the installations included in this Programmatic Draft EA undergone Phase I surveys for archaeological resources? This must be completed prior to any transfer or disposition or property.

Point of contact is Mr. Benjamin Hoksbergen, Garrison Environmental Division, Directorate of Public Works, IMSE-RED-PWE, 4488 Martin Road, Redstone Arsenal, Alabama, 35898, ben.hoksbergen@us.army.mil, 256- 955-6971.

Natural Resources
Section 3.8.2:

- Wetlands
  Check the acreage for habitat types. At Marshall Space Flight Center, the total property is listed as 1,841 acres and wetlands are listed as 122 acres. This is approximately 6% of MSFC's land, but another statement indicates wetlands account for only 3% of land type on the property. Based on this discrepancy, other calculations of habitat types may also be incorrect.

- Wildlife
  It is not necessarily true that low habitat diversity results in low wildlife diversity – the type of habitat also affects wildlife presence and diversity. MSFC is located adjacent to a large wetland complex and a unique spring, which indicates that wildlife diversity may actually be quite high.

- Protected Species
  Alabama 220-2-.92 Nongame Species Regulation lists species that are protected by the state; this list includes Tuscumbia Darter, Bald Eagle, Gray Myotis, and Indiana Bat. This regulation also protects federally threatened and endangered species. It is considered an official list.

  Alabama Invertebrate Species Regulation 220-2-.98 protects invertebrates, which are possibly found in the wetland habitats on the property.

  In Exhibit 3-45, add state listed species (species documented in Natural Heritage Inventory in 1995) to respected categories:
  Birds:
      Bald Eagle (*Haliaeetus leucocephalus*) – SP
      Green-backed Heron (*Butorides striatus*)- S2
      Solitary Vireo (*Vireo solitaries*) - S2
  Mammals:
      Northern Long-eared Myotis (*Myotis septentrionalis*) - S2
      Prairie Vole (*Microtus ochrogaster*) - S2

  Reptile:
      Eastern Box Turtle (*Terrapene carolina*) - SP/ S5
  Amphibian:
      Green Salamander (*Aneides aeneus*) - SP/S3
  Plants:
      Featherfoil (Hottonia inflate) - S2
      Limestone Adders Tongue
          (*Ophioglossum engelmannii*)-S3
      Southern rosinweed (*Silphium asteriscus*) - federal candidate for listing

Though none of the MSFC facilities are located in the Ecologically Sensitive Area for Williams Springs, there are facilities adjacent. More information on how to prevent contamination of the spring, where the state ranked and protected Tuscumbia Darter is found, should be included in the EA.

Portions of MSFC property falls within the RSA-defined groundwater protection buffer zone for the federally endangered Alabama cave shrimp. Precautions must be taken to prevent negative impacts to groundwater. Spill mitigation kits must be kept on site during construction and construction BMP's for fence installation and construction in order to prevent/minimize soil erosion and run-off. Prevent limewater seepage into storm drains by conducting concrete pours on a non-rainy day. The use of milled-up asphalt on this property is not permitted.

Point of contact is Ms. Shannon Allen, Garrison Environmental Division, Directorate of Public Works, IMSE-RED-PWE, 4488 Martin Road, Redstone Arsenal, Alabama, 35898, shannon.l.allen@us.army.mil, 256- 876-3977.

Any questions, comments, or suggestions may be submitted to Christine Fisher, Redstone Arsenal - Directorate of Public Works – Environmental Management Division at (256) 842-0019 or e-mail address: Christine.fisher2@us.army.mil

Response to the Department of Army Redstone Arsenal

Editorial and technical comments have been reviewed and incorporated as appropriate.

Common parts include items such as nuts and bolts and other fairly standard and commonplace parts. The EA will be modified to reflect this definition.

NASA agrees that while "minor" and "moderate" impacts would not be individually "significant," in combination they may be significant. "Substantial" impacts may be significant either individually or cumulatively. Changes have been made to the EA to reflect this view. However, NASA normally refrains from the use of the term "significant" in its EAs and environmental impact statements (EISs) because it is the NASA decision-maker who ultimately makes the decision as to whether the totality of identified impacts is significant. That decision ultimately is memorialized in a finding of no significant impact or record of decision. NASA would use the term "significant" to describe impacts in an EA or EIS only when it is clear on its face to everyone or nearly everyone that the impact is of such magnitude.

The estimates in Section 4.1.2.1 of the Space Shuttle Program's (SSP's) current economic footprint in the regions surrounding the major Centers is intended only as background information, to provide context. This Programmatic EA evaluates NASA's decision about how to disposition the SSP's real and personal property assets. Therefore, the socioeconomic impact analysis and finding of "less than significant impacts" refers only to the impacts of NASA's discretionary actions regarding disposition of the SSP's real and personal property. The EA does not evaluate significance of the broader socioeconomic impacts of the President's decision to discontinue the SSP, because the Presidential decision to discontinue the SSP has already been made and is not subject to NEPA.

The term "NEPA Significant" refers to effects that are significant (per NEPA, 40 CFR 1508.27), as referenced in the preceding subsection 4.12.2.4.

NASA currently complies with Redstone Arsenal (RSA) Regulation 200-7 for planned activities on Army environmental sites within the MSFC or RSA boundary. In addition, NASA has developed a similar requirement for environmental site access control within the MSFC boundaries in MPR 8500.1; the NASA contact is Mr. Farley Davis, (256) 544-6935.

The archaeological survey conducted for MSFC is now referenced in the EA and the report information will be added to the references.

NASA is aware that Executive Order (EO) 11593 directs federal agencies to locate, inventory, and nominate all potentially eligible sites, buildings, districts, and objects under their control to the Secretary of the Interior for listing on the NRHP. Federal agencies must also take precautions to prevent the sale, transfer, or demolition of historic properties. Not all Centers addressed in this EA have completed a base-wide Phase I Archaeological survey. Some disposition options, such as reutilization would not require a Phase I survey. However, if a disposition option requires a Phase 1 survey, NASA is committed to meeting the requirements of EO 11593 and all federal regulations and requirements before taking any action. If an archeological site is discovered that meets the criteria to be eligible for listing in the NRHP, NASA will complete the NHPA Section 106 process before taking any action that would affect such property.

The most recent wetland delineation was conducted by the U.S. Army Corps of Engineers in October 2005. Nine acres that were included in the previous 1994 survey are now considered as uplands, resulting in a total of 113.2 acres (45.8 hectares) of jurisdictional bottomland wetlands, which accounts for 6.15 percent of MSFC total land.

There are no planned shuttle property disposition activities in the areas of MSFC that contain diverse wildlife. Property disposition activities are only planed in the industrial areas of MSFC where there is a low wildlife diversity.

The protected species list for MSFC in the EA has been reviewed and updated according to the Alabama 220-2-92 list of species that are protected by the state, including Exhibit 3-45.

NASA currently requires that the proper best management practices (BMPs) and storm water pollution prevention measures be in place during construction activities. These requirements are detailed in the specifications for each construction project, MWI 8550, and MPR 8500.1. These two Marshall documents will be added to the final EA, along with a reference indicating that they contain construction BMP such as preventive measures for soil erosion and storm water runoff into sensitive areas.

Comments from Boeing

```
-----Original Message-----
From: Vanlandingham, Wayne R [mailto:wayne.r.vanlandingham@boeing.com]
Sent: Thursday, March 20, 2008 3:12 PM
To: HQ-NASA-SSPEA
Cc: Mcclay, Scott L. (JSC-MV8)
Subject: PalmdaleEADocument3-08Update.doc

Donna L Holland,
Palmdale has made many changes to our section of this document..that
correct many inaccuracies...

Please cut and paste the whole section of this attachment into the current
document that was sent out for final review....


   <<PalmdaleEADocument3-08Update.doc>>

Thanks

Wayne VanLandingham
Environment, Health & Safety
Palmdale/EAFB
```

Response to Boeing

Boeing submitted several editorial changes to the Palmdale section, which were incorporated.

Comment from Joshua Jeffery

April 17, 2008

AS10/Environmental NEP Coordinator
SSP Transition and Retirement Program
NASA Marshall Space Flight Center
Building 4249/100C
MSFC, Alabama 35812

To Whom It May Concern:

I recently read excerpts of NASA's Draft Space Shuttle Program Programmatic Environmental Assessment. I was especially curious about what environmental impact would take place at the Kennedy Space Center since the shuttle launches and normally lands there. I find it commendable that there would be no environmental impact on the wetlands as well as the floodplains and potable water. It is good to know that the wildlife that inhabit the wetlands in particular would not be affected in any way by the transition activities that will be and currently are being done as NASA retires the shuttle and prepares for the Constellation Program. I hope that this will remain the case once further activities at the Cape commence. I also liked what I saw out of the other NASA centers listed even though I would not have thought about any of them since a lot of what goes on with the shuttle takes place at the Kennedy Space Center. Anyway, I would like to thank you for writing this report and hope that all goes well with the retirement of the shuttle and transition to the Constellation Program. Thank you for your time.

Sincerely,

Joshua A. Jeffery
Student
Spring Arbor University

Response to Joshua Jeffery

Thank you for your input. Your comments are appreciated and noted.

Comment from Renee' Texas Lady

From: TexasLadyRenee@aol.com [mailto:TexasLadyRenee@aol.com]
Sent: Monday, March 24, 2008 4:54 PM
To: HQ-NASA-SSPEA
Subject: boldly researcher

if WE THE PEOPLE OF THE USA STOP WHAT WE'VE BEEN DOING..............god HAVE MERCY ON OUR SOULS!

Whatever it takes is a "giant leap for mankind".

We the people means 2 + more oppions I may not be able to spell well, I have disabilities. However in my not so humble opinions know we must do, what we must do.

The US Marshall coin has no "In God We Trust".   Look at where we were when our leader of the nation called for prayer when the Apollo 13 was reentering this atmosphere...... we can fit a square peg into a round hole. and YOU know what I mean.

I trust we shall continue one way or the other........ Even if the INTERNATIONAL world takes over, that's just the way it might have been meant to be...........we know better.

one of many who care!

Response to Renee' Texas Lady

Thank you for your input. Your comments are appreciated.

Comment from R. Cord

From: RCord53117@aol.com [mailto:RCord53117@aol.com]
Sent: Monday, March 24, 2008 6:13 AM
To: HQ-NASA-SSPEA
Subject: (no subject)

WHY NOT SEND THE SPACE SHUTTLE UP TO THE SPACE STATION AND USE THEM AS WORK STATION INSIDE AND OUT. THEY ALSO COULD BE USED AS EXCAPE PODS FOR THE CREW. YOU WILL HAVE INCREASE THE WORK AREA AND KEEP THE SHUUTLE S DOING A NEEDED JOB. WHY DO THEY HAVE TO COME BACK DOWN TO BE TAKEN APART . THANK YOU

Response to R. Cord

Your comment is noted. Thank you for your input and suggestions. The President has mandated that the Space Shuttle (and associated funding and personnel support) be terminated no later than 2010.

Comment from Thomas Beck

From: thomas beck [mailto:wmpa@hotmail.com]
Sent: Saturday, March 22, 2008 8:26 AM
To: HQ-NASA-SSPEA
Subject: shuttle disposition

I would like to see one of the acutual space shuttles offered to the National Museum of the United States Air Force. The USAF played a major part in the development of the program and one deserves to be on display in Dayton Ohio. They currently house one of each prior manned spacecraft.

The Space Shuttle in spite of its critics has been a great spacecraft. Being born in 1968 and not remembering the Apollo program, I am really looking forward to our return the to moon and beyond

Thomas Beck

Response to Thomas Beck

Your comment is noted. Thank you for your input and suggestions. Disposition of the Space Shuttle Orbiters will be conducted in conformance with established federal procedures.

Comment from Ron Thompson

From: ron thompson [mailto:rdthompson999@cityofbryan.net]
Sent: Thursday, March 13, 2008 10:10 PM
To: HQ-NASA-SSPEA
Subject: disposal of space shutle

I think we should store the remaining shuttles after 2010 for future emergency flights...all this money down the drain is foolish... we still have a airplane unless replaced with a more efficient space craft to repair items in space... the hubble should be kept also... great is what we have seen with this work horse.. and opened the universe to us...//

Maybe i do not understand the impact of your article as what is going to happen to the shuttle program after 2010

Congress needs to keep the programs in space funded for the future of our country.and national security items...!

Send me info if you think i have missunderstood the disposal of this great program....Rt

sincerely.... Ron Thompson,Bryan,Ohio.......

Response to Ron Thompson

Your comment is noted. Thank you for your input and suggestions. However, you should be aware that it would be extremely expensive to maintain the Space Shuttle and supporting infrastructure, even for emergency purposes. In addition, some of the existing Space Shuttle infrastructure needs to be converted for use by the Constellation Program. Retaining such Space Shuttle infrastructure (such as a launch pad) would force NASA to construct totally new facilities for the Constellation Program.

APPENDIX B-1 RESPONSES TO DRAFT EA PUBLIC REVIEW COMMENTS

Comments from Jim Barg

```
-----Original Message-----
From: Jim Barg [mailto:jimbarg@bssmedia.com]
Sent: Thursday, March 13, 2008 11:32 AM
To: HQ-NASA-SSPEA
Subject: SSP Transition and Retirement Program

To whom it may concern:

The Shuttle program should NOT be ended in 2010. The US will have no
capability to fly people into space for (at least) five years and will
need to lean on our now-unstable relationship with the Russians to keep
the ISS manned. To expect private enterprise to have a manned vehicle
ready for use during that gap is a "pipe dream". No private contractor has
demonstrated that they'll even be close to providing a manned
transportation vehicle inside that time frame.

A more realistic approach is to end the shuttle program, say, one year
from the projected completion of the Orion space vehicle. The US cannot
afford to lose such time in space!

Regarding the STS retirement itself: My feeling is in the direction of the
"no action" option. Keep STS flying for three or four more years.
But who am I to say this? I'm not George Bush.

---James Barg
```

Response to Jim Barg

Your input is appreciated. However this environmental assessment does not evaluate the impacts of retiring the Space Shuttle, because that is a Presidential mandate.

Comment from Robert Behringer

From: Robert Behringer [mailto:democrat080165@sbcglobal.net]
Sent: Thursday, March 13, 2008 12:43 AM
To: HQ-NASA-SSPEA
Subject: space shuttle(s) retirement

I am hoping to find out more information maybe by 2009 as to why it is wise to retire the fleet within the next 10 years. To me, why fix something and or improve something that's not broken?

Just for a laugh, wouldn't the shuttles make good collectors of space junk in orbit around the Earth? (nuts, bolts,etc.)

I hope to get more information from NASA web site on space shuttle uses, importances as time goes on, besides,wherever Leonard McCoy and Montgomery Scott may be, they would be proud of the great accomplishments and necessary failures...

Robert Behringer

Response to Robert Behringer

Thank you for your comment. You may find more information regarding the President's Vision for Space Exploration, including the mandate to retire the Shuttle program at NASA's website: http://www.nasa.gov/mission_pages/constellation/main/index.html. Once you access the website, click on "Exploration Vision" under "Current Missions" on the left, the scroll down to "Related Links," on the right and click on "Vision." This will explain the President's vision for space exploration.

Comment from Ahmed Mostafa El_Habbal

From: أحمد الحباك [mailto:haico_ac_7@hotmail.com]
Sent: Wednesday, March 12, 2008 7:31 PM
To: HQ-NASA-SSPEA
Subject: Space Shuttles and International Space Station in "Quraan"

I have the pleasure to send you this letter appreciating the great efforts of NASA stuff in innovating and launching Space Shuttles and constructing the International Space Station as the whole world bless this major progress.
I also wanted to tell you that God informed us about this great projects in Quraan before hundreds of years ago as follows :-
Part no. : 30
Sora name : Al_Ensheqaq .
Aya no : 19
Page no. 589 (Arabic language – madinah edition) .
God says : " La_tarkabon tabaqan an tabaq " .
The meaning : Before this Aya God swears by three obvious famous things related also to the space and ensure that " La_tarkabon tabaqan an tabaq " .
 -"La_tarkabon tabaqan": means you (the people) will make the space shuttles which are launched by us to the space.
 -"an tabaq ": means you will construct the International Space Station (Base) which will be used to launch the space shuttles to the far space.
I think this aya is one of the Quraan secrets and miracles in this century, and as you know the Quraan was sent since more than 1400 years ago to all the peoples around the world through Islam profit " Mohamed " and still includes more secrets , some of them are related to the space , but the available translated Quraan was made by the early muslims at a time there were not such space events or discoveries .
If you are interested in discover more space secrets in Quraan ,you should have to go through Arabic edition of Quraan word by word with an Arabic mother tongue muslim person, this would provide you with clues which may guide you in your future space researches, by the way , I think in Quraan also we can imagine some contents of the far space and can get the outer diameter of the earth also the distance between the earth and some places in the far space (like the seventh sky) ….etc , I'll be glad to help in this work even for free , if you accept my English because I'm not fluent .
I hope this letter meets your interest, waiting for your reply on my e.mail.

Best regards .

Name : **Eng. Ahmed Mostafa El_Habbal**

Response to Ahmed Mostafa El_Habbal

Your comment is noted. Thank you for your input and suggestions.

APPENDIX B-1 RESPONSES TO DRAFT EA PUBLIC REVIEW COMMENTS

Comment from Vinceps@aol.com

From: Vinceps@aol.com [mailto:Vinceps@aol.com]
Sent: Monday, March 03, 2008 12:17 AM
To: HQ-NASA-SSPEA
Subject: (no subject)

as our venarable shuttle program becomes decomisioned along with other programs as we move forward on our great journey through life and the knowleg we seek to help us see a clearer picture of the big picture could you guys-girls please send this tax payer a momento of yhese programs to remember them by. ex: a guage. a tile. a tire. an unused rover camera. you get the idea. I cannot thank you enough for the effort and outstanding results from your collective outstanding efforts in our ongoing qwest called life!

Response to Vinceps@aol.com

Although we are not at liberty to send mementos, we appreciate your input and support.

Comments from Jules Fraytet

From: Jules Fraytet [mailto:jlfray@ix.netcom.com]
Sent: Wednesday, February 27, 2008 1:45 PM
To: HQ-NASA-SSPEA
Subject: Proposed launch sites in wildlife refuges

To whom it may concern,

I am a frequent visitor and supporter of the National Wildlife Refuge system in the United States.

I am asking that your agency not choose any sites that will affect the national wildlife refuges, i.e., Merritt Island NWR and Canaveral National Seashore, nearby that have been selected as possible launch pads for NASA. These areas have been set aside to protect wildlife including bird species at risk and should not be damaged or compromised by activities and facilities that are not compatible with the mission of the national wildlife refuge system. It is in my opinion that the activities and construction that your agency is planning will seriously jeopardize the wildlife safety and "refuge" that the FWS is charged with maintaining.

Thank you

Response to Jules Fraytet

Thank you for your review and suggestion. The environmental impacts associated with the locations selected as possible launch pads are described in "Final Environmental Assessment for the Construction, Modification, and Operation of Three Facilities in Support of the Constellation Program, John F. Kennedy Space Center, Florida," dated April 2007. If the commenter is referring to the proposed Commercial Vertical Launch Complex (CVLC), NASA will make no final decision on the CVLC until the NEPA process is completed.

Comment from Carol Toebe

February 26, 2008

AS10/Environmental NEP Coordinator
SSP Transition and Retirement Program
NASA Marshall Space Flight Center
Building 4249/100C
MSFC, Alabama 35812

Re: Comments on NASA Space Shuttle Program Programmatic Environmental Assessment

To Whom It May Concern:

I am writing in reference to the above program for transition and property disposition at NASA facilities. Buildings can be adapted as educational facilities or adapted to research and technology for new technology and job training sites. Use this entire process as an opportunity and challenge to incorporate real science into the lives of our citizenry, especially the youth. Vibrant learning from history is crucial to building the future. Could it all get transformed to some real advances to benefit of the entire planet Earth? More monuments to peace and transformation are needed. NASA can help lead the way.

Recycle, re-use, reinvigorate the economy but above all make sound science and safe, reasoned environmental considerations paramount to all aspects of the transition and program property disposition at NASA facilities.

Very truly yours,

Carol A. Toebe

Response to Carol Toebe

Thank you for your suggestions. NASA is dedicated to advancing and communicating scientific knowledge and understanding of the earth, the solar system, and the universe to benefit the quality of life on earth. NASA excels within the federal government in waste prevention, recycling, and affirmative procurement, a program that requires federal agencies to buy recycled-content and other environmentally preferable products. Environmentally preferable purchasing benefits the environment and demonstrates our commitment to environmental stewardship. NASA also has an extensive outreach program for educators. The details of this program can be found at:
http://www.nasa.gov/audience/foreducators/index.html.

APPENDIX B-1 RESPONSES TO DRAFT EA PUBLIC REVIEW COMMENTS

Comment from Kokosrose@aol.com

From: Kokosrose@aol.com [mailto:Kokosrose@aol.com]
Sent: Monday, February 25, 2008 7:41 PM
To: HQ-NASA-SSPEA
Subject: (no subject)

What kind of space shuttles are you going to use when these shuttles are retired in 2010? I really want too know what the new space shuttles are going too look like!!!!

Response to Kokosrose@aol.com

Thank you for your comment. You may find more information regarding the proposed successor to the Space Shuttle on NASA's website: http://www.nasa.gov/mission_pages/constellation/main/index.html. Once you access the website, click on "Constellation Program" under "Current Missions." on the left. From there you will be able to read about the proposed new vehicle and its proposed missions.

Comment from Peter Lima

```
-----Original Message-----
From: PETER LIMA [mailto:plima@patmedia.net]
Sent: Monday, March 31, 2008 10:51 PM
To: HQ-NASA-SSPEA
Subject: aviation safety report

For over 3 years, you used our money to conduct an assessment report on
aviation safety.
Now, you feel that the findings may be obsolete or in-conclusive, well let
me be the judge of that.
Make the findings public, it is your responsibility to provide this
information since it was publicly funded.
If privately funded, would you respond the same way to your investors.

Who is responsible for initiating and directing such a report? Where is
the accountability.
You must reimburse the taxpayer about $11 million dollars of our money.

If not, just disclose the report for public evaluation. Its your moral
obligation.
```

Response to Peter Lima

Your comment is noted. Thank you for your input and suggestions.

This page intentionally left blank.

Appendix C: Criteria Used to Determine Historic Property Eligibility for the Space Shuttle Assets

**EVALUATING HISTORIC RESOURCES
ASSOCIATED WITH THE SPACE SHUTTLE PROGRAM: CRITERIA
OF ELIGIBILITY FOR LISTING IN THE
NATIONAL REGISTER OF HISTORIC PLACES (NRHP)**

Purpose

A "new era for the U.S. Space Program" began on February 13, 1969, when President Richard Nixon established the Space Task Group (STG). The purpose of this committee was to conduct a study to recommend a future course for the U.S. Space Program. Three years later, on January 5, 1972, the Space Shuttle Program was initiated in a speech delivered by President Nixon. During this speech, Nixon outlined the end of the Apollo era and the future of a reusable space flight vehicle, which would allow the U.S. to construct Space Station by carrying cargo to and from outer space. Subsequently, the end of the Space Shuttle Program was announced in a speech delivered by President George W. Bush in January 2004. Although plans for space exploration would advance, the technology of the Space Shuttle and its associated facilities would change or end by 2010. The significance of the Space Shuttle was noted by the National Park Service (NPS) in the 1998 National Register Bulletin, *Guidelines for Evaluating and Documenting Historic Aviation Properties*. The following excerpt is from that bulletin.

> The Space Shuttle was the U.S. space program's next generation. Key aspects of the Shuttle's design and performance were based on a rocket-powered space plane, the X-15, the world's first transatmospheric vehicle. The Space Shuttle provided a new method of space flight, taking off like a rocket and landing like an airplane. The Space Shuttle Columbia, the first reusable manned spaceship, initiated the Space Shuttle flight program in April 1981, and a new era for the U.S. Space Program (Milbrooke 1998:12).

The historic values of this program, like the Apollo-era program which preceded it, are embodied in the facilities, that is; the buildings, structures and objects within the NASA centers. The purpose of this study is to identify the NASA-controlled facilities of local, state, and/or national significance in the historic context of the U.S. Space Shuttle Program, circa 1969 to 2010. Such facilities may include, but are not necessarily limited to, those used for research, development, design, testing, fabrication, and operations. NASA will also look at certain types resources that are not facilities and are considered "personal property" under federal regulations. These resources are typically large and while they may be mobile, are also usually associated with a geographical location, An example of this type of resource are the Mobile Launch Platforms at the Kennedy Space Center.

The evaluation of facilities within the context of the Space Shuttle Program will, in part, proceed from earlier studies of the Apollo-era resources at various NASA centers. The first step in evaluating these facilities at the NASA Centers was to establish and describe the applicable historic contexts and subcontexts. The key reference relating to the Apollo program used in this assessment was the *Man In Space Theme Study*, completed in 1984 by the National Park Service. According to the study, the purpose was to evaluate:

> *All resources which relate to the theme of Man in Space and to recommend certain of those resources for designation as National Historic Landmarks.*
>
> *The Man in Space Theme Study considered resources relating to the following general subthemes:*
>
> *A. Technical Foundations before 1958*
> *B. The Effort to Land a Man on the Moon*
> *C. The Exploration of the Planets and Solar System*
> *D. The Role of Scientific and Communications Satellites*
>
> *The Theme Study considered the Space Program in an integrated fashion. In any given space mission thousands of scientists, technicians, and other support personnel were necessary to insure success. These support personnel performed vital work in a variety of ways using support facilities in many parts of the country. None of these personnel in all likelihood comprehended all aspects of each space mission, yet all were vital to the success of the program. Since individual missions lasted over many years and involved a wide variety of resources and people only a few managers at the National Aeronautics and Space Administration (NASA) were able to see all of the facets of the space program. It was this coordination, cooperation, and collaboration that enabled NASA to successfully manage the American Space Program. The theme study follows this same approach and attempts to identify, inasmuch as is possible, the surviving resources of those that were necessary to accomplish the goals of landing a man on the moon and exploring the earth, planets and solar system* (Butowsky 1984).

The National Register of Historic Places (NRHP) Criteria for Evaluation and Criteria Considerations

The significance of a cultural resource is evaluated in terms of the eligibility criteria for listing in the NRHP. The National Register Criteria for Evaluation, as described in 36 CFR Part 60.4, are as follows:

The quality of significance in American history, architecture, archeology, engineering and culture is present in districts, sites, buildings, structures and objects that possess integrity of location, design, setting, materials, workmanship, feeling and association and:

A. *That are associated with events that have made a significant contribution to the broad patterns of history; or*

B. *That are associated with the lives of persons significant in our past; or*

C. *That embody the distinctive characteristics of a type, period, or method of construction, or that represent the work of a master, or that possess high artistic values, or that represent a significant and distinguishable entity whose components may lack individual distinction; or*

D. *That have yielded, or may be likely to yield information important in prehistory or history.*

The significance of historic buildings, structures, objects and districts is usually evaluated under Criterion A (association with historic events); Criterion B (association with important persons); or Criterion C (distinctive design or distinguishing characteristics as a whole). Often, more than one criterion will apply to historic resources.

Some types of cultural resources are not typically considered eligible for the NRHP. These resources are religious properties (A), moved properties (B), birthplaces and graves (C), cemeteries (D), reconstructed properties (E), commemorative properties (F), and properties that have achieved significance within the past fifty years (G). As a result, a resource may meet one or more NRHP criteria and still not be eligible unless special requirements are met. These requirements are called Criteria Considerations and are labeled A-G. Of relevance to the Space Shuttle Program study are Criteria Considerations B and G, as follows:

Criteria Consideration B: Moved Properties - *A property removed from its original or historically significant location can be eligible if it is significant primarily for architectural value or it is the surviving property most importantly associated with a historic person or event.*

Criteria Consideration G: Properties that have Achieved Significance within the Past 50 Years – *A property achieving significance within the last fifty years is eligible if it is of exceptional importance.*

The Space Shuttle Program: Proposed NRHP Criteria for Evaluation and Criteria Considerations

In order to qualify for listing in the NRHP under this study, resources must meet all of the following general registration requirements:

- Is real or personal property owned or controlled by NASA ;

- Was constructed, modified or used for the Space Shuttle Program between the years 1969 and 2010 (or the actual end of the Space Shuttle Program);

- Is classified as a structure, building, site, object, or district;

- Is eligible under one or more of the four NRHP Criteria. All properties considered eligible for listing under;

    Criterion A - Events

- Must be of significance in reflecting the important events associated with the Space Shuttle Program during the period of significance (1969-2010); or,
- Must be distinguished as a place where significant program-level events occurred regarding the origins, operation and/or termination of the Space Shuttle Program; or

Criterion B - Significant Persons

- Must be associated with a person whose individual significance to the goals, missions, development and design of the Space Shuttle Program can be identified and documented; or
- Must be distinguished as a place where persons of significance to the Space Shuttle Program worked or trained; or
- Best represents the important achievements or the cumulative importance of prominent persons; or
- Has consequential association with a person who gained prominence relative to the Space Shuttle Program during the period of significance.

Criterion C – Design/Construction

- Was uniquely designed and constructed or modified to support the pre-launch testing, processing, launch and retrieval of the Space Shuttle and its associated payloads; or
- Reflects the historical mission of the Space Shuttle in terms of its unique design features without which the program would not have operated; or
- Reflects the distinctive progression of engineering and adaptive reuse from the Apollo-era to the Space Shuttle-era

Criterion D – Information Value

- As this criterion is primarily used for archeological sites and this document is focused on historic properties, it is inappropriate to use this criterion as a discriminator, therefore, it will not be a valid criterion for surveys used as part of the Space Shuttle Transition activities.

- Meets appropriate Criteria Considerations - Certain kinds of property that are not usually considered eligible for listing in the NRHP, although they may meet the NRHP Criteria stated above, will require special considerations. Such properties which might fall into this category are those that have been moved (Criterion Consideration B) or properties that have achieved significance within the past fifty years (Criterion Consideration G)

    - *B: Moved Properties* – Some historic resources of significance in the context of the Space Shuttle Program may meet Criteria Consideration B since they were designed to be moved. Thus, it is not required that they, or their integral components, be at their original location in order to retain integrity. These resources are generally significant for their engineering or are significant for their association with events or persons integral to the

Space Shuttle Program. However, objects removed from their original setting and that are now located within a museum are typically excluded from NRHP-listing as the change in setting and location diminishes the resources' historic integrity (NPS 1998:36).

- *G: Properties that have Achieved Significance within the Past 50 Years* – The entire Space Shuttle Program is less than 50 years old. Therefore, Criterion G cannot be a discriminator for determining eligibility, as some properties utilized by the Space Shuttle Program may be over 50 years old. Properties that are determined to possess exceptional significance in the context of the Space Shuttle Program that are less than 50-years old must meet Criteria Consideration G.

- Retains enough integrity to convey its historical significance. The NRHP recognizes seven aspects or qualities that, in various combinations, define integrity: location, setting, materials, design, workmanship, feeling, and association. However, many original NASA Apollo-era facilities, for example, have undergone major modification and are in active use supporting the Space Shuttle Program. As a general rule, in the case of highly technical and scientific facilities, "there should be continuity in function, and thus in integrity of design and materials, and there may always be integrity of association" (ACHP 1991:33).

Criteria of Eligibility by Property Type

The following twelve property types, and the associated National Register eligibility criteria, may be used in the evaluation of all NASA owned and controlled facilities at all NASA centers. Use of these categories will help narrow the list of eligible properties to those that have true significance in the overall context of the Space Shuttle Program. Many of the facilities may have already been designated as eligible under the Apollo program. The use of these criteria on those properties in no way negates their previous designations. Rather it adds to the historical context of those properties.

1. Resources Associated with Transportation: A variety of transportation resources were constructed and/or modified to support mission and launch operations in support of the Space Shuttle Program. These resources include roadways, bridges, Crawlerways, runways and landing facilities, helipads, and waterways. Special-use vehicles also are part of the transportation network. These include Payload Transporters, Crawler Transporters, Multi-use Mission Support Equipment (MMSE) Transporters, 747 Carrier Aircraft, the astrovan, External Tank barge and recovery vessels. In order to qualify for NRHP listing, transportation resources must meet one or more of the following criteria:
- Have been used for the transportation of unique objects, structures, or significant persons associated with Space Shuttle missions;
- Have been an essential component to the Space Shuttle missions, such that the program could not function without it;
- Clearly embody the distinctive characteristics of a type or method of construction specifically designed for the transportation of the Space Shuttle or its payloads;

- Have a direct historical association with the Space Shuttle (including the Orbiter, external tank and solid rocket boosters), or a significant person associated with the Space Shuttle Program;
- Must be examples of one of the identified subtypes: road-related resources, water-related resources, rail-related resources, and air-related resources.

2. Vehicle Processing Facilities: Vehicle processing facilities include those resources which are vital to the preparation of the launch vehicle for its mission. NASA vehicle processing facilities administer such operations as assembly, testing, checkout, refurbishment, and protective storage for launch vehicles and spacecrafts. Those processing facilities which are eligible for the NRHP were essential in support of the Space Shuttle Program and include but are not limited to the "Tile Shop", the Vehicle Assembly Building, the Orbiter Processing Facility, and Hangar AF. To be considered significant, the resources must have been essential to the successful completion of Space Shuttle missions. Vehicle processing facilities were specifically designed for processing the launch vehicle and, therefore, played a major role in nationally significant events related to space exploration. In order to qualify for listing, resources must:

- Have been an essential component to the processing of the Space Shuttle;
- Clearly embody the distinctive characteristics of a type or method of construction specifically designed or modified for the processing of the Space Shuttle for launch;
- Have a direct historical association with the Space Shuttle, or a significant person associated with the Space Shuttle Program.

3. Launch Operation Facilities: Launch Operation Facilities support all activities which occur after the launch vehicle has been processed up to the point of launch. These facilities provide a base and support structure for the transport and launching of the vehicle, service the launch vehicle at the launch pad, control pre-launch and launch operations, and launch the vehicle. These facilities include but are not limited to launch pads, Launch Control Center (LCC) Mobile Launch Platforms (MLPs), the Rotating Service Structure (RSS), and the Fixed Service Structure (FSS). Such facilities function as the primary resources integral to the launch of the Space Shuttle. In order to qualify for listing, resources must:

- Possess engineering importance and have facilitated nationally significant events associated with space travel;
- have been integral in pre-launch and launch preparation or the launching of the Space Shuttle;
- Clearly embody the distinctive characteristics of a type or method of construction specifically designed for the Space Shuttle;
- Have a direct historical association with the Space Shuttle, or a significant person associated with the Space Shuttle Program;

4. Mission Control Facilities: Support the design, development, planning, training and flight control operations for Space Shuttle flights. These facilities provide the infrastructure that allow the planning, training and flight operations processes necessary to support the Space Shuttle from the inception of requirements through the flight execution process. In order to qualify for listing, resources must have:

- Developed integrated flight crew and flight control plans, procedures, and training;
- Established simulators and flight control ground instrumentation;
- Configured Orbiter flight software;

- Contributed to the development and integration of spacecraft and payload support system.
- Provided onboard portable computer hardware and software for the Space Shuttle.

5. News Broadcast Facilities: Press facilities provide a primary site for news media activities at NASA-owned facilities. These broadcasting facilities were essential for relating to the American public news of the Space Shuttle Program to the nation and the world. In order to qualify for listing, resources must:
- Have been an integral facility in the dissemination of information about the Space Shuttle missions to the public;
- Clearly embody the distinctive characteristics of a type or method of construction specifically designed to broadcast information;
- Be associated with a significant person associated with the broadcast of Space Shuttle events;

6. Communication Facilities: Communication facilities in support of the Space Shuttle Program provide a vital site for instrumentation to receive, monitor, process, display and/ or record information from the space vehicle during test, launch, and/or flight. Significant communication facilities were designed specifically to house computers and computer-related technology vital to the Space Shuttle mission. In order to qualify for listing, resources must:
- Have been integral to the mission of the Space Shuttle;
- Clearly embody the distinctive characteristics of a type or method of construction specifically designed for the Space Shuttle missions;
- Have a direct historical association with the Space Shuttle, or a significant person associated with the Space Shuttle Program.

7. Engineering and Administrative Facilities: Engineering and Administrative Facilities include those resources which are essential to the administrative, scientific, and engineering work of the Space Shuttle Program. Engineering and Administrative Facilities administer such operations as research and development, testing, fiscal matters, procurement, planning, central management, and facilities engineering and construction, as well as providing offices for associated contractors and laboratories for engineers and scientists. These facilities which qualify for listing under the Space Shuttle context must:

- Be places, such as test facilities, that are directly associated with activities of significance which were associated with the development, component testing, implementation and termination of the Space Shuttle Program or missions;
- Be places where persons who made lasting achievements to the Space Shuttle Program worked or convened;
- Should clearly embody the distinctive characteristics of a type or method of construction.

8. Space Flight Vehicle (or Space Shuttle): This property type includes resources that comprise and/or facilitate the space flight vehicle or Space Shuttle. These include, but are not limited to, the Orbiter, Solid Rocket Booster (SRB), and External Tank (ET) as well as mockups of these components that were used for flight tests or other important development activities. In order to qualify for listing, resources must:
- Have been an integral component of the Space Shuttle Stack in its completed form, ready for space flight;
- Have been essential to the Space Shuttle missions and should clearly embody the distinctive aspect of reusability which reflects the goals of the Space Shuttle Program;
- Have been developed and used as test components used in preparation or evaluation for flight or flight tests;
- Have a direct historical association with the Space Shuttle, or a significant person associated with the Space Shuttle Program.

9. Manufacturing and Assembly Facilities: This property type includes facilities where major flight components were manufactured or assembled. These would include the manufacturing plants where the major components of the Space Shuttle vehicle were fabricated and assembled. In order to qualify, these facilities must:

- Have been an essential component to the manufacturing or assembling of the Space Shuttle;
- Have been constructed or modified to house this manufacturing or assembly facility exclusively;
- Embody a design that is unique to the Space Shuttle requirements;
- Have a direct historical association with the Space Shuttle, or a significant person associated with the Space Shuttle Program.

10. Resources Associated with the Training of Astronauts: This property type includes resources constructed or modified for the purpose of astronaut training and preparation for Space Shuttle missions. These facilities may include but are not limited to: processing facilities, neutral buoyancy tank, flight simulators and training aircraft. In order to qualify for listing, resources must:

- Have been designed and constructed, or modified, for the unique purpose of astronaut training and be directly associated with preparing astronauts for the completion of a Space Shuttle mission;
- Clearly embody the distinctive characteristics of a type or method of construction specifically designed for aeronautical training;
- Have a direct historical association with the Space Shuttle, or a significant person associated with the Space Shuttle Program.

11. Resources Associated with Space Flight Recovery: This property type includes resources that facilitate the recovery of the Space Flight Vehicle or Space Shuttle and its significant components after its return to Earth. These include, but are not limited to, runways, the Mate/De-mate Facility(s) and equipment, the Solid Rocket Booster Retrieval Ships (*Liberty* and *Freedom*), the Transporter and Wash Building, and the flume that brings the SRB to the building from the ships. These resources are essential to the recovery and subsequent reuse of the Space Shuttle and are therefore a significant resource to the program as a whole. In order to qualify for listing, resources must:

- Have been integral to the recovery of the Space Shuttle and/or its significant components;
- Clearly embody the distinctive characteristics of a type or method of construction specifically designed for the recovery of the Space Shuttle;
- Have a direct historical association with the Space Shuttle, or a significant person associated with the Space Shuttle Program.

12. Resources Associated with Processing Payloads: This property type is limited to facilities where fully assembled payloads are readied for insertion in the Space Shuttle Orbiter. In order to qualify for listing, resources must have been used in the processing of payloads for the Space Shuttle. Eligibility is restricted to resources which:

- Represent outstanding achievements in technological, aeronautical or scientific research which would otherwise not have been attainable without the use of the Space Shuttle;
- Clearly embody the distinctive characteristics of a type or method of construction, and which reflect the distinctive aspect of reusability unique to the goals of the Space Shuttle Program;
- Have a direct historical association with the Space Shuttle, or a significant person associated with scientific and/or technological advancements of national significance made as part of the Space Shuttle Program.

Archaeological Resources – Environmental Impacts

Archaeological resources can be affected adversely by the demolition and/or removal of buildings and structures that are present on top of an archaeological site. Intrusive ground disturbance can often upset the original stratigraphy of the archaeological deposits, thereby dislocating important artifacts and features from their original context. In many cases, intrusive ground disturbance will damage or destroy such artifacts and features. Where known or recorded archaeological sites are present, ground disturbing demolition and/or removals should be avoided. Where this is not feasible and intrusive ground disturbance is required, the underlying archaeological resource must be evaluated for NRHP eligibility and, if found to be NRHP-eligible, then appropriate mitigation measures would be required by the DHR. Often, archaeological sites are mitigated through scientific, controlled archaeological excavations and subsequent analyses and reports.

Disposition of real property may have minimal to no impact on archaeological resources because no ground disturbance would take place. However, if an NRHP-eligible archaeological site were to be transferred out of federal ownership and protection, such a transfer would be considered an adverse effect under Section 106 or 110 of the NHPA, because a future owner could damage or destroy the archaeological resource during future development. In such cases where NRHP-eligible archaeological sites must be transferred out of federal control, a data recovery program would be conducted before the transfer. Alternatively, the property could be transferred with appropriate deed restrictions to protect the archaeological resource from future harm.

In the event that archaeological sites are discovered in the course of demolition or construction, state and federal laws regarding inadvertent discoveries would be followed.

The other alternatives for the disposition of real property, including storage and reutilization, would also have no impact on archaeological sites, because they would not involve surface or subsurface disturbance

Historic Resources – Environmental Impacts
It is expected that many of the buildings used in the SSP would be reused for other NASA projects with the same or similar functions. If additions or alterations to NRHP-eligible or listed facilities are required as part of the transition, then KSC would be obligated to follow standard federal and state procedures regarding modifications to NRHP properties. NASA's compliance with these procedures would be accomplished by adherence to the Section 106 process on the federal level and the DHR's historic preservation compliance review program.

Mothballing the resource; that is, maintaining its functionality for reuse by NASA at a later time, also would have a minimal impact on the buildings or facilities. This scenario assumes there would be no alterations to the building before or during the low-maintenance mothball period and that the mothballing would not lead to destruction of the resource through neglect.

If a historic property were demolished or removed, there would be major impacts to that building or structure. Section 106 procedures and consultation with the DHR would be required before demolition of the property could begin. The conveyance of a property to another federal agency, through release to the GSA, probably would have minimal impacts to the structure itself, but would cause considered an adverse effect on the historic significance of the building because it would no longer be used by NASA for the space-related activities from which it gained NRHP significance. The building also could be transferred to the private sector, instead of to another federal agency. If the building were transferred to the GSA or another entity, the new use would have to be evaluated for potential impacts to the significance of the historic resource.

References Cited

Advisory Council on Historic Preservation (ACHP)
1991   *Balancing Historic Preservation Needs with the Operation of Highly Technical or Scientific Facilities.* Washington, D.C.

1995   *Consideration of Highly Technical and Scientific Facilities in the Section 106 Process.* Washington, D.C.

Butowsky, Dr. Harry A.
1984   *Man In Space National Historic Landmark Theme Study.* U.S. Department of the Interior, National Park Service, Washington D.C. May.

Milbrooke, Anne
1998   National Register Bulletin, *Guidelines for Evaluating and Documenting Historic Aviation Properties.* U.S. Department of the Interior, National Park Service, Washington, D.C.

National Park Service (NPS)
1998   National Register Bulletin, How to Apply the National Register Criteria for Evaluation. U.S. Department of the Interior, National Park Service, Washington, D.C.

Appendix D: Applicable Regulations and Laws

Regulation/Law	Citation	hyperlink
The Wild and Scenic Rivers Act of 1968	16 USC 1271-1287	http://www.access.gpo.gov/uscode/title16/chapter28_.html
The Marine Mammal Protection Act of 1972, as amended	16 USC 1371 et seq.	http://www.access.gpo.gov/uscode/title16/chapter31_subchapterii_.html
The Coastal Zone Management Act of 1972	16 USC 1451-1465	http://www.access.gpo.gov/uscode/title16/chapter33_.html
The Endangered Species Act (ESA) of 1973	16 USC 1531–1544, as amended	http://frwebgate3.access.gpo.gov/cgi-bin/waisgate.cgi?WAISdocID=848306967+0+0+0&WAISaction=retrieve
Archaeological and Historic Preservation Act of 1974 (AHPA)	16 USC 469-469c	http://www.access.gpo.gov/uscode/title16/chapter1a_subchapteri_.html
National Historic Preservation Act (NHPA) of 1966, as amended	16 USC 470 et seq.	http://www.access.gpo.gov/uscode/title16/chapter1a_subchapterii_.html
Archaeological Resources Protection Act of 1979 (ARPA)	16 USC 470aa-mm	http://www.access.gpo.gov/uscode/title16/chapter1b_.html
The Fish and Wildlife Coordination Act of 1958	16 USC 661 et seq.	http://frwebgate2.access.gpo.gov/cgi-bin/waisgate.cgi?WAISdocID=846802344874+0+0+0&WAISaction=retrieve
The Migratory Bird Treaty Act (MBTA) of 1918	16 USC 703-712, as amended	http://www.access.gpo.gov/uscode/title16/chapter7_subchapterii_.html
The Antiquities Act of 1906	16 USC 431-433, Stat 225	http://frwebgate3.access.gpo.gov/cgi-bin/waisgate.cgi?WAISdocID=8489621395+0+0+0&WAISaction=retrieve
Native American Graves Protection and Repatriation Act of 1990 (NAGPRA)	25 USC 3001 et seq.	http://www.access.gpo.gov/uscode/title25/chapter32_.html
OSHA	29 CFR Parts 1910 and 1926	http://www.access.gpo.gov/cgi-bin/cfrassemble.cgi?title=200629
The Clean Water Act (CWA) of 1977	33 USC 1251 et seq. [40 CFR 100-135]	http://www.access.gpo.gov/nara/cfr/waisidx_03/40cfrv19_03.html
Curation of Federally-Owned and Administered Archaeological Collections (36 CFR 79)	36 CFR 79	http://www.access.gpo.gov/nara/cfr/waisidx_06/36cfr79_06.html
Resource Conservation and Recovery Act of 1976	40 CFR 260 - 299	http://www.access.gpo.gov/uscode/title42/chapter82_.html
Comprehensive Environmental Response, Compensation, and Liability Act	40 CFR 300 - 349	http://www.access.gpo.gov/uscode/title42/chapter103_.html
Emergency Planning and Community Right-to-Know Act (also known as Title III of SARA)	40 CFR 350 - 399	http://www.access.gpo.gov/uscode/title42/chapter116_.html
The Clean Air Act (42 United States Code [USC] 7401-7671q)	40 CFR 50-99	http://www.epa.gov/air/caa/

APPENDIX D: APPLICABLE REGULATIONS AND LAWS

Regulation/Law	Citation	hyperlink
General Conformity Rule	40 CFR 51.850-860 and 40 CFR 93.150-160	http://ecfr.gpoaccess.gov/cgi/t/text/text-idx?c=ecfr&sid=7cb786046065df4b42d809b5816cf314&rgn=div6&view=text&node=40:2.0.1.1.2.19&idno=40
Toxic Substances Control Act (TSCA) of 1976	40 CFR 700 - 799	http://www.access.gpo.gov/uscode/title15/chapter53_.html
National Emission Standard for Asbestos (40 CFR Part 61, Subpart M)	40 CFR 61, Subpart M	http://ecfr.gpoaccess.gov/cgi/t/text/text-idx?c=ecfr&sid=280000c0abda6e4aaea0882a6000fcf7&rgn=div6&view=text&node=40:8.0.1.1.1.13&idno=40
National Pollution Discharge Elimination System	40 CFR 122	http://www.access.gpo.gov/nara/cfr/waisidx_06/40cfr122_06.html
The Safe Drinking Water Act (SDWA)	40 CFR 141	http://www.access.gpo.gov/nara/cfr/waisidx_03/40cfr141_03.html
Prevention of Significant Deterioration (PSD) / New Source Review (NSR)	40 CFR 51.166 / 40 CFR 51.165 (in CAA)	http://ecfr.gpoaccess.gov/cgi/t/text/text-idx?c=ecfr&sid=5f2b25d1de7e11a0da1dbe1ebd0ce9a1&rgn=div8&view=text&node=40:2.0.1.1.2.6.8.7&idno=40 http://ecfr.gpoaccess.gov/cgi/t/text/text-idx?c=ecfr&sid=5f2b25d1de7e11a0da1dbe1ebd0ce9a1&rgn=div8&view=text&node=40:2.0.1.1.2.6.8.6&idno=40
American Indian Religious Freedom Act of 1978 (AIRFA)	42 USC 1996-1996a	http://frwebgate1.access.gpo.gov/cgi-bin/waisgate.cgi?WAISdocID=849258272031+0+0+0&WAISaction=retrieve
National Environmental Policy Act of 1969 (NEPA)	42 USC 4321-4347	http://www.access.gpo.gov/uscode/title42/chapter55_.html
U.S. Department of Transportation Hazardous Materials Regulations and Motor Carrier Safety Regulations	9 CFR 107, 171-180 and 390-397	http://www.access.gpo.gov/nara/cfr/waisidx_99/49cfrv2_99.html
Eastern and Western Range (EWR) 127-1, Range Safety Requirements (U.S. Air Force, 1995)	EWR 127-1	http://snebulos.mit.edu/projects/reference/NASA-Generic/EWR/EWR-127-1.html
Protection and Enhancement of the Cultural Environment	Executive Order (EO) 11593	http://www.gsa.gov/Portal/gsa/ep/contentView.do?contentType=GSA_BASIC&contentId=12094
The Federal Emergency Management Agency (FEMA)	Executive Order (EO) 11988, Floodplain Management	http://www.fema.gov/plan/ehp/ehplaws/eo11988.shtm
Indian Sacred Sites	Executive Order (EO) 13007	http://www.achp.gov/EO13007.html
Consultation and Coordination with Indian Tribal Governments	Executive Order (EO) 13175	http://www.epa.gov/fedrgstr/eo/eo13175.htm
Preserve America	Executive Order (EO) 13287	http://www.gsa.gov/Portal/gsa/ep/contentView.do?P=PLAE&contentId=16910&contentType=GSA_BASIC
Emergency Preparedness Program (NASA directive)	NPD 8710.1	http://nodis3.gsfc.nasa.gov/displayDir.cfm?Internal_ID=N_PD_8710_001D_&page_name=main
NASA Occupational Health Program Procedures (NASA directive)	NPR 1800.1B	http://nodis3.gsfc.nasa.gov/displayDir.cfm?Internal_ID=N_PR_1800_001B_&page_name=main
NASA Procedural Requirements 8715.3B (NASA directive)	NPR 8715.3B	http://nodis3.gsfc.nasa.gov/displayDir.cfm?Internal_ID=N_PR_8715_003B_&page_name=main
Florida Endangered and Threatened Species Act of 1977	Title XXVIII Ch. 372.072	http://www.flsenate.gov/Statutes/index.cfm?App_mode=Display_Statute&Search_String=&URL=Ch0372/SEC072.HTM&Title=->2007->Ch0372->Section%20072#0372.072
The Preservation of Native Flora of Florida Act	Title XXXV Ch. 581.185	http://www.leg.state.fl.us/Statutes/index.cfm?App_mode=Display_Statute&Search_String=&URL=Ch0581/SEC185.HTM

APPENDIX D: APPLICABLE REGULATIONS AND LAWS

Regulation/Law	Citation	hyperlink
Chapter 267, Florida Statutes (F.S.) Florida Historical Resources Act	Title XVIII Ch. 267	http://leg.state.fl.us/statutes/index.cfm?App_mode=Display_Statute&URL=Ch0267/titl0267.htm&StatuteYear=2007&Title=%2D%3E2007%2D%3EChapter%20267
Section 872.05, F.S.	Title XLVI Chapter 872	http://leg.state.fl.us/statutes/index.cfm?App_mode=Display_Statute&URL=Ch0872/titl0872.htm&StatuteYear=2007&Title=%2D%3E2007%2D%3EChapter%20872
State of Florida (62-730, Florida Administrative Code [F.A.C.]): Hazardous Waste	F.A.C. 62-730	https://www.flrules.org/gateway/ChapterHome.asp?Chapter=62-730
F.A.C Chapter 62-257: Asbestos	F.A.C. 62-257	https://www.flrules.org/gateway/ChapterHome.asp?Chapter=62-257
Florida storm water regulations	F.A.C. Chapter 40C	https://www.flrules.org/gateway/ChapterHome.asp?Chapter=62-40
Florida Permitting of Consumptive Uses of Water	Section 373.216, F.S.	http://leg.state.fl.us/statutes/index.cfm?App_mode=Display_Statute&URL=Ch0373/part02.htm&StatuteYear=2007&Title=%2D%3E2007%2D%3EChapter%20373%2D%3EPart%20II
Brevard County noise standards	Chapter 46-126	http://www.municode.com/resources/gateway.asp?pid=10473&sid=9
Operation and closure of solid waste landfill facilities in Florida	F.A.C. 62-701	https://www.flrules.org/gateway/ChapterHome.asp?Chapter=62-701
Department of Transportation Regulations	F.A.C. Chapter 14	https://www.flrules.org/gateway/department.asp?id=14
FL Vehicle Code	Title XXIII–Motor Vehicles, F.S.	http://leg.state.fl.us/statutes/index.cfm?App_mode=Display_Index&Title_Request=XXIII#TitleXXIII
FL Traffic Regulations	State Law Chapter 18–Traffic	
Brevard County Code	Chapter 106–Traffic and Vehicles	http://www.municode.com/resources/gateway.asp?pid=10473&sid=9
2002 Quality/Roadway Level of Service (QLOS) Handbook	Highway Capacity Manual [HCM]), per F.A.C. Chapter 14-94	https://www.flrules.org/gateway/ChapterHome.asp?Chapter=14-94
Antiquities Code of Texas	Chapter 442, Government Code of Texas	http://www.thc.state.tx.us/rulesregs/RulesRegsPDF/AntiqCode.pdf
Texas Restricted Cultural Resource Information	Texas Government Code, Chapter 24	http://www.thc.state.tx.us/rulesregs/RulesRegsPDF/Chapter24.pdf
Texas Asbestos regulations	Title 25, Chapter 295	http://info.sos.state.tx.us/pls/pub/readtac$ext.ViewTAC?tac_view=5&ti=25&pt=1&ch=295&sch=C&rl=Y
Health and Safety Handbook, Johnson Procedural Requirement (JPR)	JPR 1700.1	http://jschandbook.jsc.nasa.gov/
Houston Noise Ordinance	City of Houston Code of Ordinances Section 30-6, 30-9	http://www.municode.com/resources/gateway.asp?pid=10123&sid=43
Texas Solid Waste Disposal Act	Title 30 TAC Chapter 335	http://info.sos.state.tx.us/pls/pub/readtac$ext.ViewTAC?tac_view=5&ti=30&pt=1&ch=335&sch=A&rl=Y

APPENDIX D: APPLICABLE REGULATIONS AND LAWS

Regulation/Law	Citation	hyperlink
"Prevention, Abatement, and Control of Environmental Pollution"	NASA Management Instruction 8800.13A JSC Management Directive 8800.3L	
Houston General Plan 2025 (proposal)		http://www.houstonplan.org/proposals/proposalfinal.pdf
Houston Traffic Ordinances	Houston Code of Ordinances Chapter 45–Traffic	http://www.municode.com/resources/gateway.asp?pid=10123&sid=43
TX Vehicle Code	TX Statutes Title 7–Vehicles and Traffic	http://tlo2.tlc.state.tx.us/statutes/tn.toc.htm
Texas Department of Transportation [TxDOT]) Regulations	TAC Title 43–Transportation Part I–TxDOT	http://info.sos.state.tx.us/pls/pub/readtac$ext.ViewTAC?tac_view=4&ti=43&pt=1&ch=26
Harris County Road Law		http://www.eng.hctx.net/Permits/pdf/road_law.pdf
Federal Aviation Regulations – Objects affecting navigable airspace	Title 14 Part 77	http://www.faa.gov/airports_airtraffic/airports/regional_guidance/central/construction/part77/
Federal Aviation Regulations – Airport Noise	Title 14 Part 150	http://www.faa.gov/airports_airtraffic/airports/environmental/airport_noise/
JSC "Procedures for Control of Hazardous Materials"	JSC Management Directive 1710.9B	
JSC "Hazardous Waste Minimization"	JSC Management Directive 8800.4A	
State of Texas Waste Reduction Policy Act (WRPA) of 1991	30 TAC 335 Subchapter Q	http://info.sos.state.tx.us/pub/plsql/readtac$ext.ViewTAC?tac_view=5&ti=30&pt=1&ch=335&sch=Q&rl=Y
City of El Paso noise ordinance	Title 9 Chapter 9.40	http://ordlink.com/codes/elpaso/_DATA/TITLE09/Chapter_9_40_NOISE.html
El Paso Vehicles and Traffic Regulations	El Paso Municipal Code Title 12–Vehicles and Traffic;	http://ordlink.com/codes/elpaso/_DATA/TITLE12/index.html
Antiquities Law of Mississippi	Title 39 Chapter 7	http://www.mscode.com/free/statutes/39/007/
SSC Health and Safety Procedural Requirements	SSC Procedural Requirements (SPR) 8715.1	https://ssctdpub.ssc.nasa.gov/smweb/dtsisapi.dll?FIELD1=SM_DOC_NUMBER&TEXT1=SPR_8715.1&JOIN1TO2=AND&FIELD2=SM_TEXT&TEXT2
Mississippi Nonhazardous Solid Waste Management: Regulations and Criteria	Mississippi Commission on Environmental Quality Regulation SW-2	http://www.deq.state.ms.us/newweb/MDEQRegulations.nsf/c70604500020692b86256e12005858cb/cd70fb66996fc55f86256bd100530dad?OpenDocument
Mississippi Vehicle Code	State Code Title 63–Motor Vehicles and Traffic Regulations	http://michie.com/mississippi/lpext.dll/mscode/12b42?fn=document-frame.htm&f=templates&2.0#
Local Coastal Resources Management Act	Act 361, La. R.S. 49:214.21 *et seq.*	http://www.legis.state.la.us/lss/lss.asp?doc=103626

APPENDIX D: APPLICABLE REGULATIONS AND LAWS

Regulation/Law	Citation	hyperlink
Archeological Treasure Act	R.S. 41:1601-1613	http://www.legis.state.la.us/lss/lss.asp?doc=99032
Louisiana Historic Preservation Program	R.S. 36:208 *et. seq.*	http://www.legis.state.la.us/lss/lss.asp?doc=92707
Louisiana Unmarked Human Burial Sites Act	R.S. 8:673	http://www.legis.state.la.us/lss/lss.asp?doc=106448
MSFC Emergency Plan	Marshall Procedural Requirement (MPR) 1040.3	http://foia.msfc.nasa.gov/docs/NAS8-01121/2_MSFC_Directives/1040.3_MPD.pdf
Louisiana Administrative Code (LAC) Title 33 Section Part VII – Solid Waste Regulations	Louisiana Administrative Code (LAC) Title 33 Section Part VII	http://www.deq.louisiana.gov/portal/LinkClick.aspx?link=planning%2fregs%2ftitle33%2f33v07.doc&tabid=1674
New Orleans Traffic and Vehicle Codes	New Orleans Municipal Codes of Ordinances Chapter 154–Traffic and Vehicles	http://www.municode.com/resources/gateway.asp?pid=10040&sid=18
LA Vehicle Code	LA Revised Statutes Title 32–Motor Vehicle and Traffic Regulations	http://www.lmvc.state.la.us/PDF/LSA-R.S.%2032-1251.pdf
LA Administrative Code for Motor Vehicles	LA Administrative Code Title 55–Public Safety Part III–Motor Vehicles	http://doa.louisiana.gov/osr/lac/51v01/55v1-17.doc
Alabama Historical Commission (AHC) (the SHPO) "Policy for Archaeological Survey and Testing in Alabama"		http://www.preserveala.org/DOCUMENTS/PDF/revisedsurveyguidelines.pdf
Alabama Historical Commission (AHC) "Guidelines for Historic Architectural Resources in Alabama"	Section 106	
City of Huntsville's noise ordinance	Ordinance 99-766	http://www.hsvcity.com/NatRes/noise1.htm
ADEM Solid Waste Program	Chapter 335, Division 13	http://www.adem.state.al.us/Regulations/regulations.htm
Huntsville City Codes on Traffic and Vehicles	Huntsville City Law Code of Ordinance Chapter 25–Traffic and Vehicles	
AL Vehicle Code	AL Code Title 32–Motor Vehicles and Traffic	http://law.justia.com/alabama/codes/22786/22786.html
AL Highway Department Regulations	AL Administrative Code Title 450–Highway Department	http://www.alabamaadministrativecode.state.al.us/docs/dot/index.html
New Mexico Wildlife Conservation Act of 1974	Chapter 17-2-37-46, New Mexico Statutes [NMSA], 1995	http://www.animallaw.info/statutes/stusnm17_2_37.htm
Cultural Properties Act	Sections 18-6 through 18-6-23, NMSA, 1978	http://nxt.ella.net/NXT/gateway.dll?f=templates$fn=default.htm$vid=nm:all
Prehistoric and Historic Sites Preservation Act of 1989	Sections 18-8-1 through 18-8-8, NMSA 1978	http://nxt.ella.net/NXT/gateway.dll?f=templates$fn=default.htm$vid=nm:all

APPENDIX D: APPLICABLE REGULATIONS AND LAWS

Regulation/Law	Citation	hyperlink
State of New Mexico occupational health and safety regulations	Title 11, Chapter 5	http://www.nmcpr.state.nm.us/NMAC/_title11/T11C005.htm
New Mexico Solid Waste Management Regulation	Title 20, Chapter 9, Part 1	http://www.nmcpr.state.nm.us/NMAC/_title20/T20C009.htm
NM State Vehicle Code	State Statutes Chapter 66–Motor Vehicles	http://nxt.ella.net/NXT/gateway.dll?f=templates$fn=default.htm$vid=nm:all
NM Transportation and Highway Regulations	NMAC Title 18–Transportation and Highways.	http://www.nmcpr.state.nm.us/NMAC/_title18/title18.htm
	California Public Resources Code (PRC). Section 5024	http://www.parks.ca.gov/pages/1071/files/public%20resources%20code%205024.pdf
California Department of Industrial Relations Division of Occupational Health and Safety (DOSH)		http://www.dir.ca.gov/dosh/dosh1.html
California Highway Patrol Regulations	Division 2, Chapter 6, of the California State Regulations	Article 1, Article 2, Article 3, Article 4
California Integrated Waste Management Board (CIWMB)		http://www.ciwmb.ca.gov/
California Vehicle Code		http://www.defend-me.com/California-Vehicle-Code/
Department of Transportation *Manual on Uniform Traffic Control Devices* (MUTCD)		http://mutcd.fhwa.dot.gov/
Kern County Code Title 10–Vehicles and Traffic	Kern County Code Title 10–Vehicles and Traffic	http://municipalcodes.lexisnexis.com/codes/kerncoun/_DATA/TITLE10/index.html

www.ingramcontent.com/pod-product-compliance
Lightning Source LLC
Chambersburg PA
CBHW081232180526
45171CB00005B/406